Decision Science and Social Risk Management

In addition to its French and Italian origins, the English word risk *has two other possible derivations which are interesting to contemplate: the Vulgar Latin word* resecum, *meaning danger, rock, risk at sea, and the Greek word* rhiza, *cliff, through the meaning "to sail round a cliff." Though unattested, these classical sources are appealing to the editors for the visual image they evoke: Ulysses and his men making their perilous voyage between Scylla and Charybdis in the Strait of Messina.*

Technology, Risk, and Society

An International Series in Risk Analysis

Miley W. Merkhofer

Applied Decision Analysis, Inc.

Decision Science and Social Risk Management

A Comparative Evaluation of Cost-Benefit Analysis, Decision Analysis, and Other Formal Decision-Aiding Approaches

D. Reidel Publishing Company

A MEMBER OF THE KLUWER ACADEMIC PUBLISHERS GROUP

Dordrecht / Boston / Lancaster / Tokyo

Library of Congress Cataloging-in-Publication Data

CIP

Merkhofer, Miley W., 1947–
Decision science and social risk management.

(Technology, risk, and society)
Bibliography: p.
Includes index.
1. Policy sciences. 2. Decision-making. 3. Cost effectiveness. 4. Social choice. I. Title. II. Series.
H97.M45 1986 361.6′1′068 86–25966

ISBN-13: 978-94-010-8589-2 e-ISBN-13: 978-94-009-4698-9
DOI: 10.1007/978-94-009-4698-9

Published by D. Reidel Publishing Company,
P.O. Box 17, 3300 AA Dordrecht, Holland.

Sold and distributed in the U.S.A. and Canada
by Kluwer Academic Publishers,
101 Philip Drive, Assinippi Park, Norwell, MA 02061, U.S.A.

In all other countries, sold and distributed
by Kluwer Academic Publishers Group,
P.O. Box 322, 3300 AH Dordrecht, Holland.

Contents

Preface

Economists, decision analysts, management scientists, and others have long argued that government should take a more scientific approach to decision making. Pointing to various theories for prescribing and rationalizing choices, they have maintained that social goals could be achieved more effectively and at lower costs if government decisions were routinely subjected to analysis. Now, government policy makers are putting decision science to the test. Recent government actions encourage and in some cases require government decisions to be evaluated using formally defined principles of rationality. Will decision science pass this test? The answer depends on whether analysts can quickly and successfully translate their theories into practical approaches and whether these approaches promote the solution of the complex, highly uncertain, and politically sensitive problems that are of greatest concern to government decision makers. The future of decision science, perhaps even the nation's well-being, depends on the outcome.

A major difficulty for the analysts who are being called upon by government to apply decision-aiding approaches is that decision science has not yet evolved a universally accepted methodology for analyzing social decisions involving risk. Numerous approaches have been proposed, including variations of cost-benefit analysis, decision analysis, and applied social welfare theory. Each of these, however, has its limitations and deficiencies and none has a proven track record for application to government decisions involving risk. Cost-benefit approaches have been extensively applied by the government, but most applications have been for decisions that were largely risk-free. Various decision analysis approaches have been used to aid risky business decisions, but not often for social decisions. Although numerous applied social welfare theory approaches have been proposed, only in a few instances have they been tried in real decision-making situations.

The lack of a single, well-tested, and universally appropriate method for analyzing a decision makes the selection of a decision-aiding approach a question of considerable importance. Because the main approaches differ fundamentally in their strengths and weaknesses, the value of analysis depends on the degree to which the selected approach matches the needs of the situation. Unfortunately, many analysts fail to understand fully the technical, political, and social factors important to a decision problem and lack familiarity with the wide range of approaches that are available. As a result, the method of analysis may be ill-suited to the problem at hand. In my opinion, the selection of an inappropriate approach is the single most frequent cause of the failure of an analysis to produce a positive effect on the decision process. To derive the greatest value from decision science, decision makers and analysts must resist the temptation to apply only those approaches which are most familiar: they must recognize features of risk problems relevant to determining the effectiveness of various types of analysis and select from the available options an approach that best matches the needs of the problem under consideration.

Several difficulties complicate the analyst's selection of a decision-aiding approach. One problem is the lack of agreement over methodology and terminology. Analysts use a variety of labels to describe their approaches, and no consensus exists for the definitions or minimal requirements of even commonly used terms. What one researcher describes as a decision analysis another may characterize as a cost-benefit analysis. A second difficulty is the absence of objective, knowledgeable comparisons of available methodologies and procedures. As Cumming (1981) notes, because risk analysis is a new and rapidly expanding field, much "jostling for position" is occurring. The need to establish one's credentials and the competition for funds create strong incentives for analysts to over-emphasize and exaggerate claims when comparing their methodological specializations with others.

This book presents an initial attempt to distinguish, compare, and evaluate altenative decision-aiding approaches. To focus the effort, the spectrum of approaches considered has been limited in several ways. First, consideration is restricted to approaches applicable to social decisions involving risks, particularly risks to health, safety, and the environment, which represent the area of greatest public concern and are the subject of the vast majority of government risk regulations. Second, consideration is limited to formal approaches that provide quantitative decision criteria, theoretical rationales, and practical procedures for identifying a "best" alternative. Third, only those approaches are considered for which there is some practical applications experience. Included within this scope are approaches that have been referred to as Pareto cost-benefit analysis, decision-maker cost-benefit analysis, clinical decision analysis, social decision analysis, supradecision-maker decision analysis, multiattribute utility theory, and applied social welfare theory. The evaluation considers questions of practicality, acceptability, accuracy, logical soundness, and completeness.

In the course of the comparative evaluation of decision-aiding approaches, this book presents a threefold conceptual framework to aid selection: (1) a taxonomy for identifying features of risk problems relevant to approach selection, (2) a means for characterizing and distinguishing approaches, and (3) criteria for approach evaluation. Conceptually, the comparision of approaches can be viewed as being conducted in a matrix whose rows correspond to different types of risk problems and whose columns correspond to the various approaches that are available for application to those problems. Each cell of such a matrix would represent the evaluation of a specific approach for use in a particular type of risk problem, the assumption being that the relative value of an approach depends on the problem to which it is applied.

The intended audience for this book is broad and not merely limited to the expanding community of professionals who analyze and manage risks; it includes technology promoters, legislators and regulators, students, and those members of the public who would like to see social risk decisions improved. Important material is provided for the analyst and for those for whom analysis is conducted. It is my opinion that analyses of social decisions could be improved if analysts achieved a better understanding of the social risk-management problem and decision makers attained a better understanding of available analytic approaches. Furthermore, both groups would profit from knowledge of the criticisms that have been directed toward quantitative approaches. Consequently, more detailed discussions of these topics are provided than would be the case were it not hoped that both analysts and nonanalysts interested in the use of formal approaches for social risk management might benefit from this effort.

The four chapters in this book describe the conceptual framework for comparing decision-aiding approaches and the results of its application. Chapter 1 discusses the social risk-management problem. Chapter 2 describes available decision-aiding approaches and their differences. Chapter 3 presents the principal criticisms and alleged limitations of approaches. Chapter 4 summarizes the basic framework for comparing approaches, evaluation criteria, and conclusions.

To provide additional descriptions and explanations, three case-study applications have been included as appendices. Appendix A describes a version of cost-benefit analysis applied to mobile source emissions control. Appendix B illustrates a version of decision applied to the evaluation of public policy for commercial cryptography. Appendix C presents an applicaton of several versions of social welfare theory to the selection of a vehicle trajectory for a space mission. Although a few of the details have been altered to simplify the exposition, each case study represents an actual professional application of a decision-aiding approach to a real social decision problem.

I wish to thank the Division of Policy Research and Analysis of the National Science Foundation for supporting initial research that provided the basis for the development of this book. Acknowledgment and thanks are also due to individuals who generously reviewed an early draft of this work and provided many helpful comments and suggestions, especially James Dyer, Baruch Fischhoff, Max Henrion, Ralph Keeney, Lester Lave, Thomas McCurdy, Harvey Richmond, and Detlof von Winterfeldt. Special thanks are due to Vincent Covello of the National Science Foundation for his assistance, support, contributions, and counsel throughout this effort.

This book was prepred with the partial support of the National Science Foundation under Grant PRA—8007238. The opinions, findings, and conclusions expressed are those of the author and do not necessarily reflect the views of the NSF.

Abbreviations

CBA	cost-benefit analysis
CBT	cost-benefit theory
CPSC	Consumer Product Safety Commission
DA	decision analysis
DES	Data Encryption Standard
DOC	Department of Commerce
DOD	Department of Defense
DOE	Department of Energy
DOT	Department of Transportation
DT	decision theory
E.O.	Executive Order
EPA	Environmental Protection Agency
FAA	Federal Aviation Administration
FDA	Food and Drug Administration
FIFRA	Federal Insecticide, Fungicide, and Rodenticide Act
GAO	General Accounting Office
JPL	Jet Propulsion Laboratory
LNG	Liquefied Natural Gas
NASA	National Air and Space Agency
NBS	National Bureau of Standards
NEPA	National Environmental Protection Act
NPV	net present value
NRC	Nuclear Regulatory Commission
OMB	Office of Management and Budget
OSHA	Occupational Safety and Health Administration
PKC	public-key cryptography

Chapter 1

Social Risk Management

The risk problems facing society today have many characteristics that limit and otherwise complicate the application of formal analysis. Since not all problems possess these complicating qualities to the same degree, some problems may be addressed more effectively by certain decision-aiding approaches than by others. To identify the most useful approach, the analyst must clearly understand the decision problem being addressed. A comparative evaluation of approaches must therefore begin with a characterization of social risk management decisions.

The purpose of this chapter is to describe the nature of social risk management and to identify some of the important dimensions along which risk decisions differ. Discussed first is the nature of the risks that create public concern and pressures for government action. This is followed by a discussion of possible government roles in risk management and an exploration of the factors that make government risk decisions difficult. Next, the institutions and mechanisms that have evolved for social risk management are summarized, and finally, a preliminary taxonomy is provided for distinguishing among risk problems. The taxonomy, which is synthesized from the discussion throughout this chapter, provides one of the elements necessary for the development of a conceptual framework for comparing decision-aiding approaches.

Nature of Risk

What is risk? What causes risk? How much risk are we currently facing? These questions are more difficult to answer than one might expect. To begin with, risk is not an easy word to define.

Meaning of Risk

People speak of business risk, social risk, economic risk, safety risk, investment risk, military risk, political risk, and so on. Depending on context, risk can mean different things. A dictionary lists several definitions, including "the possibility of suffering harm," "the amount an insurance company stands to lose," and "the possibility and degree of loss or injury." Our perspective in defining risk is consistent with the broadest and most general of the definitions presented by risk analysts: risk is defined as an uncertain situation in which a number of possible outcomes might occur, one or more of which is undesirable. With this definition, uncertainty is clearly fundamental to the concept of risk. If you know for certain that you will bear the burden of some specific undesired outcome, we might feel sorry for you, but we would not say that you are experiencing risk.

If the focus is decision making, estimating the magnitude of risk so as to permit a comparison of the risks associated with alternative actions is the major concern. In the case of a risk associated with an event, experiments show that people's perceptions of the magnitude of the risk depend on how likely they think the event is and how serious they consider the effect to be. Following this line of reasoning, risk analysts argue that the level of risk should be measured in terms of the probability (relative likelihood) of the possible outcomes (in a given time period) and measures of the magnitude (seriousness) of the consequences of those outcomes. Fundamentally, then, risk may be represented as a probability distribution over adverse consequences.[1]

Figure 1 illustrates several of the most common ways to display a risk that has been quantified as a probability distribution. Figure 1(a) applies to a case in which there are only two possible outcomes to a risk, for example, death or no death. In this case the probability of the bad

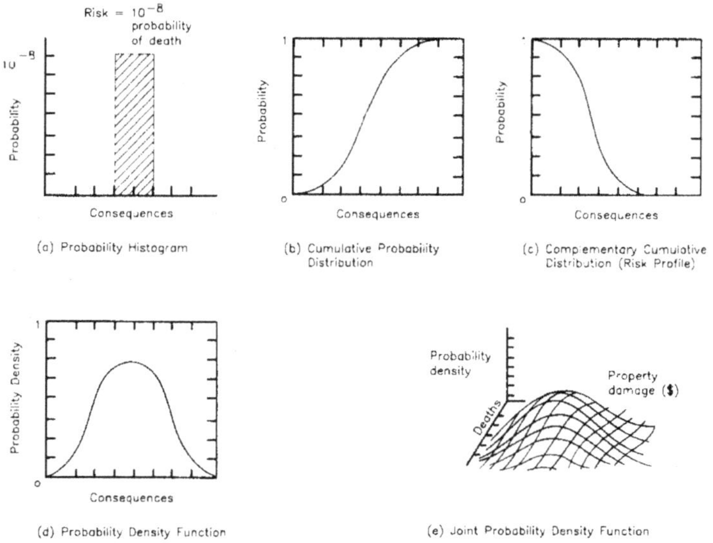

Fig. 1. Common ways to display quantified risks.

outcome, probability of death, is sufficient to describe the probability distribution. Figure 1(b) shows a probability display applicable to the case where the risk involves a range of possible consequence levels. The curve is called a cumulative probability distribution and its height at any consequence level shows the probability that actual consequences will be less than or equal to that level. Figure 1(c) is the reverse of the cumulative probability distribution. Its height at any consequence level indicates the probability that actual consequences will be greater than or equal to that level, and it is called a complementary cumulative probability distribution. It is also sometimes called a risk profile. Figure 1(d) is an intuitive but less easy to use display called a probability density function. The height of the curve at any given consequence level is proportional to the relative likelihood of that level of consequence occurring. The probability of consequence levels between any two values may be obtained as the area under the curve between those values. Figure 1(e) illustrates the sort of display that may be needed to describe risks that produce more than one type of consequence, for example, multiple deaths and property damage. The figure shows a joint probability density function. The curve defines a surface whose height at any point is proportional to the relative likelihood of simultaneously obtaining the combination of consequences associated with that point.

The consequences represented in probability distributions that quantify risks might be adverse effects to human health, plants, animals, materials, or other items of value, and might be measured in terms of fatalities, injuries, days of disability, man-hours of labor lost, incidence of cancer, property lost, or fish killed. Just how probability distributions for these sorts of consequences might be computed or estimated as part of a decision-aiding approach is described in Chapter 2 and the Appendices. The purpose of introducing such technical issues at this point is to clarify the concept of risk that is adopted throughout the discussion.

Notice that conceptualizing risk as a probability distribution allows risks to be altered either by a change in probabilities or by a change in possible consequences. This two-dimensional characteristic makes it impossible to find a completely satisfactory single number for measuring the level of risk. Common single-number summary statistics, such as the probability of loss, the maximum credible consequence level, the expected number of fatalities per year, the probability of fatality per exposed person per year, and so forth, fail to distinguish adequately among the different forms that risk can take. For example, the most commonly used summary measure is expected value of risk — the sum (or integral) of the products of probabilities and consequences. This measure fails to distinguish risks that involve a large probability of minor consequences from those that imply a small probability of a major catastrophe.

Psychologists would argue that even the two-dimensional technical definition of risk fails to capture all of the considerations that are important from a social perspective. Studies have found that in addition to probabilities and magnitude of possible consequences, other considerations influence how people "feel" about risk (e.g., Milburn and Billings 1976; Rowe, 1977b). For instance, attitudes tend to be influenced by the extent to which the potential consequences are concentrated in space and time (e.g., 200 people dying in a commercial airline crash compared with 200 unrelated deaths from automobile accidents), the degree of personal control over those risks (e.g., the risk of injury from skiing compared with that from nuclear war), and whether the individuals exposed to the risk share in the perceived advantages associated with the source of the risk (e.g., smoking risks compared with the health risks of air pollution).[2]

Keeney *et al.* (1979) suggest that, at a minimum, the analyst would have to compute the following measures to characterize mortality risk in a way that is sensitive to social concerns:

(1) Total expected fatalities per year (to measure aggregate societal risk);
(2) Probabilities of fatality for each exposed individual (to permit comparing that risk with others, such as the risk from smoking, or driving a car);
(3) Probabilities of fatality for individuals grouped by occupation, geographic location, etc. (to allow equity comparisons to be made); and
(4) Probabilities of exceeding specific numbers of fatalities per year (to allow for sensitivity to catastrophe).

Even the multitude of numbers that would be produced by such an analysis would fail to account for all the characteristics that influence risk perception. Furthermore, these measures fail to address the risk of injury and property damage. Completely measuring risk is thus a difficult, if not impossible, task. As will be described later, the social acceptability of a decision-aiding approach may be diminished to the extent that it fails to reflect important social as well as technical aspects in its representation of risk.

Character of Existing Risks

How risky is life today? According to one important statistic — the death rate — the American public has never been safer. Mortality rates have declined significantly since 1930, with the largest gains in the youngest age groups (Bailey 1980). The overall accident death rate has decreased 20 percent in the past ten years and 40 percent since 1912. Except for deaths from motor vehicle accidents, which rose prior to 1970 because of increased driving, deaths from other accidents have dropped sharply as a result of improved medical care, job safety, and a reduction in the work force engaged in dangerous occupations (such a farming). Mining accidents occur far less frequently, many industrial accidents have been reduced, and automobile fatalities show signs of decreasing. Life expectancy in the United States is now more than twenty years greater than it was in 1920 (Dodge and Civiak 1981).

Furthermore, major accidents and disasters are occurring less frequently. In the past one hundred years, eight events have occurred in the U.S. in which one thousand or more persons were killed (excluding wars and epidemics), but none has occurred since 1928 (National Safety Council 1979). Prior to 1928 an accident of this size happened roughly every eight years. Similarly, the frequency of accidents causing a hundred or more fatalities has dropped sharply since the 1940s. Historical frequency versus magnitude data show that frequency falls rapidly with magnitude for all U.S. accidents and natural disasters.[3]

Despite increased longevity and the reduction in the frequency of disasters, public concern about risk is increasing. According to a Harris poll (Harris and Associates 1980) most Americans believe life is getting riskier: 78 percent of the public surveyed agreed that "people are subject to more risk today than they were 20 years ago," and 55 percent indicated that "risks to society stemming from various scientific and technological advances will be somewhat greater 20 years from now than they are today."

Is increasing public concern in the face of a statistical decline in death rates a contradiction? Not necessarily. As noted above, risk is a multidimensional concept having more attributes than simply the age-specific average rate of death. Modern risks are perceived by the public as possessing characteristics that psychological studies show are of special concern — the possibility of catastrophe, inequities because those at risk do not directly benefit from the processes that generate the risk, lack of control by the individuals exposed, and possible long-term, irreversible consequences. Crime, an especially inequitable form of risk, is a source of considerable worry to many Americans, particularly those living within the inner cities. The risk of nuclear war is a major concern because of its catastrophic consequences. Yet, this risk is not at all reflected by increased life expectancy. Environmental risk, which is perceived as being inequitable and having potential long-term consequences, is another category that polling data indicate is of increasing concern to Americans (Harris and Associates 1980). Examples of environmental risks with possible long-term consequences include the risk of ozone depletion due to emissions of fluorocarbons, risk of radioactive leakage from nuclear accidents or from the disposal of nuclear wastes, and the risk that experiments with recombinant DNA will create new diseases for which there are no known cures or vaccinations.

There are a number of other differences between today's risks and those of the past. Risks of the distant past were primarily caused by natural events, while more recent risks, beginning with the industrial revolution, are largely the result of man-made technological developments. Along with the greater degree of enfranchisement of today's electorate has come the belief that individuals have a right to shape their environment. Because technological risks are to an extent self-imposed at the societal level, the debate over which risks are acceptable and which are not has become much more relevant. The recent increase in the federal government's role in technology regulation and its acceptance of responsibility for identifying and controlling health and safety contributes to increased public awareness and concern over risk. Finally, technology not only creates new risks but also produces greater awareness of those risks by introducing improved measurement techniques that enable even low-level risks to be identified.

Components of the Risk-Generation Process

The concept of risk may be clarified by exploring its essential components. For many risks, including those affecting humans, plants, animals, materials, and the environment, three conditions must be met (or thought to exist) before a risk can occur. First, there must be a source of risk; that is, a hazard. The hazard might be a nuclear power plant that may release radioactive material, a dam that might break, the sharp blades of a power lawn mower, the poorly understood chemical in a new drug, or the burning rays of a hot summer sun. Second, there must be an exposure process by which people, animals, plants, or materials of value may be brought into contact with the hazard. The exposure might occur as a result of wind dispersing radioactivity vented from a nuclear power plant, people living below the dam that might break, a homeowner reaching into the spinning blades of a lawn mower, doctors prescribing a new drug to their patients, or a sunbather spending too much time at the beach. Third, there must be a process by which exposure produces adverse effects. This process may consist of increased incidence of cancer resulting from elevated exposure to radioactivity, drowning from the water released in a dam failure, amputated fingers resulting from an accident with a lawn mower, unexpected side effects from a new drug, or sunburn (or skin cancer) from failing to cover up at the beach.

A hazard, an exposure process, and an effects process define a risk in the sense that they determine the level and probability of consequences. People's perceptions of risk involve an additional process that includes an evaluation by individuals or society that the severity, importance, or inequity of the effects is sufficient to be of concern. Figure 2 summarizes these basic components of risk. Together they form the "risk chain." Additional discussion of the first three links of the chain — hazards, exposure, and effects — is provided below. The valuation of risks and other aspects that influence public perception are discussed subsequently.

Hazards

Hazards are present across virtually the entire range of modern human activity. Technological health and safety hazards include substances used in growing, processing, and distributing food; drugs and medical procedures; pollutants discharged into the air or water from homes and commercial or industrial enterprises; occupational dangers, such as accidents from heavy machinery; contagious diseases; potentially harmful materials used in clothing and the construction of homes; products consumed for recreational purposes, such as tobacco and alcoholic beverages; and so on, *ad infinitum*. These and other hazards not only

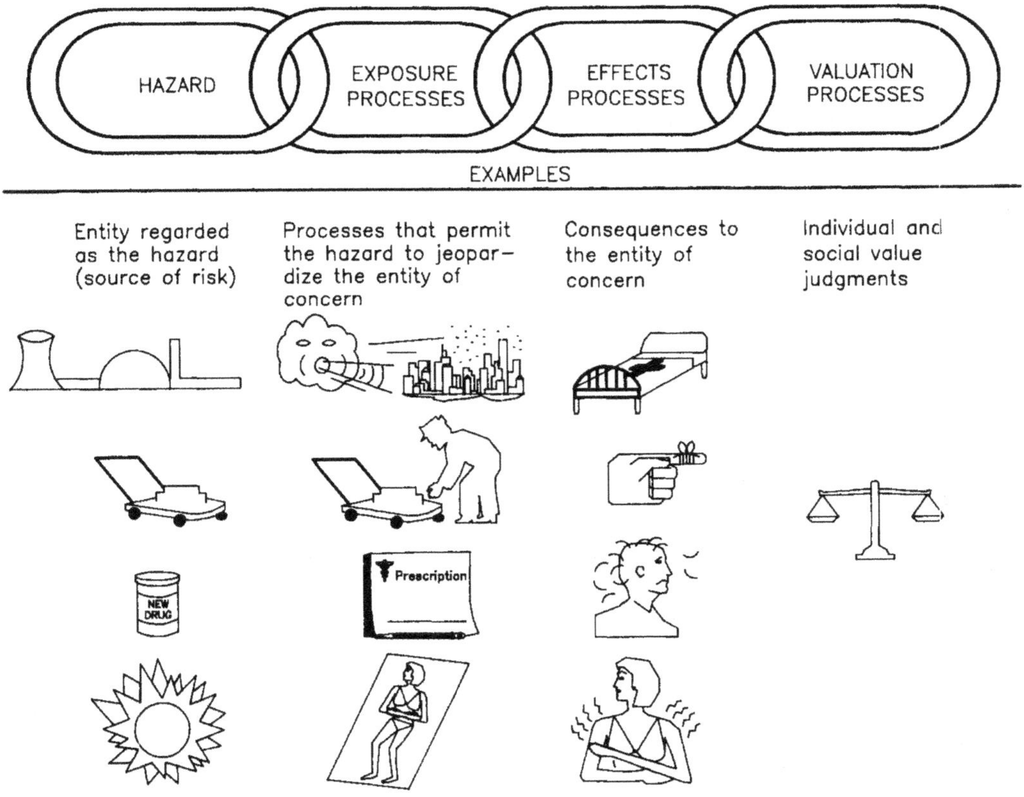

Fig. 2. The risk chain and examples.

create the potential for adverse effects to human health, they may also threaten plant and animal life and create property loss and material damage.

Several risk analysts have attempted to arrange the vast array of hazards into organizing categories. Because systems or processes may be hazardous if one or more of their parts are hazardous, one approach is to consider the most fundamental sources of risk, what some authors have referred to as risk agents. Sagan (1984), notes that fundamental agents for health, safety, and enviornmental risks are

1. Chemical (e.g., atmospheric pollutants, potentially toxic ingredients of food, drugs, and cosmetics)
2. Biological (e.g., viruses, bacteria)
3. Physical (e.g., mechanical force, acoustic energy, ionizing and non-ionizing radiation, heat, pressure)

Rowe and Broder (1979) group fundamental sources of risk into three categories:

1. Toxic materials (e.g., chemicals such as cyanide and mercury, carcinogens, radioisotopes, pesticides, DNA derivatives)
2. Kinetic energy (e.g., motor vehicles, airplanes, missiles, debris from satellites and spacecraft)
3. Stored potential energy (e.g., stored petroleum and natural gas, water in dams and tanks, combustibles)

The most significant sources of risk, Rowe and Broder argue, are systems that contain combinations of these conditions (such as an overheating, fully pressurized nuclear reactor).

For a technology, product, process, or system to be a health, safety, or environmental hazard, it must contain risk agents in sufficient quantities or intensities to be of concern. Furthermore, any natural or engineered systems that contain or limit the dangers must be less than completely effective. Because safety systems are rarely 100 percent reliable, technological systems that involve significant amounts of toxic substances or high levels of energy are generally regarded as hazards.

Exposures

The fact that a hazard exists does not guarantee that it will produce risk. For example, solar prominences — the great arches of glowing gases that erupt from the sun — create little if any risk on the earth. Some mechanism must exist by which people or the things they value may be exposed to the hazard or its risk agents. Seven mechanisms have been identified by which exposures may occur (National Research Council 1982):

1. *Self-hazardous behavior*: Situations in which individuals voluntarily expose themselves to a hazard (e.g., smoking, alcohol abuse, not wearing seatbelts, hang gliding)
2. *Hazardous behavior*: actions by individuals that expose other individuals to hazards (e.g., crime, speeding, drunken driving, child abuse, smoking in public places)
3. *Cogeneration*: Cases where the combined voluntary actions of two or more parties expose one of them to a hazard (so that an incentive exists for the nonrisk-bearing party not to fully disclose information to the risk-bearing party; e.g., risks imposed on workers by employers and risks imposed on consumers through their purchase of products supplied by producers)
4. *Production externalities:* Exposures to hazards produced or released to the environment as unwanted by-products of the production or consumption of goods (e.g., air pollution, water pollution, nuclear accidents, hazardous wastes)
5. *Natural processes:* Exposures to hazards produced by nature (e.g., earthquakes, droughts, floods, diseases, genetic mutations)
6. *Economic processes:* Exposures incurred as a result of economic conditions (e.g., poverty-induced disease and unemployment-induced stress)
7. *Government policies:* Exposures resulting from the actions of government (e.g., nuclear war, contamination of the earth from collecting and returning soil samples from other planets)

As implied by several of the mechanisms identified above, exposures often occur as an unavoidable consequence of the existence of a hazard. If, for example, the hazard is a product or a technology, then its distribution and use will bring people and the things they value into contact with it. If the hazard is a process or a system that releases risk agents into the environment, then several natural pathways will exist that may transport the risk agent to the physical proximity of humans. For example, exposures may occur through air and water transport (as is the case with many industrial effluents) or transport through the food chain (such as occurs with mercury).

Effects

Exposure to a hazard will produce risk provided that the exposure has the potential of producing adverse consequences. These consequences can include early fatalities and injuries, latent cancer fatalities, genetic effects, environment degradation, and economic losses. The magnitude or severity of such effects depends on the conditions of exposure and characteristics of the people and items exposed. In some cases, exposures to significant hazards may result in no adverse effects. Safety equipment, protective clothing, antibiotics, or vaccines may be quite effective in limiting or mitigating the effects of exposures to hazardous substances or situations. In the case of toxic chemicals, the health status of the exposed individuals can have a significant impact on the effects produced. Particularly sensitive groups might include pregnant women, very young and very old people, and persons with impaired health. Some differences in health effects can be attributed to differences in "effective" rather than "administered" dosages. If dose is interpreted as what is adsorbed or absorbed at a target organ, then different individuals might experience different doses from the same exposure, as a result of differences in physiology, personal habits, and so forth. This is of particular importance, for example, in determining the health effects of cumulative toxicants (such as lead) for which several exposure pathways may exist (e.g., breathing air, eating food, and drinking water).

Similarly, the specifics of the effects processes are critical to determining the risks to materials, plants, and animals resulting from exposure. For example, coatings such as paint often dramatically reduce the damage to materials caused by environmental contaminants. The environmental damage caused by acid rain depends on the existing *p*H of lakes and soil. Whether domestic livestock will succumb to a contagious disease may depend on the general health of the animals and whether their feed contains antibiotics to improve their resistance to the disease.

The Role of Government in Risk Management

The responsibilities and limits of government are important considerations for placing risk regulation into perspective. Discussed below are the basic economic and social arguments for government intervention to control risk and roles that government might adopt in risk regulation.

Arguments for Regulation

Economists who argue for regulation tend to focus on externalities — situations in which free markets do not produce a desired level of health and safety because the prices consumers face do not reflect true social values. According to economic theory, an individual whose initial desire for a commodity exceeds its price will continue to purchase the commodity until the benefit derived from the last amount purchased equals the price paid for that amount. An externality exists if either the production or consumption of the product produces costs or benefits to others that are not reflected in the prices consumers face.

Lave (1972) identifies several types of externalities involving risks that create problems for a free market. One consists of accidents that injure people other than the person in control. The decisions made by an individual who is driving an automobile, for example, concern all passengers. If the driver ignores the preferences of his passengers (for reaching their destination quickly and safely), he may "purchase" too little (or too much) safety. Another sort of externality revolves around the financial support given a disabled individual or the dependents of someone who is killed. The cost to society of assuming these financial burdens can be extremely large. A third sort of externality is the personal loss faced by family, friends, employer, and community when someone dies.

Lave points out that the market may also produce an insufficient level of safety as a result of imperfections in the marketing of products. For example, because of economics of scale, the cost of mass inoculation for a disease may be only pennies per inoculation. Although almost everyone would agree to mass inoculation, few people may be willing to bear the much larger costs of private inoculations. Major capital purchases can also present a problem for the free market. Once an automobile or house with insufficient safety is purchased, that item is likely to continue to have insufficient safety throughout its life, because the unrepresented preferences of second-hand purchasers may not be well reflected in the market and because the costs to upgrade the safety of such purchases may greatly exceed the costs of building in more safety in the first place. Finally, many safety decisions involve what economists refer to as "public goods." A public good is something that collectively affects individuals. Unlike an ordinary good, which may be appropriated by an individual for his own personal use, one person's satisfaction from a public good is not diminished by the satisfaction gained by others. For example, a dam for flood protection is a public good because it benefits the entire community; no individual can gain the protection on his own without freely extending it to everyone else. Many risk agents are public goods (perhaps "public bads" would be a better descriptor) because they simultaneously affect more than a single individual. Lack of consensus and other difficulties associated with organizing those who might benefit from establishing or eliminating public goods may prevent such actions from being taken through normal market processes.

Externalities and market imperfections, economists argue, often result in a level of production and consumption of goods that improperly balances risks, costs, and benefits. When externalities exist, therefore, the government may be justified in intervening to force a level of safety that is more socially desirable than the inappropriate one reached through the market. Thus failures of the marketplace provide an argument for the need for government regulation.

A more pragmatic answer to the question "When does government have a right to regulate?" is "Whenever public pressures build to make it." According to Lowrance (1976), "The demand for regulation seems to be a crying out at the futility of dealing with the multiplicity of hazards confronting us these days. People call for regulation almost reflexively, as though it were either a panacea or the last resort." Consumers tend to demand government action in situations where individual action is not possible or is very costly and where existing institutions do not appear to be lowering the risk or protecting them against the loss as inexpensively as they believe it can be done. Empirically, risk regulation tends to be associated with trends showing an increase in the incidence and magnitude of adverse effects on particular interests from technological activities or developments, growing public dissatisfaction with industry responses, and loss of confidence in the corrective efforts of the pre-regulatory mechanisms for risk management (Rowe and Broder 1979). In some cases, the impetus for regulatory solutions has come from the industrial sector, as a result of concern about tort liability or other economic consequences, technical uncertainties, lack of insurance at reasonable rates, or the vagaries of state and local laws. Some of the support for increased regulation undoubtedly reflects a general belief that an increasingly wealthy society should trade off more private goods for improved public health and safety.

Paternalism sometimes seems to be a force behind the call for regulation. Hirshleifer *et al.* (1974) note that "some social critics object to . . . the idea that individuals are capable of making their own decisions effectively." Goodin (1980), for example, argues for paternalism over individual risk decisions because, among other things, individuals are incapable of imagining how badly they would feel if the risk became reality. Historically, American society has had a strong preference for individual freedom over paternalism. Nevertheless, a variety of existing laws relating to "victimless" crimes are designed to prevent individuals from voluntarily endangering themselves, especially the young and others who are viewed as not fully capable of weighing the consequences of their actions. Regulation is used to protect those especially vulnerable; for example, stiff fire codes have been established for nursing homes. The government also tends to intervene when appraisal and control of risks require technical expertise that few people possess (Lowrance 1976).

Even though good reasons may exist for considering government intervention, the question of whether or not government should regulate is, as Morgan (1981) observes, ultimately an ethical issue. The difficulty is that regulation generally results in an involuntary incremental cost being imposed on some citizens. If this cost is small compared with perceived benefits of making people a little safer, then most people would probably be in favor of government action. This is essentially a utilitarian philosophy. Other perspectives, however, disagree with this point of view. For example, the libertarian philosophy, which places greater emphasis on individual rights, takes the position that it is wrong for society to take any action that makes some people worse off regardless of whether total social welfare is in some sense advanced. The important point is that the decision to rely on government regulation, as opposed to free market mechanisms (such as contract liability and insurance), is an ethical decision about which reasonable individuals might well disagree.

Regardless of philosophical debates, government regulation of risk is likely to be with us a while longer. Governmental authority and responsibility to intervene to reduce risk have been firmly institutionalized in the form of a complex system of regulatory agencies and processes which are accepted by the vast majority of the public as legitimate and proper means for promoting a safer and more equitable society. Despite the slow and laborious process by which government risk decisions are made, the inevitable challenges in the courts, the lack of clear evidence that risks are in fact being reduced, and the campaign promises of popularly elected officials to reduce and eliminate regulatory agencies, the regulatory effort continues to grow. The relevant question, therefore, is how to enable government to regulate risk more effectively and efficiently.

Government Roles and Alternatives for Risk Regulation

One way to explore the available options for risk regulation is to identify opportunities for altering each element of the risk chain. As noted above, for a risk to exist there must be a hazard, together with a process by which people, animals, plants, or materials may be exposed to the hazard, and a process by which exposure produces adverse effects. Furthermore, if people are concerned about risk they must judge the severity, importance, or inequity of those consequences to be significant. Thus, for example, for the risk of health effects from air pollution to be a concern, a source of air pollutants must exist, humans must have contact with the pollutants, the exposure must produce health consequences, and a value judgment must be made that those consequences are undesirable. Generally speaking, it is possible to reduce a risk by modifying any one of these essential components. In the case of air pollution, the emission may be reduced; the emissions may be prevented from reaching inhabited areas, or people may be prevented from entering the areas where pollution is the highest; people exposed can avoid strenuous activity so as to reduce the adverse effects of exposure; or those exposed can be otherwise compensated to offset the displeasure they feel for having to suffer the health effects.

Another way to explore options for reducing risk is to consider the mechanisms that are available for altering human behavior. Because people are intrinsic to the various components of the risk chain, risks may be modified by using any number of ways of motivating different behavior. Behavior may be modified directly by encouraging individuals to select existing safe alternatives, or indirectly by encouraging the generation of new alternatives that may prove to be safer and more attractive. The Clean Air Act Amendments of 1970 have been cited as an example of a government action designed to produce new technological alternatives. Lave

(1980) notes that some Congressmen who voted for passage of the act remarked that they knew that the emission requirements placed on the automobile were beyond the capability of existing technology. The standards were judged to be of practical value because they would force industry to be innovative in developing new technology.

The most direct way for government to change human behavior is through the implementation of mandatory standards and regulations designed to force individuals to take actions that lessen risk. For example, to reduce injury from automobile accidents, government could pass laws requiring the use of seatbelts. Alternatively, government could require the installation of automatic restraint systems (such as airbags) in all new cars. This illustrates that when risks are cogenerated, government can focus risk-reduction actions on either party to the cogeneration process.[4]

Another approach is to motivate human behavioral changes through incentives. For example, when risks are associated with production externalities, government can attempt to cause producers to bear the social costs of externalities so that the prices consumers pay will reflect such costs. For example, although this is not popular in the United States, Germany has regulated some forms of pollution through the sale of pollution licenses and effluent fees. Other approaches to internalizing the social costs of risks into the prices people pay include legal liability and compulsory insurance. Liability influences the incentives of people and firms to engage in risk-generating activities. Insurance converts uncertain but potentially catastrophic financial losses into relatively small and certain financial payments.

A third approach to modifying behavior is to provide information. The government can provide warnings, such as the label required for cigarette packages reminding the smoker that his behavior is hazardous to his health. It can also provide better data on the risk itself (e.g., possible levels of loss or their probabilities) and ways of protecting oneself (e.g., descriptions of available insurance). If people contribute to or engage in risky behavior because they are not fully aware of their options or of the consequences of what they do, then providing information can have a significant impact.

To illustrate the many options available, Table 1 provides examples of the various possibilities for applying different risk management strategies to different elements of the risk chain. Rows in the table correspond to three distinct processes for generating risks — self-hazardous behavior, cogenerated risks, and externalities — and three different strategies for dealing with each — mandatory requirements, incentives, and information dissemination. The four right-hand columns of the table correspond to the various elements of the risk chain — the source, the exposure process, and the valuation process. The entries in these columns illustrate intervention actions based on the different strategies and directed at different elements

Table 1. Alternative risk-management strategies and examples

		Examples targeted at modifying various links in the risk chain			
Generation process	Risk-management strategy	Modify risk source	Modify exposure process	Modify effects process	Modify consequent utility
Self-hazardous behavior Smoking Alcohol abuse Mountain climbing Skin diving Etc.	Mandatory requirements Legal restrictions Product bans Automatic protection strategies	Ban saccharin	Establish minimum age for consumption of alcohol	Require motorcyclists to wear helmets	Require personal health insurance
	Incentives Insurance premium adjustments Taxing hazardous behavior	Provide public transportation on New Year's Eve (to prevent drunken driving)	Offer reduced-cost health insurance to nonsmokers	Reduce registration fees for cars with passive restraint systems	Offer subsidized insurance for inhabitants of flood plains
	Information dissemination Educational campaigns	Publication of safe operating instructions for hazardous products	Post hazard warnings at national parks	Sponsor advertisements designed to increase seatbelt use	Provide counseling services for the terminally ill
Cogenerated risks Risks to employees from employment Risks to consumers from product purchases	Mandatory requirements occupational safety regulations Product safety standards Regulation of advertising Product bans	Prohibit use of certain pesticides in agriculture	Establish maximum daily radiation dose for workers	Require public restaurants to provide diners with water (in order to reduce dangers of choking)	Enforce product safety liability/compensation laws
	Incentives Wage premiums for risky jobs Liability rules Taxing occupational risks	Provide tax incentives for replacing unsafe equipment	Tax hazardous consumer products to reduce consumption	Provide citrus drinks for government employees working in conditions promoting dehydration	Offer supplementary worker insurance programs
	Information dissemination Product labels Warnings	Provide manufacturers with product safety information	Require safety warnings on consumer products (e.g., step ladders)	Require antidote instructions on poisonous products	Provide information on employment opportunities for the handicapped
Externalities Air, water pollution Radiation Toxic chemicals Noise pollution Nuclear war Auto accidents Etc.	Mandatory requirements Standards Regulations	Establish new source performance standards for emissions of air pollutants	Require nuclear power plant operators to develop community evacuation plans	Require auto makers to install collision-resistant bumpers	Provide emergency relief (e.g., housing) for victims of toxic chemical spills
	Incentives Pollution licenses Effluent fees Insurance Compensation	Tax polluters in proportion to quantity of pollutant emitted	Subsidize mass transit to attract drivers off highways	Offer inexpensive air filtration masks to urban commuters	Offer government-subsidized liability insurance to manufacturers of potentially toxic substances
	Information dissemination Government research Warnings	Government-sponsored R&D on air pollution control technology	Post public warnings at hazardous sites	Report pollution index to warn citizens to reduce activity on high-pollution days	Provide advice and information for horticulture in acid rain areas

of the risk chain. An important observation from Table 1 is that a wide variety of options exists for reducing any particular risk.

Complexity of Social Risk Decision Making

Selecting an appropriate alternative for government risk management is extremely difficult because of the complexity of deriving estimates of risk levels and determining the acceptabilities of these levels. Estimating risk levels is complicated by limited technical and scientific knowledge. Establishing or justifying acceptability is complicated by faulty perceptions and personal value systems (individuals often perfer statistically higher risks) and by the need (in a nonauthoritarian society) to justify decisions with which not all members agree by appeal to some principle of social consent. These three sources of complexity — limited knowledge, disfunctions in human perception of risks, and social consent — are discussed below.

Limited Knowledge

Because of limited knowledge, assessing the risks associated with proposed actions is beset with complications. Each link of the risk chain creates difficulties for decision making: hazards are often hard to identify and characterize, exposures are frequently difficult to estimate, and the damage caused by those exposures is generally difficult to assess. In addition, the impact of proposed actions on any of these links is hard to predict. Finally, as addressed in the next section, a host of problems relating to risk perception creates difficulty for valuing risks.

Hazards

One reason hazards are hard to identify stems from the difficulty of obtaining empirical data which conclusively link adverse effects to products, processes, or technologies. Chance observations of increased disease or injury may lead to a search for causal factors, but an absence of historical exposure data often makes the search difficult. Furthermore, latency (time lag) between initial exposure and appearance of harm means that indications of adverse consequences may not be apparent until the potential harm is great and irreversible. For example, statistically significant increases in lung cancer for asbestos workers were not observed until many years after the workers were first exposed.

Since ethical considerations generally prevent deliberate human experimentation to identify potentially dangerous substances, investigators must rely on epidemiological studies involving uncontrolled human exposures and animal studies. Neither epidemiological studies nor animal bioassays are very powerful, however, unless they are tested hypotheses about specific effects. Thus, the process often requires hypothesizing one of many possible effects and then hunting for it.

Animal studies are inevitably a controversial means for identifying hazardous substances because of the inherent differences in susceptibility among animal species. Exposures to the same substance may produce a significant adverse effect in one species and no apparent effect in another. For example, a chemical may produce cancer in mice because a metabolite (an intermediate product resulting from the application of the animal's metabolic processes) is a carcinogen, not the administered chemical. Because of differences in metabolic processes, the same effect may or may not show up in man. Although some 1,500 substances have reportedly been shown to be carcinogenic in animal tests, fewer than 30 have been definitely linked with cancer in humans (Tomatis *et al.* 1978).

Epidemiological studies are not only time consuming and expensive, they are also a controversial means of hazard identification. A positive result can be due to a statistical fluke or a failure to control for confounding factors. The population selected for an epidemiological study is likely to be exposed to a range of intervening, potentially causal factors, and it is frequently not possible to isolate the effects of any one of them. For example, the imposition of the 55-mile-per-hour speed limit complicates an assessment of the effectiveness of motor vehicle safety standards implemented during the past decade, and the movement of people from place to place obscures the long-term effects of local air pollution on chronic lung disease. Even in those situations where there is strong evidence that a hazard exists, its characteristics (potency, intensity, probability of releasing toxic substances) may be very difficult to determine. This is especially true for the extreme cases that are often of greatest concern; namely, massive harm resulting from catastrophes and continuous low-level exposures leading to chronic diseases. In the first instance, because there are few or no actual occurrences of the feared catastrophic accident, its probability and consequences cannot be obtained from historical data. In the second, it is virtually impossible to measure directly the magnitude of the effect because the levels are generally masked by higher background levels of ongoing, natural incidences of the effects.

Exposures

Gauging exposure is usually difficult because of the complexity of exposure processes and because information is often incomplete. Only rarely is there adequate information about who is exposed and the degree of their exposure. Large chemical manufacturers may try to keep records on workers exposed to potential hazards, but data on the level of exposure to particular chemicals usually are not precise. The situation with respect to the general population is even worse, because of the lack of effective monitoring mechanisms. Mean contamination levels at specific sites can be monitored; but individuals move from location to location, and contamination levels generally vary dramatically over relatively short distances. Even if accurate exposure data are available, those data have direct applicability only for the assessment of past risks or future risks assuming that exposure levels do not change. Since contemplated actions almost always alter exposure patterns, existing exposure data have only limited applicability for comparative analyses.

Exposure estimation is complicated by the many different "pathways" by which a risk agent might achieve proximity to humans or the things they value. Figure 3, for example, illustrates exposure pathways for a hazardous waste disposal site. If the site fails to contain its toxic chemicals, pollutants may contaminate the air, surface water and sediment, soil, groundwater, and the food chain. Any or all of these pathways may produce significant exposure to humans or the things they value.

If exposure is through the air or water, then the level of risk depends on the fate and transport of risk agents. In many cases, however, the mode of transport, the length of time involved, the distance the substance might travel, and the transformations that it undergoes as it moves are all unknowns. For example, acid precipitation is thought to arise when sulfur and nitrogen oxides emitted during industrial processes and fuel combustion combine with water in the atmosphere. Currently, there is little understanding of how these compounds move through the atmosphere or how they react chemically as they do. Such unknowns can be extremely important in predicting impacts. In the case of air pollution, for example,

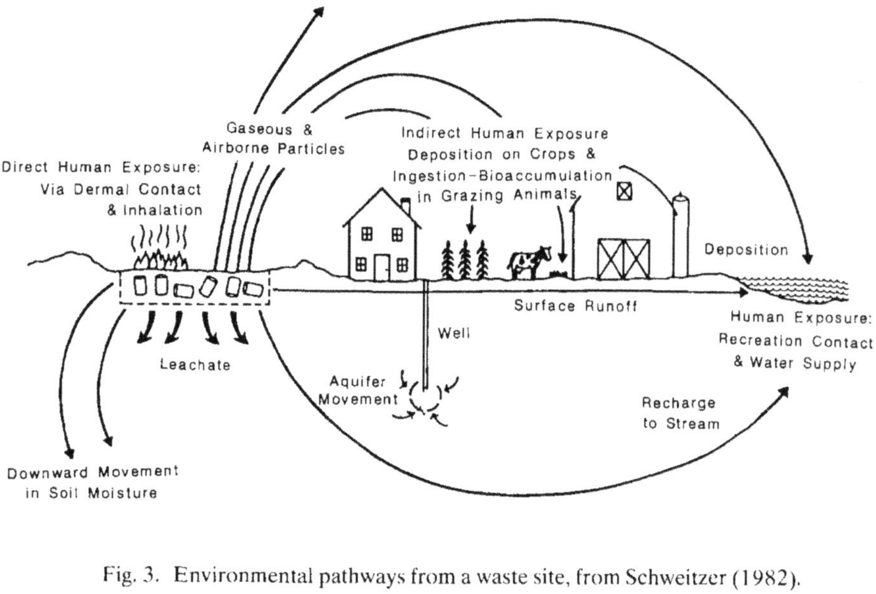

Fig. 3. Environmental pathways from a waste site, from Schweitzer (1982).

order-of-magnitude changes in one-hour average ambient concentrations at a given location are common due to changes in atmospheric conditions.

Even in the simpler cases in which the hazard is associated with consumer products, differences in personal habits make exposure difficult to assess. For example, the amount of a risk agent present in food may be measurable, but differences in food storage practices, food preparation, and dietary habits lead to wide variations in the amount of the agent that individuals ingest. To estimate the risks associated with the use of a pesticide applied to citrus fruit, for example, it would be necessary to know not only how many oranges are consumed by a typical person, but also how many people are fond of eating large numbers of oranges, because the latter group would be at greatest risk. If the risk agent is absorbed when a consumer product is used, patterns of use affect exposure. A solvent whose vapor is potentially toxic, for example, may be used outdoors or in a small, poorly ventilated garage.

Effects

Assessing the effects of exposure to a risk source is complicated because the analytic relationship between dose and response is inescapably tied to the mechanisms by which exposure causes harm, and these mechanisms are rarely understood. Harm to human health may include injury, illness, death, and genetic effects. Dose-response data are ordinarily available only for relatively high exposure levels. Extrapolation of these data to more frequently occurring low levels generally requires theoretical models. The simplest model would be linear, with the magnitude of the effect depending on the total accumulated exposure. In cases where evidence is available, however, dose-response curves usually show substantial nonlinearities. Sometimes exposure have thresholds below which no effects are observed. In the case of living organisms, for example, this may result because the organisms are able to completely metabolize low levels of some materials, converting them into less harmal materials or excreting them. Furthermore, exposures to a given hazardous substance often produce a response that depends critically on synergistic or antagonistic cofactors. For example, human response to a specific air pollutant is strongly affected by the other air pollutants present, activity level (e.g., exercise rate), the distribution of exposure with respect to time (duration, cumulative exposure, and latency periods), and other environmental conditions (e.g., temperature and humidity).

Extrapolating dose-response relationships from animals to humans is highly uncertain. Differences in size and metabolic rates between man and laboratory animals require that doses used experimentally be converted to infer equivalent human doses. Even under equivalent doses, however, the inherent susceptibility across different species can vary dramatically. For example, the response of different species to equivalent doses of a specific carcinogen may vary by a factor of 100,000 (Gori 1980). Moreover, because the human population is in many respects highly variable, the susceptibility of individuals exposed to harmful substances varies tremendously. Human susceptibility to a carcinogen, for example, may vary by a hundred fold or more among individuals, depending on sex, age, hormonal balance, health, general genetic predisposition, and so forth (Leape 1980).

Estimating the response of ecological systems is often even more difficult than estimating individual human health responses. The state of an ecological system is a delicate balance derived from the interaction of many factors; altering one of them can produce a sequence of changes and an ultimate impact that is much greater than might be expected. Because of the complexity of each ecosystem, estimating its response to a given exposure level is extremely difficult. In the case of acid rain, for example, the rate and extent to which lakes might become acidified or forests damaged through continuation of current rates of acid precipitation is a major source of uncertainty in assessing the significance of this problem.

Impact of regulatory actions

In situations where the relationship between exposure and harm is hard to establish, it will be difficult to project the degree to which a regulatory action might lessen the harm. Furthermore, the effectiveness with which a decision is implemented must be considered. Although analyses generally assume that there will be 100 percent compliance with a regulation, the actual extent may be much less than complete, depending on the nature of the hazard, the cost of compliance, the ease of enforcement, the availability of agency resources, and the degree of cooperation of the regulated parties. Compliance is likely to be a problem in situations where the benefits from an action accrue largely to a single party but the costs are mostly borne by the community at large; for example, disconnecting automobile emission controls for the purpose of improving acceleration. Such situations require an enforcement mechanism to prevent private decisions from countervailing the public interest.

Risk Perception

Critical to determining social response to risk is how individuals perceive the risk of potentially hazardous situations. Psychological research shows that individuals typically over- or underestimate the probabilities associated with certain dangers compared with their statistical frequencies Slovic *et al.* (1983), for example, note that individuals typically believe familiar hazards, such as home appliances and cigarettes, are considerably less dangerous that historical data indicate. Furthermore, people's concern over risks depends on much more than just the perceived likelihood of fatalities.

Risk perception, according to behavioral psychologists, is "an idiosyncratic process of interpretation which involves a subjective probability judgment about the occurrence of an unpleasant event and an interpretation by the individual that reflects how he or she defines and feels about the outcome" (Thomas 1981). The basic research method consists of presenting subjects with lists of hazards. The subjects' task is to scale the hazards in order of severity or to evaluate each hazard according to some set of dimensions. The results point to the importance of external events such as past experiences, as well as internal dynamics such as discussions with others, in determining people's perceptions of risk. As Otway (1980) observes, risk perception "depends upon the information people have been exposed to, what information they have chosen to believe, the social experiences to which they have had access, the dynamics of stakeholder groups, the legitimacy of institutions, the vagaries of the political process, and the historical moment in which it is all happening."

The empirical basis assembled to date is insufficient to support convincingly any firm conclusions about either the specific factors influencing risk perception or the nature of the effects. However, recent research suggests certain plausible relationships. Slovic *et al.* (1981) have found that ratings of risk closely relate to feelings of dread and the likelihood of the mishap being fatal. Other characteristics found to influence risk perception were the degree to which risks are voluntary, controllable, known to science, known to those exposed, familiar, catastrophic, and immediately manifested. They found that judged characteristics generally correlate; for example, risks faced voluntarily were typically judged well known and controllable. Rowe and Broder (1979) explain this effect by suggesting that familiarity through experience (as in the case of relatively frequent events such as storms and earthquakes) leads to the perception of a degree of control over one's survival. For more unfamiliar events (such as the re-entry of Sky-Lab or a nuclear meltdown), there is little experience, and the result is a perception of a lack of control. Similarily, where it is perceived that society is controlling disasters to some degree (e.g., early warning systems for hurricanes coupled with storm shelters and evacuation plans), public anxiety is assuaged. The influence of such characteristics on perception has been found to be much stronger for lay people than for experts, with experts' perceptions lying closer to statistical or theoretical expectations (Fischhoff *et al.* 1981a).

Table 2, based on Otway and von Winterfeldt (1982) and Covello (1985), summarizes the factors that psychological studies suggest have some bearing on risk perception. Although most of the studies on which these findings were based involved small, highly specialized and unrepresentative groups, the results serve to demonstrate that describing risk in terms of a simple probability distribution over adverse consequences may miss many of the factors that people intuitively feel to be relevant.

Table 2. Associations observed in studies of public risk perception

Factor	Conditions associated with increased public concern	Conditions associated with decreased public concern
Severity of consequences	Large numbers of fatalities or injuries per event	Small numbers of fatalities or injuries per event
Probability of occurrence	High probability of adverse consequences	Low probability of adverse consequences
Catastrophic potential	Fatalities or injuries grouped in time or space	Fatalities distributed randomly in time and space
Reversibility	Irreversible consequences	Consequences appear reversible
Delayed effects	Somatic effects that are delayed in time (e.g., cancer)	All effects immediately realized
Impact on future generations	Risk posed to future generations	All risks borne by current generation
Impact on children	Children specifically at risk	Risks threaten adults only
Victim identity	Identifiable victim (e.g., yachtsman lost at sea)	Statistical victims
Familiarity	Unfamiliar risks (e.g., ozone depletion due to emissions of fluorocarbons)	Familiar risks (e.g., household accidents)
Understanding	Lack of personal understanding of mechanisms or processes involved	Mechanisms or processes involved personally understood
Scientific uncertainty	Risks appear to be unclear or uncertain to scientists (e.g., when scientists disagree)	Risks relatively known to science (e.g., actuarial data on automobile accidents)
Dread	Risk evokes fear, terror, or anxiety (e.g., toxic waste dumps)	Risk not dreaded
Voluntariness	Involuntary exposures (e.g., air pollution)	Risks taken at one's own choice (e.g., skiing)
Controllability	Little personal control over risk (e.g., travelling as a passenger in an airplane)	Some personal control perceived (e.g., driving an automobile)
Clarity of benefits	Benefit or need for activity generating risk is questioned (e.g., nuclear power)	Clear benefits
Equity	Those at risk from an activity do not directly gain its benefits	Distribution of risks and benefits appears equitable
Institutional trust	Lack of trust in institutions responsible for risk management (e.g., the Nuclear Regulatory Commission with its perceived ties to industry)	Responsible institutions well trusted (e.g., management of recombinant DNA by universities and the National Institutes of Health)
Personal stake	Individual personally at risk	Individual not personally at risk
Attributability	Risk caused by human failure	Risk caused by nature
Media attention	Much media coverage (e.g., airline crashes)	Little media attention (e.g., on-the-job injuries)

Adapted from Otway and von Winterfeldt (1982) and Covello (1985)

Social Consent

According to the American Declaration of Independence, "... Governments are instituted among Men, deriving their just powers from the consent of the governed." Thus, while individual freedom is a central tenet of our society, freedom is acknowledged to be constrained by an obligation to others, and this social obligation is derived from consent. Individual differences prevent unanimous consent, and few would argue that everyone must agree to every social action. Nevertheless, when government acting as the agent of society attempts to reduce risk by placing constraints on freedom, legitimization demands the availability of ideological arguments that demonstrate that society consents to these constraints.

Douglas and Wildavsky (1982) introduce social consent as a factor contributing to the complexity of social decision making by describing a four-way classification scheme for categorizing risk problems. Both the nature of the risk problem and the appropriate solution depend on whether knowledge is certain or uncertain and on whether consent is contested or complete:

1. In the first case, knowledge is certain and consent complete. This is a situation in which objectives are agreed on and all alternatives and their consequences are well understood. Here the problem is technical, and the solution is one of calculation.
2. In the second case, knowledge may be certain, but consent is contested. Here, the problem is one of disagreement about how to value consequences, and the solution is either more discussion or more coercion.
3. In the third case, where consent is hampered only by uncertain knowledge, the problem is essentially one of insufficient information; hence, the solution is seen as further research.
4. In the last situation knowledge is uncertain and consent is contested. This is not only the most difficult, but also the situation that most commonly is associated with risk. It does not, according to Douglas and Wildavsky, have any obvious solution.

MacLean (1982a) describes three models for justifying social decisions in situations where the explicit consent of those affected cannot be obtained. The first involves justification through an appeal to a concept of implicit consent. This model argues that centralized decisions are allowable as long as the tradeoffs between safety and other benefits imposed by the decisions resemble the tradeoffs that individuals would have made on their own if the situation permitted it. Under these conditions it might then be claimed that individuals implicitly consent to the imposed decisions. An important example would be to select a level of safety for a social decision that is identical to the level that people would select on their own were it not for externalities. Thus, a decision-making mechanism that arrives at social decisions from the preferences for risk and safety revealed by the decisions of individuals in the marketplace might be justified, not so much because market solutions are efficient, but because in a properly functioning market people are assumed to consent to their transactions. MacLean does, however, point out two important difficulties with this model. First, rights and liabilities must be assigned so as to distinguish compensation from blackmail. Second, preferences and values differ: preferences are claimed to reflect what an individual wants, whereas values explain what kind of person one aspires to be. Market prices may not necessarily reflect social values.

MacLean's second model, hypothetical consent, is based on a division of decision making into two stages — establishing agreement on the principles and procedures for making decisions — followed by applications to specific cases. If people consent to the process by which social decisions are reached, then it might be claimed that they accept the decisions that result. For example, a hypothetical consent theorist might argue for principles that ideally rational people would consent to and then deduce further principles for designing "fair" decision-making institutions. In this situation, the consent would be hypothetical because a normative argument would be used for justifying the institutions and principles. The hypothetical consent model provides a comfortable basis for sanctioning fromal decision-aiding approaches based on concepts of rationality or fairness. However, the model requires the legitimization of the acceptable decision-making principles and procedures. In MacLean's view, it seems reasonable to insist that hypothesized consent to procedures be unanimous.

MacLean's last model, nonconsent, is based on the proposition that political and social decisions may be justified by arguments that do not appeal to the consent of the parties affected. This model maintains that some decisions might be legitimitized by a direct appeal to the social values that we seek to secure, even though this might entail endorsing and ranking the values through philosophical arguments. Thus, for example, we may need to interfere with liberty to promote our interest in equality, security, or posterity, because at this particular time we view one value as having more weight or urgency. Support of the nonconsent model requires acknowledging that not all individual preferences need to be given equal weight in the social decision-making process. For example, appeal might be made to arguments of nonconsent to justify locating a hazardous-waste facility in a particular community against the wishes of the local citizens.

As will become clearer in Chapter 2, the decision-aiding approaches appeal for their justification with varying degrees to the different models of consent. The controversy associated with MacLean's three models illustrates that the complexity of social decisions associated with lack of knowledge and disagreements in risk perception is compounded by the difficulty of establishing social consent.

The Existing Risk-Management System

Consideration of the relevant institutional structure and processes for risk management is important for evaluating and selecting a formal decision-aiding approach. This subsection describes government institutional structure for risk management and the coping strategies that have evolved, including the extent to which quantitative approaches may be used.

Participants

The most important institutional structures for managing risk are tort and common law, particularly those laws related to negligence, liability, nuisance, and trespass; insurance and compensatory plans, including self-insurance pools, true self-insurance, workers' compensation, bonding and escrow, and restoration funds (which might be offered either by private companies, through joint private/government arrangements, or directly by government programs); voluntary standards-setting organizations (e.g., the Underwriters' Laboratory, the National Fire Protection Association, the American National Standards Institute, the American Society for Testing and Materials); and mandatory government standards or regulations. Government can also manage risk to an extent through the dissemination of information and publicity, government procurement, and issuing advanced notice of an intent to regulate. In view of this study's focus on government decision making, a more detailed discussion of the principal participants in the government risk-management system is presented in the following sections.

Congress

Congress enacts the statutory laws for promoting national environmental and health goals and creates the agencies that produce the regulations and standards for achieving these goals. Furthermore, through the laws it passes, Congress permits the courts to regulate conduct by judicial imposition of liability and injunction. Congress also sometimes enacts legislation that directly establishes standards of conduct or prohibits certain activities. It can dictate the factors to be included in and excluded from regulatory decision making (e.g., requiring the banning of any food additive causing cancer in animals), and it can pass special legislation to preempt agency discretion (as it did in acting to prevent the removal of saccharin from the market).[5]

The politics of elective office has a strong influence on congressional decision making. The qualities needed to win an election are not necessarily those required to govern effectively. The vast majority of Congressmen are heavily trained in law and business, with very little formal training in science and technology. Most are white males in their late forties or early fifties. Contenders for political office offer competing ideas and programs; those who are "out" criticize the "ins," pose alternatives, and invite shifts of public allegiance. Congressmen must continually be conscious of the risk of defeat at the polls.

Congress is by no means autonomous in choosing its tasks and ends; it is more reactive in character than active. Issues with near-term significance dominate longer-term concerns. With "elections to win, wars to fight, dams to build, and hungry mouths to feed, it is hard for any politician . . . to concentrate funds and attention on a problem which is multidimensional and seemingly long-term in nature" (Bogen 1981). The seniority system and safe election districts combine to produce a regional and mature age bias in committee chairmen who play a major role in controlling and shaping legislation. The congressional decision to spend extensively for cancer research, for example, has been interpreted as a result of the personal value system of aging decision makers (Starr 1978).

Because of majority rule, the political process for congressional decision making is pursuit of consensus. Obtaining consensus is often difficult and requires compromise. Policy issues written into legislative and funding bills are extensively screened in committees, through staff analysis, hearings, and "markup" sessions. Noncontroversial bills readily establish consensus; but when issues are highly controversial and sharply drawn, the contest for votes is critical, and a single vote may make the difference between passage and defeat. In contests dealing with broad national issues and large public outlays, partisan politics becomes less important than coalition politics. Members of Congress vote party positions when they can, but constituency interests win out when the issue bears heavily on the member's political security (Roback 1972).

As a result of such political realities, the bills that win passage almost never explicity promote actions that would be offensive to a locally powerful political faction. Statutes therefore frequently contain glowing declarations of congressional purpose but evade critical issues through ambiguous wording. To express objectives for dealing with health and environmental risk, vague criteria such as "as low as practicable," "adequate margin of safety," "best practicable technology," and "best available control technology" are often used. Although such phrases create difficulties for regulatory agencies charged with implementing an unclear "congressional intent," they provide necessary flexibility to permit laws to evolve incrementally through the regulatory and legal process.

Although most members of Congress have a limited awareness of formal decision-aiding approaches, a few Congressmen are intensely interested in the subject. This interest is reflected in recent bills designed to promote greater use of formal quantitative analysis (discussed below) and the activities of several congressional committees, especially the House Committee on Science and Technology, which has held hearings on risk-benefit analysis and comparative risk assessment, and the House Committee on Interstate and Foreign Commerce, which has held hearings on the use of cost-benefit analysis. The contrast between the intense interest of a few members of Congress and committees and the general lack of understanding on the part of the broader body is not unusual. For a variety of reasons, including personal interest and training, a small number of congressional policy activists often dominate institutional policy making in a complex, technical area, and a committee's legislative activity can be largely accounted for by the initiative and energy of a few members (Kraft 1982a).

Regulatory agencies

Congress created the present regulatory agencies in the late 1960s and early 1970s. The major agencies have been charged by the legislature with broad national goals dealing with the protection of health and the environment. The Environmental Protection Agency (EPA) has responsibilities that include the development of standards for environmental quality and motor vehicle emissions, the approval of abatement plans for curtailing individual industry pollution emissions, emergency cleanup response, and the funding of waste water treatment plants. The Consumer Product Safety Commission (CPSC) sets product safety standards, and the Nuclear Regulatory Commission (NRC) licenses the construction and operation of civilian nuclear power plants and other uses of nuclear energy. The Department of Energy (DOE) has responsibilities for national energy policy, including fossil, nuclear, and renewable forms of energy and conservation. Within the Department of Labor the Occupational Safety and Health Administration (OSHA) sets and enforces workers' safety and health regulations to reduce work-related accidents and disease, while the Mine Safety and Health Administration has responsibility for safety and health in the nation's mines.

A variety of agencies within the Department of Transportation also have responsibilities for controlling risks. For example, the Materials Transportation Bureau issues regulations for transporting hazardous materials. The Federal Aviation Administration (FAA) operates the air-traffic control system and sets safety standards for aircraft and airports. The National Highway Traffic Safety Administration sets safety standards for automobiles, while the Federal Highway Administration has responsibilities for the safety of the highway system. The Food and Drug Administration (FDA), within the Department of Health and Human Services, controls the labeling and content of foods and drugs.

Other agencies concerned with health, safety, and environmental risks include the U.S. Coast Guard and the Federal Railroad Administration (both within the Department of Transportation); the Packers and Stockyards Administration, the Agricultural Marketing Service, and the Animal and Plant Health Inspection Service (Department of Agriculture); the Mining Enforcement and Safety Administration, the U.S. Fish and Wildlife Service, the U.S. Geological Survey, the Bureau of Land Management, and the Office of Surface Mining (all in the Department of Interior); the Drug Enforcement Administration (Department of Justice); the National Bureau of Standards and the National Oceanic and Atmospherics Administration (Department of Commerce); and the Army Corps of Engineers.[6]

As might be expected from the nature of the legislative process, regulatory legislation and regulating agencies have evolved largely as a reaction to public pressure; they are products of specific concerns dominant at the time of their genesis. As a result, agency responsibilities frequently overlap. For example, a host of federal agencies, including CPSC, EPA, FDA, and OSHA, share the task of regulating environmental carcinogens. In some cases a regulatory agency's ability to achieve results is limited by its having jurisdiction over only a part of the hazard causal chain (Fischhoff *et al.* 1978a). The incremental manner of dealing with health and safety concerns has even caused a few important risks to miss regulation altogether. Indoor air pollution is an example. No agency has a clear mandate to regulate indoor air, but this microenvironment is the most important exposure setting for many air pollutants that are otherwise subject to regulation.

Whereas the legislature has virtually unlimited power to consider and dispose of risk issues as it sees fit, the power of the administrative agencies is limited by the charter the legislature granted to them. Sabatier (1977) identifies a number of factors that constrain regulatory agencies. Most critical are the statutes, court decisions, and executive orders which determine the agency's legal resources and constraints. Important variables in this category include the nature and clarity of policy directives, the scope of agency authority, the ability of the agency to apply sanctions to modify behavior, and administrative structure. Of particular importance is the opportunity provided by an agency's originating statute for citizen participation. Most recent legislation offers explicit avenues for citizens or activist groups to sue to enforce legislative mandates.

The procedures used by Federal agencies when establishing regulations may be formal or informal, as distinguished in the Administrative Procedures Act.[7] Most agencies involved in risk management use relatively informal procedures in regulatory decisions. Such procedures generally provide for public notice of proposed and final rules (published, e.g., in the Federal Register) and an opportunity for affected parties to comment. The agency must respond to all substantive comments by outside groups in the preamble of its final rule. The entire process, including hearings, often takes several years. An example of a regulatory decision requiring formal procedures is the cancellation by the EPA of registrations for pesticides under the Federal Insecticide, Fungicide, and Rodenticide Act (FIFRA). Formal rule making involves a courtlike adjudicatory hearing at which competing claims are heard before an administrative law judge who issues a decision based on the oral and written evidence presented. This decision becomes the basis for the agency's final action.

Because of the open and adversarial character of regulatory proceedings, areas of scientific uncertainty generally tend to become highlighted and extensively debated in public. Administrative proceedings provide a favored forum for these discussions. Thus the adversarial character of the proceedings tends to encourage expert witnesses to take definite public positions on controversial issues, and the pressures to compile a detailed record promote considerable exploration of issues by the regulatory agencies (Brickman *et al.* 1985).

Within the boundaries defined by statutory provisions and court decisions, an agency's policy is influenced by the interplay of competing political pressures, organization policy, and institutional philosophy. Naturally, major decisions such as regulation setting and large resource allocations (budget, facilities, or manpower) are made at the highest levels of an agency (by agency administrators, department heads, and program managers), while operational decisions are usually made at the lower levels. The right to make policy decisions is restricted at lower levels through mechanisms such as budget approval, limitations on the amount of money that can be committed on any one decision, or the imposition of special review committee structures. Decisions at lower levels also tend to be prestructured by custom or by written procedures. Agency decision makers may act individually, but more often than not they involve their staff and advisors. Agencies also frequently make use of advisory committees composed of relatively stable memberships of individuals who are not regular employees of the agency. Some of these do little more than provide window-dressing for decisions made elsewhere, while others have a virtually determining voice.

The personal preferences and leadership abilities of agency officials obviously affect their discretionary decisions. Lave (1972) suggests that agencies that regulate risk naturally attract individuals most concerned about risk and then "socialize these individuals to become even more concerned." Kelman (1980b) claims that it is the ideology of the professionals on OSHA's staff that has played the most important role in determining the agency's regulatory direction. Sabatier (1977) cites the example of Clarence Klassen, long-time administrator of Illinois' state air and water pollution control programs, to illustrate the influence of personal views on decision making. Viewing the passage of the Environment Protection Act of 1970 as a threat to the cautious policies he had been pursuing, Klassen bitterly opposed more aggressive regulatory actions. After six months he was asked to retire so that he could be replaced by someone more attuned to the new mandate.

Notwithstanding organization policy and personal philosophies, political realities and limited resources restrict and temper agency actions. The funds available to regulatory agencies are subject to political control, and their actions are further restricted by the manpower recommendations of the Office of Management and Budget (OMB), by the oversight of Congress, and by the impact of congressional investigations. Although the administrator has ultimate authority for any single agency decision, the position is filled by presidential appointment and the administrator continues at the President's pleasure.[8] Regulatory actions must therefore be compatible with political reality if for no other reason than the political survival of the agency's officials. Dissatisfied constituency groups (regulated firms, consumers of regulated products, etc.) have the option of appealing directly to those officials or institutions that control the agency's legal or budgetary resources (e.g., Congress, the president, the federal courts). Illustrations would be the tobacco industry's successful effort to have Congress overturn the Federal Trade Commission's prohibition of cigarette advertising and the court suits brought by environmental groups against Atomic Energy Commission siting decisions (Sabatier 1977).

Aggressive enforcement of regulations and a resort to legal coercion are contingent on a credible threat to prosecute violators successfully. This requires a staff of sufficient size to monitor compliance and enough expertise to prove the agency's case in court in the face of counterclaims by the regulated industry and other parties. The technical complexity of the problems confronting an agency will itself be a factor limiting the stringency of its regulations. For example, the number of chemicals in the jurisdiction of Federal regulatory agencies is enormous. Of the more than 70,000 chemicals in commercial use, only about 7,000 have ever been tested for carcinogenicity (Office of Technology Assessment 1981). Most agencies do not have the resources to do a thorough analysis for priority setting, much less to analyze individually the many problem chemicals within their jurisdiction. To the extent permitted by their budgets, agencies must contract to external consultants for much of their analysis. In 1980 alone, EPA's Office of Pesticides and Toxic Substances spent over $10 million on outside contractors (Environmental Protection Agency 1980).

The relationship between regulatory agencies and business contains elements of both mutual antagonism and dependence. Litigation produces delay and creates enmity between the hazard manager and the technology sponsor. Yet regulatory agencies typically have strong ties to the industries they regulate. Government agencies often depend on the technical expertise of industry for information and analysis and generally recruit their staffs from the ranks of the regulated.

That American institutions are federal rather than unitary ensures that policies made at one level must frequently be implemented at others. Because regulatory decisions often must be administered either by regional offices that are headed by independent appointees or by autonomous state and local governments, an agency's decisions must be acceptable to a broad range of political interests and must anticipate practical problems of implementation. Without general congressional support and a degree of business and consumer cooperation, regulatory agencies can make little progress toward achieving their goals.

To date, there has been relatively little use by regulatory agencies of formal decision-aiding approaches. However, as a result of recent executive orders and legislative proposals (described below), agency interest and applications are increasing significantly. The extent and type of analysis used varies from agency to agency and office to office, reflecting the different types of policy problems dealt with, statutory differences, and differences in the internal characteristics of the agencies. Kraft (1982b) concluded from a survey of regulatory agencies that the following factors tend to influence agency use of quantitative methods for assessing and evaluating risks: internal capabilities (especially scientific and technical skills), the extent of familiarity with relevant analytic procedures, the degree to which such procedures are well established within the organization, the financial resources to support the work, and the attitudes of top policy makers toward the use of quantitative analysis. The EPA, NRC, and DOE are among the agencies that are most active in the development and application of quantitative approaches.

Courts

Adjudication by the courts is playing an increasingly important role in social risk management. By hearing charges of misrepresention, mislabeling, negligence, and fraud and by awarding damages the courts create strong incentives for behavioral change. When manufacturers know that they are liable for defects in their products, they are motivated to improve safety. The courts' increased role in adjudicating scientific and technological controversies at the same time places severe strains on the legal system (Markey 1980). Basic rules of adjudication, such as those for determining liability, have largely evolved out of a framework of two-party litigation. These procedures tend to break down in modern technology-related litigation because the complexity of issues often makes if difficult to determine precisely those who are affected, how they have been injured, and who is responsible.[9]

In the judicial review of regulatory decisions, the courts have evolved three important roles. Sometimes judicial review of agency activities is directed at forcing the agencies to meet statutory requirements. As indicated above, statutes often provide specific deadlines for action and liberal citizen suit authorization. A second role is clarifying and interpreting statutory language. Judical review is often invoked to resolve issues of statutory interpretation in cases where legislative guidance is incomplete or vague. Federal courts, for example, have frequently reviewed EPA's interpretations of statutory provisions, such as the definition for "adequately demonstrated" technology used in the Clean Air Act.[10] The courts have tended to defer to an agency's expertise when the agency presents a sound interpretation of an ambiguous directive but have occasionally overturned some agency interpretations. A third role for courts is monitoring and checking to ensure that regulatory decisions are made according to proper procedures and that they are reasonably and rationally based on appropriate, accurate, and relevant information. The courts' involvement in procedural aspects of regulation has tended to promote establishment of administrative procedures in which decisions are based on clearly established facts and fully articulated policies. When reviewing substantive aspects of agency decisions, courts consider whether the decision was based on a consideration of relevant factors and whether any clear error in judgment has been made (National Research Council 1977).

The judicial process consists of three steps (Sinden and Worrel 1979). The first is an attempt to determine what previous decisions or judgments have a bearing on the case at hand. In this step the court is trying to find out how society has expressed its attitude toward similar problems in the past. The second step is an advocacy proceeding in which representatives of various alternative points of view present supporting evidence. An important part of the advocacy proceeding is an attempt to offset or discredit evidence represented by the opposition. The final step is a judgment or finding about comparative values. The court must weigh the evidence presented by the parties and decide the matter according to the preponderance of the evidence or according to which version of the facts is the more probable. This has a subjective element, and various devices such as juries, panels of judges, and hearing boards are used as means of reducing individual biases. As a further safeguard, the judicial process provides for appeals to higher courts or authorities if the judgment of the first court is questioned.

On several occasions the courts have sought to establish legal ground rules for the use and application of formal analysis (Lawless *et al.* 1984; Morris and Duvernoy 1984; Ricci and Molton 1981). In *Reserve Mining Co. v. EPA,* for example, the defendant's processing plant was discharging asbestos fibers into Lake Superior and the atmosphere. The court concluded that the decision about whether the allegedly harmful activity should be banned must be based on a weighing of the gravity of the disputed potential harm against the benefits of allowing the activity to continue.[11] An unusually explicit recommendation of how this should be accomplished was provided by a judge presiding over a case involving damage caused by an adrift and unmanned ship:

> Since there are occasions when every vessel will break from her moorings, and since, if she does, she becomes a menace to those about her, the owner's duty, as in other similar situations, to provide against resulting injuries is a function of three variables: (1) the probability that she will break away; (2) the gravity of the resulting injury, if she does, (3) the burden of adequate precautions . . . if the probability be called P; the injury, L; and the burden, B; the liability depends on whether B is less than L multiplied by P: i.e., where $B < PL$.[12]

Numerous court rulings have established that some risks are too slight to be the subject of regulation, a position derived from the centuries-old maxim of *de minimis noncurat lex* — the law does not concern itself with trifles. For example, in *Environmental Defense Fund v. EPA,* the court concluded that EPA has the power, "inherent in most statutory schemes, to overlook circumstances that in context may fairly be considered *de minimis.*" Should the agency wish to invoke this concept to justify a decision not to regulate, however, the court said that the agency must first find "the concentration at which there are only trivial benefits to be derived from regulation."[13] Similarly, in *Ethyl Corp. v. EPA,* the court sanctioned precautionary regulation to control "a significant risk" and "the quasi-legislative powers of the administrative rulemaker to exercise his own judgment in areas of scientific uncertainty."[14] In another case, however, the court rejected an industry contention that EPA must support "effluent limitations with a mathematically exact cost-benefit analysis" (Baram 1980).

Only recently have analyses based on formal decision-aiding approaches been introduced as evidence in adjudicatory proceedings. In such situations, the emphasis has been placed on risk estimates, and the reaction of the courts has been mixed. For example, to support regulation of uranium mill tailings under the Uranium Mill Tailings Radiation Control Act, the NRC estimated additional cancer deaths resulting from the release of radium from uncontrolled tailings. In *Kerr-McGee Nuclear Corp. v. NRC,* the court accepted that the NRC had appropriately exercised its powers and found its estimate of the additional risk of cancer death reasonable and indicative of a significant risk.[15] In *Scott v. Food and Drug Administration,* the FDA defended its decision to allow continued use of the color additive Green No. 5, even though it had apparently been shown to be an animal carcinogen.[16] The agency justified its position with a quantitative risk estimate showing that the contaminant was present in such small amounts as to produce only a very small incremental increase in human cancers. The court upheld the agency's use of risk assessment as a basis for its decision.

An unprecedented court critique of an agency's use of analysis occurred in a 1983 decision by the Fifth Circuit Court of Appeals striking down CPSC's ban on urea-formaldehyde foam insulation. In *Gulf South Insulation Co. v. CPSC,* the Court agreed with industry's contention that the agency's exposure estimates were flawed, because the agency's measurements were collected in large part from a biased sample of homes where the residents had previously complained about odors and various acute irritations.[17] The court also criticized as "not good science" CPSC's reliance on a single study for evidence of formaldehyde's carcinogenicity without evaluating other studies that were available. The ruling has been interpreted as a rebuke by the courts of worst-case analysis and on the extrapolation of cancer risks to man on the basis of animal studies alone (Brickman *et al.* 1985). A similar result occurred in 1984 in *Asbestos Information Association v. OSHA.*[18] In this case, the court struck down OSHA's use of an emergency temporary standard which the agency had backed up in part with an estimate of the number of lives its action would save. In siding with industry, the court questioned the reliability of the agency's estimates. The court was also concerned that the agency's risk assessment had not been previously exposed to outside scrutiny, even though under the law an emergency standard can be issued without any opportunity for public comment (Merrill 1985).

The extent to which these and other court decisions will extend to future cases is unclear. To date, court rulings have tended to be couched in the context of specific laws. Their generalizability, therefore, depends on the regulations that would be relevant in other cases.

Federal Health and Safety Regulations and Related Government Initiatives

Statutes enacted by Congress to protect public health, safety, and the environment were among the earliest of Federal regulatory laws. In 1848 Congress prohibited the importation of any drug that was "improper, unsafe, or dangerous," and in 1886 enacted the Oleomargarine Act, which prohibited the use of any ingredients "deleterious to public health" (Hutt 1984). The greatest increase in the number and scope of such laws occured in the late 1960s and early to mid-1970s. During this period, public concern for newly discovered risks, optimism over government capabilities, and sustained economic growth combined to produce widespread support for strengthening existing regulatory policies and expanding Federal regulation into new areas. By the late 1970s conditions had changed, and regulatory reform became the new fashion in Washington. Most health, safety, and environmental policies, however, have survived relatively unscathed. Table 3 lists thirty-seven of the major pieces of legislation in effect by 1980, and Figure 4 shows the growth in the number of these laws over time.

Table 3. Major health, safety, and environmental laws

Year enacted	Title of act
1877	Dangerous Cargo Act
1893	Safety Appliance Acts
1906	Federal Food, Drug, and Cosmetic Act
1906	Federal Meat Inspection Act
1954	Atomic Energy Act
1957	Poultry Products Inspection Act
1958	Federal Aviation Act
1960	Federal Hazardous Substances Act
1966	National Traffic and Motor Vehicle Safety Act
1968	Radiation Control for Health and Safety Act
1969	Federal Mine Safety and Health Act
1969	National Environmental Policy Act
1970	Poison Prevention Packaging Act
1970	Occupational Safety and Health Act
1970	Clean Air Act
1970	Federal Railroad Safety Act
1971	Lead-Based Paint Poisoning Prevention Act
1971	National Cancer Act
1972	Federal Insecticide, Fungicide, and Rodenticide Act
1972	Consumer Protection Safety Act
1972	Marine Protection, Research and Conservation Act
1972	Ports and Waterways Safety and Health Act
1972	Federal Water Pollution Control Act (Clean Water Act)
1972	Marine Protection Research and Sanctuaries Act
1972	Noise Control Act
1973	Endangered Species Act
1974	Safe Drinking Water Act
1975	Rail Safety Improvement Act
1975	Hazardous Materials Transportation Act
1976	Toxic Substances Control Act
1976	Solid Waste Disposal Act
1976	Resource Conservation and Recovery Act
1977	Surface Mining Control and Reclamation Act
1978	Uranium Mill Tailings Control Act
1978	Outer Continental Shelf Lands Act
1979	Hazardous Liquid Pipeline Safety Act
1980	Comprehensive Environmental Response, Compensation and Liability Act (Superfund)

Adapted from Dodge and Civiak (1981)

The wording used in current health, safety, and environmental laws varies greatly, since Congress has rarely selected more than once exactly the same phrases to describe the means for obtaining safety goals. Despite these differences, regulatory agencies have been left with considerable flexibility in establishing specific standards and regulations. There are

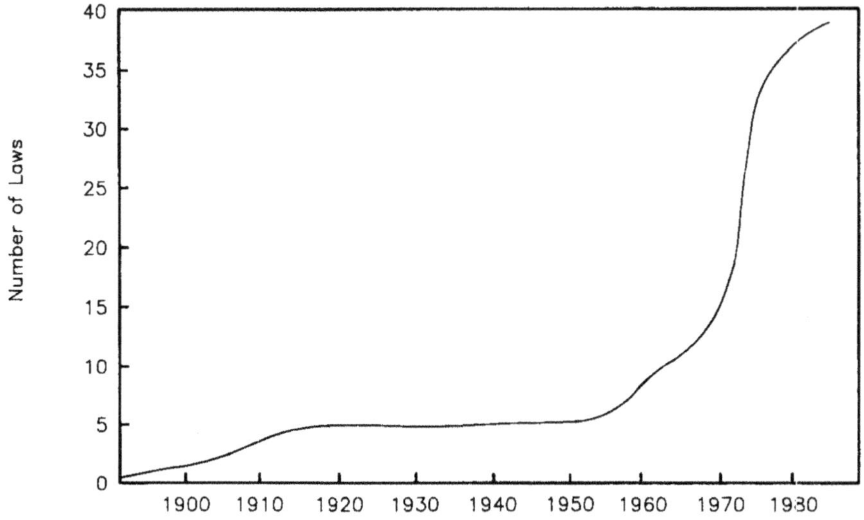

Fig. 4. Number of major federal health, safety, and environmental laws in effect over time.

several reasons for agency leeway. First, the broad and general language and avoidance of controversial issues which are necessary for the attainment of congressional consensus allows much room for agency interpretation. Second, it is generally not possible to determine whether the differences in statutory wording reflect differences in underlying congressional intent or are merely reflective of historical accident. Third, in the case of many of these statutes, there is relatively little judicial interpretation. Fourth, when the courts have wrestled with statutory language, as noted in the previous section, the unique facts of the case often result in a decision that provides no helpful precedent.

The latitude for agency discretion extends, in almost all cases, to flexibility in the use and selection of decision-aiding approaches. Statutes typically provide only very general guidance. For example, regulators are to use "latest scientific knowledge" under the Clear Air Act; "a systematic interdisciplinary approach" under the National Environmental Policy Act; and "research, demonstrations, experiments, and such other information as may be appropriate" under the Occupational Safety and Health Act. Superfund requires development and use of methods for discovering and investigating problem facilities, methods for evaluating and remedying releases, methods for determining the appropriate extent of remedial action, and so forth. An analysis by Hutt (1984) concludes, "The statutory requirements of current health and safety laws implemented by FDA, EPA, CPSC, and OSHA are sufficiently flexible to allow the adoption of whatever analytical and decision-making methodology best represents the public interest."

There are, however, several statutory differences that influence the role of an approach in regulatory decision making. One important difference is the extent to which the applicable law requires the agency to strike a balance between the benefits of risk reduction and the costs imposed by regulation, allows the agency to consider technical or economic feasibility of controls only, or allows the agency to consider only health effects when formulating regulations. Moss and Lubin (1981) estimate that roughly 75 percent of existing health and safety laws are of the balancing type, 5 percent permit only consideration of technical feasibility, and 20 percent permit only health or level of risk considerations. Depending on the requirements of the law, an agency may or may not be able to appeal directly to the results of a particular type of formal approach to justify its decisions.

Another important distinguishing characteristic is whether the enabling statute requires that agency approval be obtained prior to introducing a potentially hazardous substance or process, or whether it operates after the introduction of the hazard. The latter is important because those regulatory programs involving premarketing approval (such as those for pesticides, new drugs, and new food additives) empower an agency to require the submission of data to support a comprehensive analysis.[19] When premarketing approval is not required, the agency is forced to fend for itself in the acquisition of necessary data.

To provide a better understanding of existing regulations and their implications for the use and role of quantitative approaches in regulatory decision making, a more detailed description is presented of some of the major statutory laws dealing with the control of risk. The discussion is organized according to whether the law (or applicable portion of the law) represents a balancing statute, a technology-only statute, or a health-only statute.

Balancing statutes

The large number of statutes that require an agency to balance health benefits against the economic costs of regulation provide an obvious opportunity for quantitative, decision-aiding approaches to play major role in agency decision making. The earliest and most demanding of the balancing statutes is the Flood Control Act of 1936. This law essentially mandates the use of cost-benefit analysis for water resource development projects by requiring that projects planned and constructed by the Army Corps of Engineers, such as dams and navigation systems, produce benefits that exceed their economic costs.

Most of the more recent environmental statutes also contain a balancing requirement. The National Environment Protection Act (NEPA) mandates "balancing the environmental costs of a project against its economic and technological benefits."[20] Because NEPA does not demand that intangible values be quantified, only given "appropriate consideration" in agency procedures, it has not been construed as requiring formal cost-benefit analysis.[21] Similar interpretations have been given to the Outer Continental Shelf Lands Act, amended in 1978, which requires offshore drilling to be done with the safest technology, except when "the incremental benefits are clearly insufficient to justify the incremental costs of using such technologies."[22] Balancing requirements are also contained in the Federal Meat Inspection Act, the Atomic Energy Act, portions of the Federal Food, Drug, and Cosmetic Act dealing with drugs and contaminants to food, and portions of the Clean Air Act dealing with fuel additives.

A large category of statutory laws requiring a balancing of costs and benefits are those that mandate protecting the public from "unreasonable risk." The Toxic Substances Control Act (TSCA), for example, regulates substances that present "an unreasonable risk of injury to health or the environment,"[23] and the Federal Environmental Pesticide Control Act bans pesticides that pose "any unreasonable risk to man or the environment."[24] Other examples of such statutes include the Consumer Protection Safety Act, the Federal Hazardous Substances Act, and the Compensation and Liability Act. The courts have held that when the "unreasonable risk" language appears, a balancing of costs and benefits is required (Ricci and Molton 1981). Some statutes specifically identify the benefits that should be balanced against the risks of the activity. For example, FIFRA requires consideration of the "economic, social, and environmental" benefits of a substance that is to be regulated.

A similar situation has occurred where statutes permit costs to be considered, but in a way that is less central to the decision. This is the case in the Clean Water Act. In establishing effluent standards the EPA must consider the total cost of the standards, but it need not make a quantitative comparison of costs and benefits. A standard is to be imposed unless the marginal level of effluent reduction is "wholly out of proportion to the cost." When an individual polluter wishes a variance from the effluent standards, the economic capability of the individual plant to bear the cost of the standard may not be considered.

Thus Congress frequently specifies certain conditions for balancing but does not specify to regulatory agencies the particular approach to use in assessing benefits and balancing against costs and risks.[25] Although Congress is cognizant of agency responses through its constant oversight activities, it has generally acquiesced to agency choices in dealing with balancing requirements.[26] As a result, agencies governed by balancing statutes have great flexibility in using and choosing among formal decision-aiding approaches, but so far they have chosen not to base their decisions solely on the results of cost-benefit analysis or some related approach that explicitly and quantitatively balances benefits and costs.

Typically, an agency has met a statutory balancing requirement by simply discussing at length the benefits and costs estimated to result from the regulatory action. Thus, for example, in the numerous pesticide cases considered by EPA, the agency's quantitative analyses have focused almost exclusively on the level of risk: if the risk level was judged qualitatively to be insignificant, no action was taken; if it was regarded as substantial, the pesticide was banned. Neither EPA nor any other agency has enunciated an operational methodology for balancing costs and benefits or has purported to make a decision based solely on a quantitative cost-benefit comparison.

Technology-only statutes

Technology-only statutes, which generally contain language requiring that risks be reduced through the use of the "best available control technology" or the "best practicable control technology," permit only technological or economic feasibility as criteria in determining appropriate regulation. They apply in situations where risks can be reduced by electro-mechanical systems and are, therefore, primarily associated with agencies having responsibility involving complex technological systems, such as the EPA has over power plants and industrial process plants. Standards implementing technology-only statutes often specify numerical goals for emissions or effluents from a particular process or point source, procedures (such as labeling, permitting, or notification requirements), or necessary types of safety or control equipment.

An example of a technology-only statute is provided by some sections of the Clean Air Act, which requires the use of best available control technology (BACT) to limit pollution from motor vehicles. Portions of the Clean Water Act and the Federal Water Pollution Control Act also require BACT. Technology-only standards have the advantage of reducing risks as technology improves; however, they prohibit regulatory agencies from explicitly taking into account the cost of the technology in setting the standard or even from considering whether or not the required use of the technology will result in a net reduction in risk.[27] In practice, however, economic considerations do play a role in defining the BACT — political realities temper decisions, and engineers will implicitly consider costs when defining technical feasibility.

Another example of a technology-only statute, the Occupational Safety and Health Act, requires that "to the extent feasible . . . no employee will suffer material impairment of health." The courts have devoted considerable energy to determining whether this language requires a technology-only approach or a balancing of costs and benefits. In 1974 the Court of the District of Columbia ruled that proposed OSHA standards were feasible even if they bankrupted some individual employers.[28] A later court, however, noted that regulations that caused massive dislocation or "crippling" of an industry would not be viewed as feasible.[29] Then, in a 1978 ruling on a proposed benzene standard, the Fifth Circuit Court squarely endorsed cost-benefit balancing.[30] The proposed standard was struck down on grounds that OSHA had not performed a quantitative estimate of the risks and benefits of the standard nor had shown that the benefits bore a reasonable relationship to the costs of control. The U.S. Supreme Court affirmed the circuit court's decision in 1980, but then in a 1981 ruling regarding cotton dust safety standards it held that OSHA should not base its decisions on a balancing of costs and benefits.[31] The Court reasoned that balancing was inappropriate in view of the legislative history of the Occupational Safety and Health Act. Its position was that Congress had already performed the balancing and intended to place "the 'benefit' of worker health above all other considerations save those making attainment of this 'benefit' unachievable. Any standard based on a balancing of costs and benefits by the Secretary that strikes a different balance than that struck by Congress would be inconsistent with the command set forth [by the "to the extent feasible" language]" (Ricci and Molton 1981). A similar conclusion was reached when the Tenth Circuit Court considered "to the extent feasible" language in the Uranium Mill Tailings Radiation Control Act and the Atomic Energy Act.[32]

Technology-only statutes constrain the role of formal approaches within the decision-making process but do not prevent their use. For example, formal approaches that do not involve cost considerations (such as certain variations of decision analysis and applied social welfare theory) might be used to aid the identification of the most effective controls or the situations in which they should be required.[33] In fact, technology-only standards do not necessarily even rule out the use of decision-aiding approaches based on a cost-benefit balancing criterion. The Supreme Court OSHA decision on cotton dust was specifically based on a limited interpretation of congressional intent in the workplace and does not necessarily apply to other hazards, even those covered by statutes using the "to the extent feasible" language (Ricci and Molton 1981).[34] In any case, the Supreme Court ruling leaves open the use of cost-benefit balancing for the selection among various methods to achieve a standard. The Court noted that if two methods that achieved the same reduction of health risk were both feasible, the more burdensome method might not be "reasonably necessary." Thus, two courts of appeal have allowed employers to avoid expensive engineering controls by demonstrating that they are not cost-effective.[35] Cost-benefit analysis or other decision-aiding approaches might also be used to set priorities, to review residual risk after application of best-available technology to see if anything more needs to be done, and to provide information regarding the urgency of situations where subpopulation groups are inadvertently exposed to toxic agents (such as populations near uncontrolled waste sites).

Health-only statutes

A few statutes require that consideration be given only to health issues in the formulation of standards. Such statutes generally contain language stipulating that the agency adopt regulations "as may be necessary to protect human health," regulations that provide "an ample margin of safety," or "the highest possible degree of safety in the public interest," or outright prohibitions of the use, sale, transportation, or manufacture of certain hazards. The most dramatic example of the latter is the Delaney Clause in the 1958 amendment to the Food, Drug, and Cosmetic Act, which requires the FDA to ban any food additive that has been shown to cause cancer in animals.[36] The Resource Conservation and Recovery Act also permits the regulation of carcinogens based on health issues only. In theory at least, once tests demonstrate the carcinogicity of a substance, no consideration of its benefits or costs of removal, or even of the technological feasibility of producing the product without the additive, is relevant. Another important example of a health-only statute is provided by portions of the Clean Air Act dealing with mobile and stationary sources. For example, the Clean Air Act requires the EPA to set primary national ambient air quality standards on the basis of health considerations only; considerations of cost or technological feasibility are not relevant in the determination of these standards.

In health-only statutes, designation of a substance as "unsafe," "carcinogenic," "hazardous," or "toxic" rests largely within the discretion of the agency. Once a substance is so listed, however, regulatory response is compulsory, and the result may produce significant impacts on industry. Not surprisingly, regulatory agencies have tended to refrain from listing a substance until there is complete confidence in the decision. In practice, therefore, health-only statutes, like all other statutes, are administered in a way that permits cost-benefit comparisons, albeit in a broad brush and common sense way. Thus, for example, the FDA has understandably declined to ban essential nutrients shown to be carcinogenic in tests where had the same results been obtained on trivial ingredients there would have been an immediate ban. According to the Conservation Foundation (1984), the EPA has managed to take into account the economic impact of primary air quality standards by choosing which sensitive populations to consider and by adjusting the margin of safety so that thinner margins are set for the pollutants that require more expensive control measures.

Despite the prohibition against explicity balancing costs and benefits in health-only statutes, agencies operating under these statutes may use decisioin-aiding approaches in priority setting and frequently use quantitative methods in risk assessment. EPA's proposal for regulating airborne carcinogens, for example, includes the development of quantitative risk assessments to classify source categories as high, medium, or low priority, where the classification is based on whether there is a strong likelihood that the source category emits a carcinogen and that individuals receive a significant exposure to the substance emitted (Environmental Protection Agency 1983b). In the case of recently proposed rules for inorganic arsenic and radionuclides, the agency explicitly considered the economic effects of additional control requirements in determining appropriate control technologies for various source categories.[37] The EPA is also exploring the use of quantitative risk assessment as an aid in setting national primary ambient air quality standards (Richmond 1981). Provided that cost-benefit balancing does not become the basis for decision, agencies can also use decision-aiding approaches to provide information useful to setting standards governed by health-only statutes. EPA's administrator acknowledged that while she was "legally bound not to consider economic and technological feasibility in setting air quality standards," she considered "benefits analysis a useful tool which can be very helpful in evaluating alternative standards" (General Accounting Office 1984). According to Marraro (1982), EPA's relaxation of the ozone standard was justified in part by subjective probability assessments obtained through the application of a decision analysis approach.[38]

The Food, Drug, and Cosmetics Act prohibits the addition of any carcinogenic compounds to the feed of food-producing animals unless "no residue" is left in the ultimate human food product. Even this has been interpreted by the agency as allowing room for quantitative analysis and balancing. The FDA has proposed to define "no residue" as quantitatively specified low risk level — for example, a residue that would produce one cancer death per million lifetimes (Leape 1980). If a drug sponsor can provide a detection method capable of measuring residues posing risks at this level, the FDA would ignore any residues that could not be detected using the sponsor's method. Since agencies operating under health-only statutes already implicitly take costs and benefits into account, for example, by relying more or less heavily on the results of questionably positive studies when deciding whether to take action, there is ample opportunity for formal approaches to influence decision making.

Executive orders and legislative proposals

Regulation has produced progress toward a healthier, safer, and cleaner environment, but it has done so at a high economic cost. Some estimates of the full costs of regulation range as high as $100 billion annually (Weidenbaum 1978). Not surprisingly, the growth in regulation has brought with it increasing concern that the government is not as effective nor as efficient as it might be, and that regulators are not sufficiently sensitive to the economic consequences of their actions. These pressures for regulatory reform have produced a number of attempts to countermand the vagaries and inconsistencies of statutory direction with a uniform requirement for agencies to apply formal approaches that produce an explicit accounting of costs and benefits. In at least two cases before the Surpreme Court it has been argued that regulatory agencies, such as EPA and OSHA, be allowed to promulgate only those rules whose benefits can be demonstrated to be "reasonably commensurate" with the costs, and at least one Supreme Court Justice has concurred with this view.[39] An increasing number of executive and congressional initiatives are aimed at achieving this same objective.

Since the administration of Gerald Ford, presidents have endeavored to compel regulatory agencies to undertake a more systematic comparison of regulatory costs and benefits. President Ford's Executive Order No. 11821 was designed to reduce the inflationary impact of new regulations by requiring that "inflation impact statements" (later called "economic impact statements") be prepared for "major" agency proposals (specifically, those expected to have a national economic impact greater than $100 million).[40] The process was to be reviewed by the president's Council on Wage and Price Stability and overseen by the Office of Management and Budget. In preparing impact statements, analyses were to include an examination of the inflationary effects of the proposal, a comparison of its estimated costs and benefits, and a comparison of these costs and benefits with the costs and benefits estimated for alternatives to the proposal.

President Carter replaced the inflation impact program with Executive Order 12044, a similar directive that required agencies to prepare "regulatory analyses" for all regulations producing an annual impact of at least $100 million or causing a "major" price increase.[41] A regulatory analysis was to include a description of the alternative ways of dealing with the problem, an economic analysis of each, and a detailed explanation of the reasons for choosing one alternative over the others. Although the order did not explicitly require the use of a quantitative approach, it stipulated that the various burdens, gains, and overall economic impacts of regulations be compared. Each year ten to twenty regulatory analyses were to be reviewed by the Regulatory Analysis Review Group (RARG), which was chaired by the chairman of the Council of Economic Advisors.

Although President Ford's and President Carter's executive orders spurred agency interest in and some additional use of cost-benefit and related approaches, there is little evidence suggesting that they produced much impact. An OMB report concluded that regulatory analyses frequently contained analytical shortcomings, such as failure to consider all effects of proposed regulations and failure to deal with uncertainty (Office of Management and Budget 1980). In a study of Ford's inflation impact statement, Bernick (1982) concluded that the program had little effect on inflation or specific regulatory decisions. Reviews of impact statements and regulatory analyses tended to be conducted at the last minute and off the record (Ayres 1979). Reflecting President Carter's lack of continued enthusiasm for his order, RARG actually examined only nine out of forty proposals eligible for review (DeMuth 1980). Baram (1980) suggests that the Carter administration's regulatory analyses resulted in EPA standards that were "significantly less stringent and less protective of health" at a "presumably" lower cost, but Liroff (1982) believes they did not, "except in one or two cases, significantly affect specific regulations."[42]

In February 1981 the Reagan administration issued its own version of a regulatory analysis program in Executive Order 12291.[43] Like the orders issued by previous administrations, it required executive agencies to prepare regulatory impact analyses to assist them in selecting regulations with the property that the "potential benefits to society from the regulation outweigh the potential costs to society." President Reagan's order was, however, much more demanding. For one thing, it expanded the definition of "major rules" to incorporate a large number of regulations: in this case actions likely to have an economic impact of $100 million a year or more; likely to produce significant adverse effects on competition, employment, productivity, or other industrial concerns; or likely to produce significant increases in costs or prices in the economy. In addition, it required any set of related rules to be considered together as a major rule.

Most important, E.O. 12291 went much further in attempting to institutionalize cost-benefit balancing; it not only stipulated that agencies should apply cost-benefit analysis or a similar approach to major decisions, it demanded that agency actions follow the results. The only exceptions were those cases (such as health-only or technology-only statutes) where the agency was legally prohibited from cost-benefit balancing. Even in these cases, however, the agency was required to conduct the analysis for "informational purposes." Specifically, the order required agencies to

1. Estimate the potential costs and benefits of proposed actions
2. Identify those likely to bear the costs and enjoy the benefits
3. Compare the estimated costs and benefits with those associated with alternative policy actions
4. Select actions that maximize net benefits and minimize net costs consistent with regulatory objectives

The agencies were directed to identify for review and possible change or rescission any existing rules that did not follow a least costly approach and to postpone new rules until regulatory reviews were completed, unless a statutory or court-imposed deadline or emergency existed.

Another significant characteristic of President Reagan's E.O. 12291 was the powerful role in regulatory analysis that it provided to OMB. The OMB director was given authority to designate certain regulations as major and existing rules as subject for review. Agencies were required to refrain from publishing preliminary regulatory impact analyses or final rules in the Federal Register until OMB concluded its review (within a specified time period) and to respond to and incorporate OMB's views and their own responses in the rule-making record. OMB's role was expanded in January 1985 with the issuance of E.O. 12498, which requires agencies to submit annually to OMB a statement of planned major rules. The OMB has authority to review these statements for consistency with E.O. 12291 and to identify actions necessary to achieve such consistency.

Considering the scope of E.O. 12291, it is not surprising that its legality has been questioned. Rosenberg (1981) has claimed that the order violated the Administrative Procedures Act because cost-benefit analysis does not consist solely of "neutral, value-free" procedures, as required by the act. Sohn and Litan (1981), however, conclude that a requirement to perform cost-benefit analysis is legal provided that the underlying statute does not expressly forbid cost-benefit balancing. A court challenge occurred in 1985 when Rep. John Dingell (D-Michigan) and four other House members filed a friends of the court brief with the D.C. Court of Appeals in a case concerning the regulation of a chemical by OSHA. The brief argues that "by causing the deletion of a short-term exposure limit from the standard established for ethylene oxide under the purported authority of E.O. 12291, [OMB] unlawfully interfered with the rulemaking functions delegated by Congress to the Department of Labor . . .".[44] The Court sided in favor of the challenge for other reasons, but elected to sidestep the question of the constitutionality of the order.

Confronted with the technical difficulties and tenuous legal standing of E.O. 12291, regulatory agencies have been somewhat lax in meeting its requirements. According to the General Accounting Office (GAO), some agencies have deliberately kept annual cost estimates of proposed regulations below $100 million to avoid classifying actions as "major" (GAO 1984). In 1981 OMB reviewed nearly 2,700 regulations, but only twenty-two of these had regulatory impact analyses prepared. In roughly half of the cases in which a regulation was deemed major, OMB waived the requirement for analysis. Eads and Fix (1984) observed that it was fortunate for OMB that so few regulatory impact analyses were submitted in the initial period covered by the order: "Given the extremely small number of technically qualified analysts, a significantly larger number of 'major' rules, accompanied by detailed regulatory analyses that themselves had to be subjected to scrutiny, would have swamped the agency."

The analyses that have been conducted under Reagan's order have been criticized for failing to consider all relevant alternatives, failing to identify and compare the various costs and benefits, and inadequately accounting for uncertainty (GAO 1982). Noting that the EPA analyses emphasized single-value benefit and cost estimates or, at best, provided only a range of possible values, the GAO recommended that the EPA administrator "direct the program office performing cost-benefit analysis to use special techniques for analyzing uncertainty . . ." (GAO 1984). The GAO (1982) concluded:

> OMB does not appear to exercise its powers under E.O. 12291 . . . . The provisions of the executive order . . . imply a need for achieving consistent methodologies for measuring regulatory impacts . . . . However, problems of coordinating methodologies for estimating costs and benefits in any given industry have not been addressed.

Although E.O. 12291 does not have the force of a statutory law, the commitment of the Reagan administration to reduce the economic costs of regulation and the similar interests of the OMB have produced a degree of agency attention greater than that of previous regulatory analysis orders. Since its enactment more than a hundred regulatory impact analyses have been conducted which involve at least some characteristics of cost-benefit or related formal approaches. The majority of these have been prepared by or for EPA and OSHA, since these agencies' actions most often are classified as major. In addition to producing applications of cost-benefit analysis, another effect of the order has been the initiation of a variety of efforts by and for regulatory agencies to help them familiarize themselves with and improve available methodology for assessing and evaluating costs, risks, and benefits. For example, at the request of the FDA, the National Academy of Sciences prepared a study entitled "Risk Assessment in the Federal Government" (National Research Council 1983); and the National Science Foundation, at the request of OMB, sponsored surveys of the state of the art in risk assessment (Merkhofer *et al.* 1984) and benefit assessment (Bentkover *et al.* 1986).

In addition to steps taken by the executive branch, several congressional initiatives also reflect increasing pressures to use quantitative methods to promote regulatory reform. Several bills have been introduced into Congress that if passed would have encouraged or required agencies to perform risk analysis. For example, Senators Robert Dole (R-Kansas) and Paul Laxalt (R-Nevada) have in the past each introduced bills that would require agencies proposing regulations to describe the costs and benefits of such proposals in quantitative terms and to explain how estimated benefits are expected to justify estimated costs. These were similar to E.O. 12291 except that the bills refrained from requiring agencies to make quantitative results the basis for decisions. Laxalt's bill, which extended the cost-benefit analysis requirement to such independent agencies as the Federal Trade Commission, successfully passed in the Senate. Congressman James H. Martin (R-North Carolina) has introduced a bill "for the establishment of a Central Board of Scientific Risk Assessment as a means of improving the scientific review and evaluation of risk assessments made by Federal agencies . . . ." Congressman Donald Ritter (R-Pennsylvania) has more than once introduced bills that would require agencies to conduct demonstration projects and establish a research program to improve knowledge and use of risk assessment by federal agencies. At least one version has managed to successfully pass in the House.[45] Although none of these bills has, as of the time of this writing, yet passed both houses, efforts to approve such legislation are continuing.

Clearly, not all of the support behind these ongoing initiatives comes from a noble desire to improve the effectiveness and efficiency of achieving the goals articulated in environmental statutory law. Some is prompted by self-interested attempts by those who pay the costs of environmental controls to limit the nation's efforts to achieve these goals. The use of cost-benefit and other formal approaches as a political weapon for challenging health, safety, and other consumer values may reduce their credibility and acceptability in the long run. Nevertheless, the immediate effect of these continuing activities has been increased agency interest in and use of formal decision-aiding approaches.

A Taxonomy of Risk Problems

The presentation in this chapter has indicated some of the significant characteristics associated with social decisions involving risk. The specifics of these and other characteristics associated with risk problems are important considerations for the selection of a decision-aiding approach. To begin constructing a conceptual framework for the comparative evaluation of approaches, it is useful to consider how risk problems might be distinguished according to characteristics that affect the choice among approaches. The discussion of social risk management is concluded, therefore, with an illustration of how the concept of a taxonomy of risk problems might be used in the characterization of decision situations.

As a practical matter, the first step toward problem characterization should be diagnosis. The issues that have been identified that surround social risk management provide assistance in this respect. For example, it is useful to consider which of the sources and processes identified above in the section entitled Components of the Risk-Generation Process are responsible for the risk. The relevant characteristics of each component of the risk chain should be considered. For example, the hazard should be defined, including the substances or energies that pose the fundamental danger. The processes that generate and produce potential exposures to the hazard should be understood. For example, is the risk due to self-hazardous behavior, cogeneration, the actions of individuals? The nature of the effects process should be considered, and any special features that influence social valuation should be explicitly recognized. The various motivations for government intervention should be considered to appreciate the rationale for government action. For example, has there been an apparent failure of the free market because of externalities, imperfections in the flow of information, concentration of market power? A comprehensive set of alternative courses of government action should be identified. As indicated in that section government intervention can take many forms, and the alternatives may be broader than they first appear. Sources of problem complexity should be identified. How serious are problems of lack of knowledge, risk perception, and social consent? Finally, the existing laws and regulations that are relevant to the problem and the extent to which they define objectives or otherwise constrain the analysis should be identified and understood.

To organize these considerations into a problem taxonomy, it is helpful to examine the relationships between a decision-aiding approach and the factors that influence social decision making. These factors include the substantive characteristics of the decision problem, the decision process used by those who have authority for the decision, and the social and political pressures that create the need for a decision.

Consider first the technical characteristics of the decision problem; that is, dimensions that describe the substance of the problem as opposed to the political and social setting in which a decision must be made. As will be discussed in Chapter 2, all formal approaches entail the development of an analytic decision model. One category of relevant factors distinguishing risk problems consists, therefore, of any technical characteristics that create modeling difficulties for one or more of the approaches. For example, are there only a few distinct options, or are there many complex plans that must be compared? Is it fairly simple to estimate the consequences of any given choice or are there many complex interactions that must be considered? How much uncertainty is present? Are the relevant social values fairly straightforward or highly complex? The quality and quantity of available data are very important to the selection of a decision-aiding approach. Is there a large amount of directly applicable data available (implying that statistical procedures might be used)? Is there a large amount of indirectly applicable data available (implying that extrapolative procedures must be employed)? If statistical data are lacking, is there, nevertheless, a high degree of relevant scientific understanding (implying that credible models may exist or might be developed)? Or are both data and understanding lacking (implying that the best source of information may be procedures for systematically obtaining subjective judgments from qualified experts)?

Another characteristic that influences the choice of modeling methods relates to whether the risk is "continuous" or "discrete." Continuous risks are induced by activities that regularly produce dangers (such as pollution emitted from a coal-fired electric power plant), result in continuous exposures, or tend to produce chronic health effects (due, perhaps, to the cumulative effects of exposures). Discrete risks, on the other hand, are generally associated with the nonroutine functioning of a system (especially accidents), tend to be associated with brief periods of exposure, and frequently produce acute health effects. Risks can also involve both continuous and discrete aspects. For example, the accidental release of radioactivity from a nuclear power plant might create continuous and cumulative adverse health effects. Conversely, the continuous emission of atmospheric pollutants could produce unusually high, short-term ambient concentrations that result in acute health effects.

The second category of relevant characteristics involves aspects of the decision process. Those characteristics that influence the value of approaches include resource constraints (such as the time available for analysis), the availability of qualified personnel to apply the approach, and the availability of computational resources. Institutional and statutory requirements of the sort discussed in the section on Federal Health and Safety Regulation may influence the acceptability and applicability of an approach, as will the basic philosophy of the decision makers. Finally, because decision-aiding approaches are designed to be in some specific sense "rational," the demand for agency accountability and rationality will be a characteristic relevant to the evaluation of decision-aiding approaches.

The last category of relevant characteristics reflects the recognition that a decision-aiding approach may influence social and political demands by affecting public perceptions. Because analyses done for government will usually be in the public domain, the results not only provide informational inputs for government decision makers, they also may be scrutinized by industry, citizen groups, and the public at large. The extent to which alternative approaches influence the pressures felt by government decision makers may, therefore, be a relevant consideration for selecting among approaches. An obvious example from this category is the extent to which a decision-aiding approach may influence public risk perceptions; for example, by altering public concern over a technological risk. Another consideration is the level of controversy regarding the decision. The inherent explicitness of various approaches can, under certain situations, either resolve or fuel such controversies.

These and other considerations are represented in the sample taxonomy for risk problems shown in Table 4. Although the table contains nearly one hundred entries, there are many other distinguishing characteristics that might, in certain situations, have an important bearing on the selection of a decision-aiding approach. Although not comprehensive, Table 4 does serve to illustrate the very large number of ways in which social decision problems involving risk may differ and suggests that a careful, systematic consideration is needed to produce sufficient problem understanding to permit an informed choice among decision-aiding approaches. The use of a risk taxonomy such as Table 4 as an aid for the selection and application of a decision-aiding approach will be addressed in Chapter 4.

Table 4. A taxonomy for characterizing risk-decision problems

Category	Consideration	Dimension	Distinct cases
1. Technical characteristics influencing difficulty of analysis	1.1 Complexity of alternatives	1.1.1 Scope	1.1.1.1 Narrow (e.g., site- or situation-specific) 1.1.1.2 Broad (e.g., generic)
		1.1.2 One-shot v. strategy	1.1.2.1 Situation demands action based on existing information 1.1.2.2. Situation requires contingency plan 1.1.2.3 Situation demands dynamic strategy involving issues of timing and phasing actions
		1.1.3 Commitment v. information gathering	1.1.3.1 Situation requires substantive action 1.1.3.2 Situation involves deciding what information to collect
		1.1.4 Incremental v. fundamental	1.1.4.1 Action involves incremental change (e.g., level of existing activity or resource) 1.1.4.2 Action involves qualitative change (e.g., relocation, new technology, outright ban)
		1.1.5 Few v. many options	1.1.5.1 There are only two options (e.g., "yes", "no") 1.1.5.2 There are several competing alternatives 1.1.5.3 There are many, fundamentally different, options 1.1.5.4 Situation requires selecting a value for a continuous variable 1.1.5.5 Interrelated values must be selected for several variables
		1.1.6 Ranking v. portfolio	1.1.6.1 The single "best" option must be identified 1.1.6.2 The "best mix" of favorable options must be identified
		1.1.7 Fixed v. adjustable	1.1.7.1 Once taken, the decision cannot be easily altered 1.1.7.2 Commitments can be readily modified as results become known
	1.2 Complexity of relationships	1.2.1 Few v. many relevant factors	1.2.1.1 Few factors influence decision outcomes 1.2.1.2 Many factors influence decision outcomes
		1.2.2 Static v. dynamic	1.2.2.1 Relationships between decisions and outcomes do not change with time 1.2.2.2 Relationships are evolving
		1.2.3 Compensating v. synergistic	1.2.3.1 Predominant effects tend to cancel out 1.2.3.2 Predominant effects are additive 1.2.3.3 Predominant effects are synergistic

Table 4 (*Continued*)

Category	Consideration	Dimension	Distinct cases
		1.2.4 Continuous v. discontinuous	1.2.4.1 Incremental changes in actions produce small changes in outcomes
			1.2.4.2 Small changes in action can produce extreme changes in outcomes
		1.2.5 Separable v. dependent	1.2.5.1 Decision outcomes do not depend much on other decisions
			1.2.5.2 Outcomes depend strongly on other decisions
		1.2.6 Reversible v. irreversible	1.2.6.1 Decision consequences are irreversible
			1.2.6.2 Consequences may be reversed
	1.3 Nature of relevant information	1.3.1 Degree of uncertainty	1.3.1.1 Only moderate uncertainty exists over decision outcomes
			1.3.1.2 Extreme uncertainty exists over decision outcomes
		1.3.2 Continuous v. discrete risk	1.3.2.1 Risks derive continuously from the routine functioning of a system
			1.3.2.2 Risks occur infrequently due to failures or non-routine functioning of a system (e.g., accidents)
		1.3.3 Type of evidence available for characterizing uncertainty	1.3.3.1 Statistical data provides the best source for characterizing uncertainty
			1.3.3.2 Knowledge of system provides the best source for characterizing uncertainty
			1.3.3.3 Credible experts provide the best source for characterizing uncertainty
	1.4 Outcome valuation complexity	1.4.1 Critical issues for net evaluation	1.4.1.1 Achieving equity is of primary importance
			1.4.1.2 Achieving efficiency is of primary importance
			1.4.1.3 Balancing conflicting objectives is of primary importance
		1.4.2 Few v. many value dimensions	1.4.2.1 Very few value criteria are needed to measure decision effectiveness
			1.4.2.2 Multiple criteria are important
		1.4.3 Source of values	1.4.3.1 Impact will be primarily on marketed commodities
			1.4.3.2 Group preferences are the appropriate source of values
			1.4.3.3 Preferences of single decision maker should drive decision
		1.4.4 Timing of decision consequences	1.4.4.1 Costs and benefits accrue at comparable times
			1.4.4.2 Costs and benefits accrue at very different times
			1.4.4.3 Costs and benefits accrue to different generations
		1.4.5 Risk attitude	1.4.5.1 Risk neutrality may be assumed
			1.4.5.2 Risk aversion (or preference) must be accounted for in analysis

Table 4 (*Continued*)

Category	Consideration	Dimension	Distinct cases
2. Decision process characteristics influencing acceptability or applicability of analysis	2.1 Constraints	2.1.1 Time	2.1.1.1 Extremely limited time for analysis 2.1.1.2 Moderate time for analysis 2.1.1.3 Much time for analysis
		2.1.2 Decision criteria	2.1.2.1 Basis for decision must be cost-benefit balancing 2.1.2.2 Consideration of costs is prohibited
		2.1.3 Information sources	2.1.3.1 Decision must be based on specific information sources (which?) 2.1.3.2 No constraints exist on information sources
		2.1.4 Analytic resources	2.1.4.1 Constraints on analytic resources are severe (nature?) 2.1.4.2 No significant constraints exist on analytic resources
	2.2 Decision maker characteristics	2.2.1 Analytic skills	2.2.1.1 Decision maker(s) is (are) familiar with analytic approaches (which?) 2.2.1.2 Decision maker(s) has (have) little familiarity with analytic approaches
		2.2.2 Problem knowledge	2.2.2.1 Decision maker(s) has (have) much familiarity with problem 2.2.2.2 Decision maker(s) is (are) not very familiar with problem
		2.2.3 Role	2.2.3.1 Sponsor of analysis will make final decision 2.2.3.2 Sponsor of analysis will make a recommendation 2.2.3.3 Sponsor will provide information for others who make decisions
		2.2.4 Personality/ security	2.2.4.1 Decision maker sensitive to personal criticism 2.2.4.2 Decision maker not very sensitive to personal criticism
		2.2.5 Motivation for analysis	2.2.5.1 Motivation for analysis is choosing an alternative 2.2.5.2 Motivation is defending a position 2.2.5.3 Motivation is convincing others
		2.2.6 Authority level	2.2.6.1 Authority level of decision maker(s) is low (e.g., first-line manager(s)) 2.2.6.2 Authority level of decision maker(s) is moderate 2.2.6.3 Authority level of decision maker(s) is high (e.g., top manager(s))
		2.2.7 Accessibility	2.2.7.1 Decision maker will be easily accessible throughout the analysis 2.2.7.2 Decision maker will not be available to assist analysts
	2.3 Organizational characteristics	2.3.1 Rational actor model	2.3.1.1 Organizational decision process is well approximated by rational actor model 2.3.1.2 Decision process does not approximate rational actor model

Table 4 (*Continued*)

Category	Consideration	Dimension	Distinct cases
		2.3.2 Dispersion of decision responsibility	2.3.2.1 Centralized decision making 2.3.2.2 Decentralized decision making (dispersed horizontally or vertically
		2.3.3 Accountability	2.3.3.1 Institution highly accountable for decisions 2.3.3.2 Institutional accountability for decisions is low
3. Characteristics influencing impact of analysis on public perceptions	3.1 Sensitivity of public perceptions	3.1.1 Issue familiarity	3.1.1.1 Public is widely familiar with and concerned about decision 3.1.1.2 Issues have not achieved significant public attention
		3.1.2 Technical ambiguity	3.1.2.1 Issue understanding requires considerable technical expertise 3.1.2.2 Basic issue understanding does not require sophisticated technical expertise
		3.1.3 Risk perception biases	3.1.3.1 Experts and public perceive level of risk very differently 3.1.3.2 Experts and public do not disagree on level of risk
	3.2 Controversy	3.2.1 Stakeholder motivations	3.2.1.1 Stakeholder and decision maker motivations differ significantly 3.2.1.2 Stakeholder and decision maker priorities differ 3.2.1.3 Stakeholder and decision maker motivations similar
		3.2.2 Equity	3.2.2.1 Situation is regarded as very "unfair" 3.2.2.2 Equity concerns are not significant
		3.2.3 Social consent	3.2.3.1 Authority or right to make decision questioned seriously 3.2.3.2 Broad consensus exists that decision maker should make decision
	3.3 Expectations	3.3.1 Situation acceptance	3.3.1.1 Public accepts current situation 3.3.1.2 Considerable public pressure exists for quick action
		3.3.2 Action demanded	3.3.2.1 Symbolic action needed 3.3.2.2 Effective action needed
		3.3.3 Capacity to achieve change	3.3.3.1 Options are available for achieving change 3.3.3.2 Little can be done to change situation

Chapter 2

Decision-Aiding Approaches

A comparative evaluation of decision-aiding approaches requires that alternative approaches first be identified and differentiated. Characterizing and distinguishing among approaches, however, is not an easy task. There are two main sources of difficulty. One is the sheer quantity of available options; the number of possible approaches is essentially infinite, given the many different ways in which analytic procedures can be combined. The second difficulty is the lack of consensus about terminology and requirements.

This chapter addresses the problem of distinguishing approaches from two perspectives: differences in the concepts and theories that underlie alternative approaches and variations in the procedures that are used within an approach to implement the underlying theories. Characterizing approaches according to theories and procedures provides a practical means for differentiation and facilitates judgments of the value of applying alternative approaches in various problem situations.

The chapter begins by investigating the nature of decision-aiding approaches. This involves defining the approaches, indicating two fundamentally different interpretations that might be given to probability and risk, and describing the characteristics that are common to all approaches. Next, the principal normative theories for decision making are described. Several popular types of approaches are then summarized and compared on the basis of their procedures. Finally, a simply taxonomy is proposed for characterizing and differentiating approaches.

Nature of Decision-Aiding Approaches

A major difficulty in describing decision-aiding approaches is that in practice the development and application of an approach is largely a creative art.[46] There are certainly scientific elements to this art; indeed the mathematics can become quite sophisticated. However, reducing an application to a series of predefined steps is neither possible nor potentially useful. Because there are many different problem situations and the responses of individuals engaged in a decision are unpredictable, the analyst cannot be constrained to a hard and fast, step-by-step process.

Rather, he establishes as a goal a general idea of the information and structure that must be developed to apply some decision logic and then gradually modifies, clarifies, and realizes this goal through reliance on experience, ingenuity, and a "tool kit" of analytical procedures.

Before attempting to clarify this characterization of decision-aiding approaches, it is helpful to provide some background information. To initiate the discussion, a definition of a decision-aiding approach is presented. As noted in the Preface, the scope is confined to normative, analytical quantitative approaches that provide formal criteria for identifying a "best" alternative under situations of risk. Furthermore, only those approaches are included for which there exists a body of practical applications literature.

Definition of a Decision-Aiding Approach

What is a normative, formal, analytical, quantitative decision-aiding approach for dealing with situations of risk? In the broadest sense, a decision-aiding approach is any orderly, systematic way of supporting the process of reaching a decision. Decision-making theory distinguishes between descriptive theory (which is concerned with understanding and predicting how people actually reach decisions) and normative or prescriptive theory (which is concerned with how people ideally should reach them).[47] A normative decision-aiding approach is thus concerned with identifying which of a number of available alternatives is in some sense "best."

Conceptually, normative decision-aiding approaches can range from highly intuitive to very formal. Churchman (1961), White (1975), and others have defined formal approaches as those consisting of three identifiable phases: a problem formulation phase, an analysis phase, and a solution phase. Analytical approaches proceed through the development of a representation (model) of the decision problem based on mathematical or logical symbols. This analytical representation is used as the basis for simulating outcomes or otherwise analyzing the decision options. Quantitative approaches develop numerical measures of the important attributes of the decision — generally, measures of decision outcomes and their valuation.

Risk may be addressed either directly, through the development of quantitative measures of the likelihood and magnitudes of possible decision outcomes, or indirectly, through the use of methods such as sensitivity analysis to determine the range of possible outcomes. This book focuses primarily on approaches that develop probabilities to quantify uncertainty. Such approaches are concerned with the proper way to develop measures for probabilities and values and the proper way to combine these measures into a figure of merit.

Objective versus Subjective Probability

Although probabilities are widely accepted as the preferred means for quantifying uncertainty, there are two fundamentally different views regarding the meaning of probability. These two views are often termed the objective (or classical) definition and the subjective (or personal) definition. Because the analyst's interpretation of probabilities has a significant impact on the nature of the decision-aiding approach, a brief introduction to objective and subjective probabilities is presented. Expanded discussions are provided by many authors, including de Finetti (1968), Savage (1972), Holloway (1979), and Parry and Winter (1981).

Objectivists and subjectivists agree on many of the characteristics of probabilities. Both require probabilities to be numbers between zero and one, with zero implying impossibility and one implying certainty. The basic means for manipulating probabilities are also agreed upon: probabilities may be added to obtain the probabilities of sets of events that are mutually exclusive, and probabilities add to one for sets of events that are collectively exhaustive. The difference between objectivists and subjectivists is that the former interpret a probability as a characteristic of an identifiable physical process while the latter interpret it as a degree of belief held by a given individual.

The objective perspective allows for two means for obtaining probabilities. The means used most often involves a relative frequency interpretation: the probability of an event is the fraction of times the event would occur in a series of identical opportunities. Since the relative frequency of a particular event derived from a small number of trials can be different from that obtained with a very large number, the "true" probability of an event is taken to be that which would occur if an infinite number of trials were conducted. The second means for deriving probabilities is a theoretical approach that relies on conditions of symmetry. The idea is that if there is no information to the contrary, a symmetrical situation is one in which all possible outcomes are equally likely. The usual example is a throw of a six-sided die. Symmetry exists in this case because there is no apparent reason to believe that any one side of the die should be favored over any other. Hence, each side is regarded as equally likely and is assigned a probability of 1/6. With either interpretation, the objectivist is able to view probability as a "hard," measurable property of the real world, even though the actual measurement may be possible only in a "thought experiment."

The subjective view argues that probabilities are not intrinsic to physical objects but are rather a product of perceptions. The numbers assigned to represent probabilities are personal, corresponding to a degree of belief that depends on the information available to the individual, his experience, theories, and so forth. Subjectivists calibrate probabilities using the frequency with which an event would occur in a sequence of trials as a reference. For example, if a subjectivist says that he believes the probability of rain tomorrow is 0.75, he has the same degree of faith in that outcome as he does in randomly selecting a red ball from an urn containing three red balls and one white ball. Thus, though they subjectively generate probabilities, subjectivists have an objective measure in mind when they quote a given value for a probability. The fact that probabilities depend on the full spectrum of information forces subjectivists to consider all information relevant to an assessment, and they have developed logically consistent mechanisms (discussed later) for combining new information with old to permit the revision of probabilities.

Since the objective and subjective views of probability have advantages as well as disadvantages, there is no easy, universal answer to the debate over which interpretation should be adopted for a decision-aiding approach for social decisions involving risk.[48] As will become apparent in subsequent sections, the means for defining and measuring probabilities (as well as other fundamental characteristics of a decision-aiding approach) must be selected according to characteristics of the decision situation.

Characteristics Common to All Approaches

Many of the arguments for, as well as criticisms of, decision-aiding approaches are intimately connected with features common to all approaches. Therefore, before further distinguishing among available approaches, a discussion of their common characteristics is presented.

Decomposition

Decision-aiding approaches are all based on the concept of divide and conquer. *Analysis,* according to Webster, has a Greek etymology (*ana* + *lysis*) and means a loosening or breaking up of any whole into its parts. In the context of decision making, this means that the critical components of a problem should be sorted out, identified, analyzed separately, and then synthesized at the end. Typically, the application of an approach entails the execution of several disaggregated tasks, including structuring the problem, determining probabilities of risky event outcomes, interpreting those outcomes in terms of impacts, and valuing those impacts (Sage and White 1980).

Decomposition has several benefits. It provides a formal structure for collating expertise on different facets of a problem. If assessments of uncertainties are conducted separately from evaluation, it may be possible to tap the expertise of scientists without embroiling them in value judgments. Decomposition also makes explicit the rationale used in reasoning. The more one structures a problem and articulates the rationales for conclusions about component parts, the easier it is for others to evaluate the reasoning used.

Modeling

Decision-aiding approaches are based on the development and analysis of a decision model. Just what the model is designed to represent depends on the criteria adopted for judging the "goodness" of a decision. For approaches that view the goodness of a decision in terms of the processes used to reach it, the model represents an idealized process for reaching a decision. Exercising this sort of model yields a decision designed to approximate what would result if the idealized process could be implemented in the real world. For approaches that view the goodness of a decision in terms of its ability to achieve desired ends, the model relates alternative courses of action to possible decision consequences. In either case, the model is a proxy for the real world.

The principle of decomposition is embodied in models by the fact that the model decomposes a complicated process into more manageable components whose relationships are made explicit. For example, a model relating alternatives to outcomes often consists of submodels for physical, environmental, or market processes; each of these, in turn, is composed of variables representing important factors and their relationships. Thus, the overall decision model is a simplified representation of the important characteristics of the system that will be affected by the decision. Although simplified, the model is meant to include all essential elements and their interrelationships; simplification occurs by omitting nonessentials. Models used in decision-aiding approaches are generally composed of mathematical equations and procedures, but they need not necessarily be highly refined or complex. Anything that illuminates by simplifying and identifying key elements and their relationships, from a simple equation or graph to a sophisticated computer program, is a legitimate model.

Strauch (1980) observes that while a model attempts to match or come very close to the actual problem structure, the complexity and lack of knowledge surrounding situations of risk mean that the models developed in these situations generally differ considerably from the substantive problem. As a result, the coupling between the problem and the model is indirect; one or more layers of formulation, simplification, and redefinition must be inserted between the problem and the model, necessitating corresponding layers of interpretation between analytical results and substantive conclusions. Thus the application of an approach requires not only analysis of a model but also its formulation and the interpretation of its results. Therefore in applying a decision-aiding approach, the analyst must function as a detective ferreting out important issues, as a synthesizer of information, and as a facilitator of the decision process.

Iteration

Iteration is fundamental to all of the decision-aiding approaches considered in this book. The decision model used as the basis for analysis in the application of an approach is constructed through a process of trial and error. The model is repeatedly compared with the real world. If decision makers or others serving as clients or sponsors of the analysis are dissatisfied with the analysis, if statistical or logical evidence indicates that the model does not adequately describe the real situation, or if a previously unrecognized dimension of the problem is uncovered, then some or all of the steps in the decision-aiding approach may need to be repeated. Iteration is thus a crucial component in the formulation stage of analysis. It is also largely responsible for producing the credibility and improved problem understanding that are often the most important benefits derived from the application of a formal decision-aiding approach.

Quantification

The approaches addressed in this book all rely on quantification: data are transformed into numeric scales of measurement prior to analysis. Risk is quantified through the development of probabilities and measures of the magnitude of consequences. Consequences are equated to a common unit (e.g., dollar value). Finally, all components of the analysis are integrated to produce a bottom-line, numeric evaluation of each alternative.

Quantification is highly valued in decision-aiding approaches because it provides a precise language for communication. For example, the scientific judgment that a reduction in emissions will probably lead to reduced pollutant concentrations is much more ambiguous than the statement that scientist *X* believes that a 50 percent reduction in emissions will lead to a decrease of at least 4 micrograms per cubic meter in average ambient concentration at location *Y* with an estimated 90 percent probability. The latter statement can always be expressed as the former if it is more appropriate for a given audience. However, the reverse, of course, is not true.

Since quantification represents a specific language for expressing information, models based on quantification will have errors to the extent that holistic or qualitative elements of the problem are not captured. Decision makers may respond to aspects of the problem not captured by variables transformed to quantitative scales. The analysis, however, is based solely on the transformed data. This as well as other limitations of formal decision-aiding approaches will be addressed in Chapter 3. The remainder of this chapter presents various underlying theories and analytic procedures that distinguish available approaches.

Decision-Making Theories

To aid the characterization of decision-aiding approaches, this section describes three distinct, internally consistent theories of normative decision making: cost-benefit theory, decision theory, and social choice theory.[49] Understanding the basic theories is important for three reasons. First, the theories serve as a source of (implicit) philosophical rationales for most decision-aiding approaches. Second, the procedures used in the various approaches are to a considerable extent natural derivations from the fundamental theories. Third, the criteria by which approaches are evaluated as "right" or "wrong" must stem directly from a philosophical judgment of the appropriateness and usefulness of the theories. As will be made clear in the final section of this chapter, the basic decision-making theories described below play an important role in the framework proposed for distinguishing among approaches.

Cost-Benefit Theory

Cost-benefit theory (CBT) is the basic theory underlying cost-benefit analysis. It emerged from applications of economic theory originally conducted in the late nineteenth century directed at evaluating proposals for the construction of waste-water disposal systems. The intellectual father of CBT is said to be the French economist and engineer Jules Dupuit, who in 1844 wrote the frequently cited study "On the Measure of Utility of Public Works." Even today most authors of cost-benefit literature are economists.

The basic premise of CBT is that alternatives should be selected according to a systematic comparison of the advantages (benefits) and disadvantages (costs) that result from the estimated consequences of the choice. Thus the theory does not involve the concept of a social decision maker with special responsibility for the decision. Individuals define the perspective for establishing ultimate consequences, and individuals are assumed to be the appropriate judges for valuing consequences. More specifically, CBT identifies a "best" alternative in terms of an "efficiency criterion" which specifies both how advantages and disadvantages to individuals should be measured and how they should be aggregated to obtain an overall measure of social good. To value consequences to individuals, the concept of willingness to pay is applied; that is, an individual's valuation of impacts is what he would be willing to sacrifice to incur or avoid them. To aggregate values to the social level, individuals' values — the values of all consequences, regardless of to whom they accrue — are added together to yield a measure of the total social utility of the alternative. Thus, CBT is concerned with the maximization of the aggregate value of goods and services consumed by individuals.

According to CBT, the adoption of the principle of willingness to pay implies the use of market values for valuing marketable commodities. Hirshleifer (1975) justifies this through an "equi-marginal principle":

If a loaf of bread has a market value of one dollar, this means that it is valued (to a close approximation) at one dollar by everyone in the community. For, every consumer will have adjusted his rate of purchase of bread until the last loaf purchased is worth (has a marginal value) to him just equal to the worth to him of whatever else he could have purchased for one dollar. And similarly for production as well as consumption. Suppliers of bread will devote resources to its production only up to the point where the marginal value of a resource in bread making is equal to its yield elsewhere. So market values are not arbitrary measurements. They are a kind of social consensus, approaching unanimity, as to the worth of increments or decrements of products and resources.

The equi-marginal principle implies that a free market (one in which competition governs the exchange of all goods and services and there are no externalities or other impediments to the allocation of resources) would lead to a completely efficient allocation of resources. For a private good (where consumption by any individual eliminates the availability to others), prices resulting under such a situation would exactly measure the marginal benefit realized from each individual's consumption of it. For public goods (such as entertainment, where consumption by one individual does not diminish the amount available to others), the sum of the marginal benefits (across individual consumers) equals the marginal costs of provision. Because CBT appeals to the voluntary actions of individuals in free markets to derive aggregate preferences, it may be argued that people implicitly consent to its use as a basis for social decisions. MacLean's model of implicit consent discussed in Chapter 1 thus provides a possible foundation for the legitimization of CBT.

Although the argument for CBT is conceptually appealing, its acceptance embodies several troublesome implications. Most notably, the basing of social decisions on aggregate net benefits means that CBT considers only total societal welfare; it is insensitive to the distribution of that welfare across people or groups within society. Thus CBT could result in the redistribution of resources from the poor to the rich, so long as the rich gain more than the poor lose. The assumption that market values reflect willingness to pay also has deficiencies. It assumes that what was true for the past should be true for the future. It also applies only to people who purchase the commodity. The value of pork to a Moslem or bourbon to a Mormon may be less than the market price. Furthermore, if the market for a commodity is distorted in any way, by taxes or externalities, for example, then prices may not equal individuals' willingness to pay. In any case, individuals' willingness to pay for impacts of importance that are not traded in the marketplace must be established in some other way. (The methods for applying CBT are discussed later).

The efficiency criterion underlying CBT is sometimes referred to as a criterion of "potential Pareto optimality."[50] Pareto optimality is an economic principle that states that a decision is an improvement on the status quo if it makes one or more persons better off without simultaneously making at least one other person worse off. Potential Pareto optimality, as defined by the efficiency criterion, is not as restrictive as Pareto optimality. The efficiency criterion states that a decision is an improvement if those who are better off could potentially share some of their gains with those who are worse off in such a way that everyone would be better off. It does not, however, require that such compensation actually be made.

Hirshleifer *et al.* (1974) have defended the efficiency criterion on several grounds. For one thing, application of this decision-making rule will tend to increase the size of the total economic "pie." The redistribution of these returns can be a separate decision for policy makers. Taxes, for example, may be used to redistribute gains. Another argument is that of the many projects to be evaluated; some may benefit the rich, but others are likely to benefit the poor. In the long run, as the argument goes, everyone will be made better off.

The efficiency criterion has important implications for the values that are reflected in decision making. The value perspective implied by CBT is that of society as a whole, rather than any particular person or group within it. The theory considers the benefits and costs that occur throughout the social system, not merely the direct benefits to intended beneficiaries and direct expenditures by the implementing party. Furthermore, authority for value judgments (e.g., expressed through market prices) is decentralized to all stakeholders in the decision. Thus the specific preferences of the policy maker or the government agency responsible for the decision may not be reflected by CBT. In addition, the focus of CBT on the consequences of a decision means that the process by which alternatives are generated or selected (e.g., the extent of participation of affected parties) is not considered.

Anderson *et al.* (1981) summarize several additional assumptions embodied in CBT. Two have to do with the relationship between monetary value and welfare (utility). Implicitly, it is assumed that this relationship is linear (i.e., an individual will be twice as well off with twice the money) and that the relationship is the same for all persons (a given increment of money would provide all individuals with the same pleasure). CBT also implies a fundamentally materialistic view of individuals and society. On the individual level, CBT collapses multidimensional impacts into a single monetary equivalent, in effect assuming that "individuals can feel just as well off with greater command over marketable commodities but less of certain nonmarketable gratifications as in the reverse situation" (Rothenberg 1975). In effect, individuals can be compensated with money for any impact. The same assumption holds true at the societal level: any adverse effect with a given monetary value can be offset by any favorable effect of the same value. Thus resources do not have intrinsic value; their value is derived from their consumption as part of a mixture of resources which maximizes some objective. Another inherent assumption is that the current societal distribution of income is equitable. This is implied because market prices and willingness to pay depend on individuals' wealth as well as on their preferences. Therefore, accepting existing market prices requires accepting the existing distribution of wealth, because an individual's influence on prices is proportional to the individual's wealth.

Decision Theory

Decision theory (DT) describes how individuals should make decisions in the face of uncertainty. The origins of DT can be traced to the eighteenth-century mathematician Daniel Bernoulli, who held that choice logically depends on the probabilities of the consequences of a decision and the utility (worth) of those consequences to the decider.[51]

The criterion used by DT to identify a "best" alternative is largely a result of von Neumann's and Morgenstern's efforts (1947) "to find mathematically complete principles which define 'rational behavior.'" Their approach consisted of proposing a set of basic axioms of rationality. Most individuals find these axioms quite compelling. For example, one axiom states that choices are comparable. Offer a decision maker any pair of outcomes or lotteries, and that decision maker must be able to identify one or the other as preferable. Another deals with transitivity of preferences. It states that if a decision has three outcomes, A, B, and C, and if the decision maker prefers A to B and B to C, then the decision maker must prefer A to C.[52]

The von Neumann–Morgenstern axioms of rationality can be expressed in a variety of ways (cf. Keeney and Raiffa 1976; Hooker *et al.* 1978; Schoemaker 1980). Each statement of the axioms, however, leads to a proof of two results central to DT. The first is that a decision maker's preferences can be encoded in terms of a function called a utility function, which represents a scaling of the values the decision maker assigns to decision outcomes that captures his attitude toward accepting risk. The second result is that a decision maker's preference for various alternatives may be measured by calculating expected utility (the sum of the utilities of possible outcomes weighted by their probabilities of occurrence). The preferred alternative will have the highest expected utility. The axioms themselves do not include a notion of expected utility. Rather, this criterion is a logical consequence of the axioms.

Most modern decision theorists adopt the subjective definition of probability; that is, probability is not an objective property but exists only in the mind of the decision maker. Thus probabilities assessed by an individual are regarded as subjective and dependent on the information, experience, and theories, of that individual. Keeney and Raiffa (1976) emphasize the subjectivist view of DT by arguing that its purpose is "to elaborate the logical implications of subjective probability and utility theory and to extend them by means of mathematics."[53]

The criterion of modern DT can therefore be expressed as subjective expected utility — the sum of the utilities of an alternative's possible outcomes weighted by their subjective probabilities of occurrence (Edwards 1954). The key to DT is the construction of the "decision set": specifying the available alternative actions; defining a set of outcome variables, the outcome vector, on which the outcome will be judged; assigning probabilities on the outcome vector given each alternative; and finally establishing a utility function on the outcome vector. By the axioms of rational behavior the preferred alternative is the one with the highest subjective expected utility. The validity of DT is its mathematical logic. Once the decision set is developed, the theory stands as a logical structure for decision making. Aiding decision makers to achieve consistency is, thus, a principal justification for DT.

A number of assumptions are implied in DT. Most significant of these are that (1) all possible events and all significant consequences can be enumerated in advance, (2) meaningful probability and utility values can be obtained and assigned to them, and (3) the often disparate outcomes of concern to the decision maker can somehow be made comparable to one another. Another assumption is that the decision maker accepts the von Neumann–Morgenstern axioms of rationality. These axioms can be shown under some problem specifications to imply decisions that are counter to common behavior, particularly in cases where decision outcomes involve large amounts of money or when the probabilities or differences among the probabilities associated with the choice are small. Allais (1953), among others, has constructed a number of examples of this type. Decision theorists argue, however, that it is precisely these departures of intuitive decision making from logical theory that make the theory useful (Howard 1968a). They contend that once the axioms are carefully explained, any reasonable person would want to change his behavior to be in accord with these axioms.

One frequently cited implication of the axioms is that DT is result oriented. For example, it is sometimes argued that the decision maker must be indifferent between two uncertain situations yielding the same outcomes with the same probabilities, even if the situations involved quite different physical processes to generate these outcomes. This does not mean that process implications may not be included in outcome definitions. The decision maker may, for example, express a difference in preference over breaking a leg while skiing on a mountain as opposed to tripping on the walk back to the car. All that is required is to incorporate into the outcome definitions important process characteristics.

Finally, it must be emphasized that DT is focused on individual decision making. Using DT as a basis for reaching social decisions requires answering a difficult question: how should individual decision making be extended to collective decision making?

Social Choice Theory

Social choice theory (SCT) takes the perspective that the appropriate criterion for social decisions is not the preferences of some single decision maker but rather a rational synthesis of the preferences of all those individuals who will be affected by the decision. The theory is thus concerned with finding decision rules or procedures by which preferences specified by individuals may be incorporated into the decision process. It may be interpreted, therefore, as a "normative theory of elections and legislative choices" (Arrow 1977).

Research in SCT has taken several related directions. Harsanyi (1977a) divides the field into game theory and ethics. Game theory is concerned with individuals who are pursuing self-interest and personal values against other individuals who are doing the same. Ethics, on the other hand, involves the rational pursuit of the interests of the group as a whole. In the area of ethics, the development of normative theory has taken two courses. One line of investigation has attempted to derive ways of aggregating individual preferences into a social utility function that can then be used to extend individual decision-making theory to group decision theory. The second approach is the analysis of mechanisms by which group decisions may be reached directly through a process (such as voting) in which individuals encode and simultaneously incorporate their preferences in the decision-making process without explicitly developing a group utility function (Lind 1972).

In developing normative rules for social choices, researchers naturally have sought to minimize restrictive assumptions and difficult input requirements. A question central to the theory has been whether ordinal expressions of individual preferences are sufficient inputs to a rational process of social choice. Ordinal expressions of preference are limited to statements such as A is preferred to or indifferent to B. Cardinal expressions require statements of the strength of preferences — that is, the amount by which an individual prefers A to B must be measured. Procedures that require individuals to express the intensity of their preferences are particularly vulnerable to subversion by the strategic misrepresentation by individuals of their true preferences (Green and Laffont 1979).

The most significant results of SCT research to date are, in a sense, disappointing: whenever a minimum set of reasonable sounding conditions is imposed on the process of forming social choices from individual preferences, then it can be demonstrated that no processes will always satisfy those conditions. The central result is Arrow's impossibility theorem (Arrow 1974; Keeney and Raiffa 1976), which states that given only ordinal information on individual preferences over a set of alternatives, there is no collective ranking of those alternatives that does not contradict one of the following assumptions:

1. *Complete domain.* The procedure works for all possible patterns of individual preference rankings (for at least two individuals, three alternatives). In particular, the social orderings produced must always be transitive; if A is ranked ahead of B and B ahead of C, then A would also be ranked ahead of C.[54]
2. *Positive association of social and individual orderings.* If the group originally prefers A to B, and if any change is, if anything, in favor of A, the group still prefers A to B.
3. *Independence of irrelevant alternatives.* If an alternative is eliminated from the possible set, the group ordering for remaining alternatives is unchanged.
4. *Individual's sovereignty.* For each pair of alternatives A and B, there is some set of individual orderings such that the group prefers A to B.
5. *Nondictatorial.* There is no individual with the property that whenever that person prefers A to B, the group will also prefer A to B regardless of the others' preferences.

Arrow's theorem has been generalized by Sen (1970), who proves that there is no procedure for quantifying a group's preference structure that is consistent with the spirit of Arrow's assumptions and that does not include the interpersonal comparison of preferences. Gibbard (1973) and Green and Laffont (1979) extend this further by showing that no social choice mechanism exists that is nondictatorial and that can never be advantageously manipulated by some member of the group. Intuitively, this says that all nondictatorial social choice mechanisms will be subject to the misrepresentation of preferences if these preferences are privately held pieces of information. Even if some optimal social decision-making rule can be found, it is not clear that individuals would reveal their true tastes. Hence, in practice such a decision process may find the optimal social choice for a distorted set of preferences and result in allocations that are far from optimal.

Despite the difficulty associated with assessing cardinal preferences, the impossibility theorem has tended to force normative SCT research toward theories based on cardinal utilities. The cardinal utility functions used in SCT may be of the sort derived by von Neumann and Morgenstern, which measure preferences over uncertain outcomes, or the sort used in conventional microeconomics, which measure the strength of preferences for multiattribute consequences known with certainty. Some authors (e.g., Howard 1968a; Keeney and Raiffa 1976) distinguish von Neumann and Morgenstern utility functions from those used in classical economics by referring to the latter as (measurable) value functions (Dyer and Sarin 1978a and 1978b). Like decision-theoretic utility functions, the existence of a value function for measuring an individual's preferences may be derived using certain assumptions regarding rational choice.

Efforts in SCT to develop social utility functions generally assume that each individual's preferences are described by a utility (or value) function and proceed by expressing the group utility function in terms of the individual utility functions. To deduce the appropriate form for this expression, specific properties are postulated for the process of aggregating individual preferences. These properties are then used to derive the relationship between the group utility function and the individual utility functions describing group members' preferences.

Several different sets of postulated properties for the aggregation of individual preferences have been explored. Harsanyi (1955, 1975), for example, considers a situation in which it is assumed that group preferences satisfy the same axioms of rationality as individual preferences, and if all individuals within the group are indifferent between two alternatives defined by probability distributions over their consequences, then the group will be indifferent between them. Harsanyi shows that these assumptions imply that the social utility function must be a linear combination of the individual utility functions; that is, group utilities are weighted averages of individual utilities. The linear form has also been derived using other sets of postulated properties. Kirkwood (1979), for example, shows that the linear form is obtained if Pareto-efficiency is adopted as the principle for aggregating individual preferences (i.e., if one alternative is better than a second for at least one individual and no worse for any other, then that alternative must be preferred by the group). Harsanyi (1977a) obtains the linear form using the principle of "impersonally situated" rational decision makers. An individual is said to be impersonally situated if he faces an equal chance of ending up in the shoes of any of the citizens affected by a policy decision. Harsanyi shows that if an impersonally situated citizen is rational in the von Neumann–Morgenstern sense, it follows from the expected utility maximization theorem that the he will select the policy that maximizes the sum of the utility function associated with each individual.

Keeney and Kirkwood (1975) find that the following three assumptions are sufficient to require the group utility function to have a linear form: (1) social utilities depend only on individual utilities; (2) if everyone has the same utility function, then the social utility function must be "strategically equivalent" to it, meaning that it is a scaled version of the same utility function and hence leads to the same choices; and (3) if every group member but one is indifferent between two alternatives, then group preferences should match those of the individual who is not indifferent. Keeney and Kirkwood then extend their analysis by relaxing slightly their strategic equivalence under unanimity assumption. Their revised assumption is that group preferences between any two alternatives should depend only on the preferences of those individuals who are not indifferent between them. When this is substituted for their assumption (2) above, a multiplicative form (for which the linear form is a special case) is obtained. Thus, while the linear form is most common, not all derivations lead to the linear result.

The most obvious difficulty associated with SCT is the need to assume utility functions for each individual affected by a social decision. The assumption implies that all members of the group are "rational" in the sense that they wish to behave in accordance with a set of axioms defining rationality. Exhaustively identifying the individuals affected by a decision is likely to be controversial; obtaining their true preferences would be extremely difficult. In any real application, participants might be tempted to alter their reported utilities rather than report their true preferences so as to influence the decision. If the preferences are hypothesized rather than directly assessed, any group not in agreement with the decision would probably argue that its preferences were misrepresented.

Another major difficulty is determining how individual utility functions should be aggregated to obtain a social utility function. If, for example, the social function is a weighted average, selecting the weights will be controversial. At a more fundamental level, even though additive and multiplicative forms can be derived by making assumptions regarding interpersonal comparisons, there is no demonstrably correct procedure for making these comparisons. Should welfares be made equal for all people? Or should greater welfare be the reward of those who show more ability, work harder, or behave better? As pointed out in Chapter 1, these questions are inherent to any regulatory decision that produces increased safety at the expense of increased prices or higher taxes. That SCT makes these interpersonal comparisons in an explicit way ensures that the process will be controversial.

Differences among Decision-Making Theories

Table 5 summarizes important differences among CBT, DT, and SCT. As illustrated, the theories differ not only in their respective criterion for defining a "best" decision but also in the perspective each adopts in its interpretation of probabilities, values, and information.

An interesting parallel may be drawn between the three identified decision-making theories and Linstone's "multiple perspectives on technology assessment" (1981). Linstone distinguishes among three perspectives: technical, personal, and organizational. The technical perspective represents an impersonal worldview based on scientific objectivity. Its preferred modes of inquiry can be traced to the thinking of Locke (data-based), Leibniz (models), and Kant (data plus model- or multi-

Table 5. Comparison of decision-making theories

	Cost-benefit theory	Decision theory	Social choice theory
Intellectual roots	Engineering economics	Engineering, psychology, management science, economics	Welfare economics
Conceptual basis	Economic efficiency	Axioms of individual choice	Axioms of social choice
Method of analysis	Comparison of aggregate value of estimated consequences of alternative actions	Determination of logical implications of alternatives, information, and preferences of decision maker	Derivation of group decision from acceptable mechanisms for incorporating individual preferences
Perspective on value	Total monetary equivalent as determined by economic actors in a free market	Responsibility of decision maker, objective is consistency	Social preference derived from "equitable" synthesis of preferences of impacted parties
View of uncertainty	Objective characterization of environment	Subjective beliefs of the individual	Product of individuals coping with erratic environment

model-based). This may be associated with CBT, which takes a technical view of decision making in its aggregate measure of value and its objective view of information and uncertainty. Linstone's personal perspective focuses on the individal. It is subjective in its worldview: reality is the perception of the senses. This may be associated with DT, in which the decision-making unit is taken to be an individual, and information, uncertainty, and values are all taken as subjective. Linstone's organizational perspective is concerned with structure and the fundamental importance of social processes. This can be associated with SCT, where the focus is more on equity and the process by which individuals incorporate their preferences into decision making than on the logic or efficiency of the end results.

Decision-Aiding Approaches

Decision-aiding approaches are distinguished both by the basic theory or theories on which they are based and by the specific procedures used in their execution. A wide variety of procedures exists for dealing with virtually every aspect of analysis, and the flexibility of most of these is such that a great many variations are possible. Essentially, therefore, procedures can be defined and combined in an infinite number of ways to produce approaches.

Because it is impossible to describe every possible approach, consideration is given to several classes or "types" that are discussed in the literature. These are the Paretian approach to cost-benefit analysis, the decision-maker approach to cost-benefit analysis, the clinical art approach to decision analysis, the supradecision-maker approach to decision analysis, social decision analysis, and applied social welfare theory. Brief overviews of these basic types of approaches are provided below, and more detailed discussions of the procedures used in decision-aiding approaches follow. For more complete descriptions and applications, the reader is referred to the cited literature. Examples of several types of approaches — the Paretian approach to cost-benefit analysis, social decision analysis, and several variations of applied social choice theory — are presented in the appendices.

Cost-Benefit Analysis Approaches

Cost-benefit analysis (CBA)[55] represents the implementation of cost-benefit theory. It is, in essence, a pragmatic methodology for determining whether a proposed change in the use of resources will improve efficiency. Sugden and Williams (1978) and Watson (1979) recognize two distinct types of CBA — a Paretian approach and a decision-maker approach.

The Paretian approach to CBA is a relatively uncompromising application of CBT. To the extent possible, consequences of alternatives are estimated in an objective way and preferences are deduced from choices and prices observed in the marketplace. Some practitioners evaluate only those commodities and services that have a readily measurable market value (e.g., costs, sales, wages), while others extend the range of considerations assigned dollar values by using indirect economic valuation methods (e.g., Parish 1976).

Because controversial assessments are avoided in the Paretian approach, the analysis is recognized as being incomplete. The advantage is that there is no introduction of noneconomic consequences to "confuse the analysis" or "obscure the purely economic facts" (Watson 1979). Results are regarded as providing only one of many necessary inputs to a public policy decision: namely, a measure of the economic efficiency of an option. The conclusion is thus "an ethical prescription of a rather weak kind: 'such-and-such course of action ought to be taken, provided that it does not adversely affect social welfare in some dimension other than that of economic efficiency' " (Sugden and Williams 1978).

The decision-maker approach to CBA attempts a much more ambitious accounting of decision factors. In addition to including estimates of dollar values for decision outcomes that are not traded in the marketplace, this form of CBA may even include factors not addressed by CBT (e.g., by including personal values of decision makers and measures of the equity of distributional impacts). While the Paretian approach is meant to produce an objective model of the environment, the decision-maker approach settles for a quasi-objective representation (Watson 1979).

Because the decision-maker approach attempts to introduce explicitly all issues of importance into the analysis, including the decision maker's evaluations and judgments, it is viewed by practitioners as providing a close approximation to the decision problem. Speaking of this approach, Sugden and Williams say, "There is, then, no sense in which a decision maker could regard the outcome of a cost-benefit analysis as an independent piece of evidence on which to base a decision; it *is* his decision."

In theory, the decision-maker approach to CBA should rely on decision makers for the development of assumptions and the establishment of measures for costs and benefits. Applications, however, generally appear to be conducted without detailed participation of decision makers (Van Horn and Wilson 1976). Applications of cost-benefit analyses include such diverse areas as education programs, prison reform, disease control, labor and manpower training programs (Anderson and Settle 1977), facility siting (Flowerdes 1972; Abelson 1979), public construction (Foster and Beesely 1963), resource development and conservation (Krutilla and Eckstein 1958; Throsby and Abelson 1976; Australian Department of Environment, 1980) and regulation and standards setting (Moll *et al.* 1975; Jackson *et al.* 1976). Appendix A contains an example of a CBA approach.

Decision Analysis Approaches

Decision analysis (DA)[56] uses concepts of analysis from systems engineering to implement decision theory. Although the philosophical justification of DA thus rests on the assumption of a single decision maker, several schemes have been developed for applications to social problems.

The simplest way to apply DA to social decisions is to assume that a single individual (or group of individuals with coincident views) has the responsibility and authority for making public decisions. Thus the group decision problem is reduced to a single decision-maker case, and the justification for DA follows from the same arguments that justify its use for personal decision making. Buede (1979) uses the phrase "clinical art approach" to denote a version of DA which relies heavily on the skills of the decision analyst to elicit and structure the elements of the decision set — alternative actions, outcome variables, probabilities, and utilities — required to apply formal decision theory. The clinical art approach can be applied either in a one-to-one session with the decision maker or, in the case where the decision maker wishes to involve his staff, as a conferencing exercise. Among the applications of clinical DA described by Buede is a conferencing application involving an analysis of systems acquisition decisions conducted for the military.

A variation of the social decision-maker concept is the "supradecision-maker" model for applying DA in the social sector. The assumption of this approach is that a (supra) decision maker (i.e., a benevolent dictator) would want to synthesize a social utility function from the preferences of the individuals affected. The procedure involves assessing a utility function for each of the affected parties and then obtaining parameters for weighting the preferences from the (supra) decision maker. Loosely speaking, the weights reflect the decision maker's view of the relative importance of the affected parties' preferences.

The alternative approach to applying DA in the public sector is to base the analysis directly on group judgments and preferences without recourse to a single decision maker. Howard's "social decision analysis" (1975) and a similar approach proposed by Edwards (1977) advocate specific schemes for obtaining probabilities and utilities that represent society as a whole. With social DA, the decision analyst creates an extrapersonal, explicit model of the decision under consideration. Information on possible alternatives, uncertainties, relationships, and preferences comes from different groups and is represented in the decision model, with the implication that the basis for the decision is apparent to all. The most important distinction between the supradecision-maker approach and social DA is the source of authorization and verification of assumptions. Social DA tends to rely more on normative arguments. Applications include evaluations of automotive pollution regulations (Howard 1975), hurricane-seeding strategy (Howard *et al.* 1972), synthetics fuel programs (Tani 1978), and breeder reactor research and development (Manne and Richels 1978). Appendix B provides an example of a social DA approach.

In all DA approaches, the decision analyst relies heavily on interviews conducted with decision makers and technical experts to obtain the information on which the analysis is based. An important characteristic of these interviews is the posing of choice situations (frequently involving only two options) in which the subject must indicate either a preference or indifference among clearly defined alternatives. The process tends to be dynamic with subsequent questions determined from the subject's responses to earlier choices. A complete analysis using this method may require hundreds of judgments and decisions from a single individual. Howard *et al.* (1975) and Keeney (1977) provide examples of extensive DA interviews.

Applied Social Welfare Theory Approaches

Social welfare theory (SWT) approaches (Dyer 1978) reflect the ethical setting of social choice theory; that is, their concern is the identification of group decisions that maximize the welfare of the group as a whole. Such approaches develop a social welfare function for aggregating individual preferences to yield a preference rule for society. Because of the limitations of social welfare theories based on ordinal preferences, applications have tended to make use of cardinal preference representations — either value functions or von Neumann and Morgenstern utility functions.

Uncertainty in the consequences of a decision may be dealt with by SWT approaches through either a centralized or a decentralized process (Cox 1986). With the centralized approach, individuals provide value functions describing their preferences for different time streams of outcomes, or utility functions describing their preferences for specified probability distributions over those time streams. In both cases individuals are asked to provide preference data without regard to exactly what the outcomes to the decision might be. The estimation of decision outcomes is done centrally using models — for example, models for aggregating the subjective probabilities of individuals for possible decision outcomes or models based on the expertise of specialists. Thus preferences for consequences are separated from consequence estimation, and the combination of the two is centrally conducted to obtain a measure of the expected increase in social welfare associated with each alternative.

With the decentralized approach to SWT, each individual is asked to express his or her preferences for each alternative directly. The social welfare function is then concerned with the aggregation of individual preferences over alternatives and does not explicitly concern itself with the uncertainty in the consequences to those alternatives. Essentially the individual is assumed to determine preferences for each alternative by relying on his her own beliefs and information regarding its likely outcome. In this case a decentralized process is used to deal with uncertainty.

Decision-aiding approaches have also been proposed that reflect the results of social choice theory research in game theory. For example, if the social decision can be regarded as a cooperative game, then various game theoretic concepts and models can be applied to identify resolutions that satisfy certain normative arguments. (A cooperative game is one in which participants have the opportunity to communicate with one another and to form binding and enforceable agreements.) Game theory typically concentrates on the payoffs that a decision brings to the individual players in the game rather than on aggregate social welfare. Because applications of game theory approaches are extremely rare compared with other types of approaches, they are not dealt with in detail in this book. Appendix C, however, illustrates an approach based on cooperative game theory as well as several approaches based on the development of a social welfare function. Additional examples of SWT applications are provided by Keeney and Raiffa (1976) and Fisher and von Winterfeldt (1978).

A Comparison of the Procedures Used in Decision-Aiding Approaches

Table 6 provides a summary of the various types of decision-aiding

Table 6. Summary comparison of several types of decision-aiding approaches

	Cost-benefit analysis	Decision-maker cost-benefit analysis	Social decision analysis	Supradecision-maker decision analysis	Clinical decision analysis	Applied social welfare theory
Description	Straightforward application of cost-benefit theory	Cost-benefit theory augmented to reflect decision maker's preferences	Decision analysis from a social perspective	Decision analysis based on decision maker's "imaginative empathy"	Straightforward application of decision theory	Application of social choice theory
Decision criterion	Maximization of economic efficiency	Maximization of decision maker's definition of equitable equivalent economic value	Maximization of expected social value	Maximization of decision maker's subjective expected social utility	Maximization of decision maker's personal subjective expected utility	Maximization of social utility
Typical characteristics and procedures	Models based primarily on empirical relationships		Models based primarily on experts' judgments of causal logic		Models based on decision maker's assessments	Models based on simple empirical or theoretical relationships
	Outcomes valued in monetary units			Outcomes valued in units of utility		
	Outcome valuation based on market prices and willingness to pay	Outcome valuation based on market values, willingness to pay, and decision maker's preferences		Outcome valuation based on interpersonal preference comparisons	Outcome valuation based on decision maker's preferences	Outcome valuation based on interpersonal preference comparisons
	Limited consideration of uncertainties through frequency distributions and sensitivity analysis		Explicit treatment of expert uncertainties through probability encoding			Limited consideration of uncertainties based on subjective or objective probabilities

approaches and their general characteristics. As illustrated, the different types differ from one another both in terms of the criterion adopted for identifying a given alternative as "best" and in terms of the characteristics of the procedures they employ to implement their criterion.

To a considerable extent the procedures used in a decision-aiding approach can be traced to the theories on which the approach is based. For example, CBA approaches (following CBT) rely heavily on market arguments and procedures based on the aggregation of data. Probability is used sparingly; and when it is used, the objectivist view is normally taken (i.e., probability is interpreted as an objective property of the physical system under study), and the procedures for manipulating probabilities are derived from classical statistics. DA approaches (following DT) emphasize interviewing procedures designed to reduce cognitive biases and to minimize the influence of the analyst on the judgments expressed. The subjectivist view of probability is adopted. SWT approaches focus mainly on the representation of social preferences and emphasize procedures drawn from utility theory.

The strong connection between the procedures used in a decision-aiding approach and its associated decision-making theories is based in part on a conscious effort by the analyst to maintain a sound theoretical rationale for the analysis. In part, the selection of procedures may also be due to biases in the training and professional experiences of the analyst. For example, DA is generally taught in engineering and business schools, and CBA tends to be taught in the economics departments of universities. No doubt this helps to explain the fact that decision analyses tend to use engineering and management science procedures, and cost-benefit analyses rely heavily on procedures from economic theory.

Although the procedures of an approach tend to be associated with the primary decision-making theory on which the approach is based, most approaches also use procedures that are closely related to the other theories. For example, the decision-maker approach to CBA may use procedures derived from DT or SCT for representing preferences. Social DA generally uses willingness-to-pay and efficiency arguments from CBT to establish a social utility function. SWT approaches often rely on procedures used in DA for encoding individuals' probabilities and utilities. Procedures consistent with different theories are used for reasons of practicality; if the lack of necessary inputs, computational resources, or expertise make it difficult to apply one procedure, the resourceful analyst uses another. This borrowing of procedures tends to blur the philosophical rationale underlying approaches and creates some conflict of perspective.

Because many procedures are used in decision-aiding approaches, some means is needed to organize a discussion about them. One way of categorizing procedures is according to the elements of analysis to which each relates. All of the approaches described above involve steps that attempt to separate the knowledge relevant to determining the implications of a decision from the values that permit those implications to be evaluated. The six basic elements identified in this generic partition of a decision-aiding approach are summarized in Figure 5.[57] Knowledge is

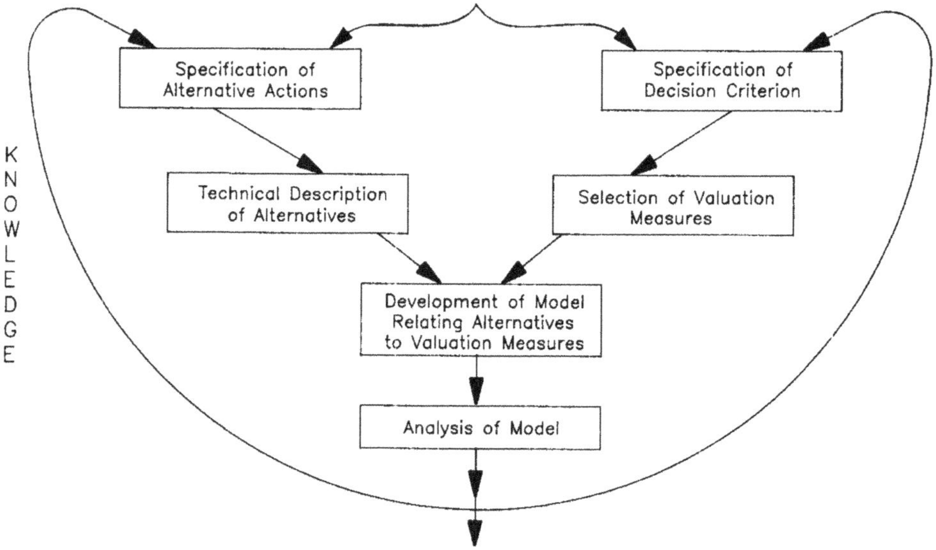

Fig. 5. Generic partition of a decision-aiding approach useful for procedure identification.

represented by (1) specifying alternative actions and (2) characterizing those alternatives in terms of technical descriptors. Values are represented by (3) defining a decision criterion and (4) specifying measures for evaluating decisions according to this criterion. To complete an analysis requires (5) developing a model that translates the technical descriptions of the alternatives to the measures needed by the decision criterion and (6) analyzing the model to identify a "best" alternative and provide measures of risk and other information useful to decision makers.[58] The sections below provide more detailed discussions of the procedures typically used by various approaches within each of these basic elements of analysis.

Specifying and Characterizing Alternatives

Selecting and characterizing alternatives is a crucial step for a decision-aiding approach, since the comparison is obviously limited to the set of alternatives analyzed. Alternative generation is essentially a creative rather than analytic process (e.g., Newell and Simon 1972; Wasson and Johnson-Laird 1972), and methodological descriptions of decision-aiding approaches rarely provide much advice about the specific procedures to be used in this step. As a result, the diversity of the alternatives represented in an analysis depends largely on the analyst's understanding of the problem and the effort devoted to the creation of options.

In general, DA approaches tend to evaluate a wider and more diverse range of options than CBA approaches. Decision analysts tend to interact most extensively with the various participants in the decision process, which often leads to a wider consideration of alternatives. Whereas a DA will typically evaluate a range of regulatory options or project configurations, CBA most often evaluates a yes-no decision about one particular project (i.e., its costs either do or do not outweigh its benefits) (Slovic and Fischhoff 1982). DA approaches also often evaluate contingency plans (policies for taking future actions as a function of anticipated information). The subjective interpretation of probability held by decision analysts allows them to account for anticipated information producing a future improvement in the ability to estimate decision consequences. Thus in recommending an initial action, DA approaches can account for the fact that some actions may be more or less desirable than others because they affect the ability to select actions in the future as more information becomes available.

Cost-benefit analysis approaches rarely evaluate contingency policies, because the objective view of probability makes it more difficult to update the estimation of consequences as new information arises. Contingency plans are also rarely evaluated using SWT approaches. To do so requires the centralized approach to assigning probabilities because the decentralized approach would result in individuals using different assumptions for future actions. Even the centralized approach, however, cannot really capture people's preferences for time streams of consequences resulting from future social decisions because those preferences are likely to depend on the processes by which those decisions are made (e.g., the degree of their participation) and on the institutional arrangements (such as insurance) that determine how those consequences affect them personally.

The means used by decision-aiding approaches for characterizing alternatives is largely a matter of selecting an appropriate level of simplification so as to achieve computational tractability without omitting critical factors. The analyst simplifies by initially screening options to eliminate possibilities that appear less favorable or by grouping alternatives that are similar into distinct classes and then selecting a "typical" member from each class to represent the spectrum of possibilities within that class. Simplification also occurs in determining how the alternatives for analysis should be specified. Because actual regulatory actions may be quite complex, analysts often characterize an alternative by an approximation to its effect rather than by the actions that would actually be taken. For example, a regulation governing the emission of atmospheric pollutants might contain different restrictions for different sources, depending on such factors as geographic location, age, and so forth. Instead of characterizing these alternatives in terms of their actual requirements, they might be specified by the net reduction in emissions that they are likely to achieve. Such differences among decision-aiding approaches are less a function of the type of approach than they are of the complexity of the problem and the computer resources available to the analyst.

Decision Criteria

Cost-benefit analysis approaches strive to implement the efficiency criterion of cost-benefit theory. This is normally accomplished by investigating whether the consequences of an alternative are likely to increase net present value; that is, the present value of all benefits less that of all costs. Frequently, benefits and costs are estimated on an annual basis, in which case the net present value, NPV, is calculated from an equation of the form

$$NPV = B_0 - C_0 + \frac{B_1 - C_1}{(1 + r)} + \cdots + \frac{B_T - C_T}{(1 + r)^T}$$

where B_t and C_t are the benefits and costs in year t, r is the discount or interest rate (assumed to be constant), and T is the length of time considered for the evaluation. If probabilities are not used to quantify uncertainties, risk is sometimes represented by increasing the value of the discount rate r. Generally, however, cost-benefit analysts advocate associating probabilities with alternative net benefit (benefit minus cost) levels and then computing expected values (weighting net benefits by probabilities and adding). In this case the criterion function is expected net present value. Any action with a positive expected net present value is desirable, but the one that maximizes this quantity is most desirable. Expressed with symbols, the CBA criterion function is

$$\text{maximize } E(NPV)$$

where E denotes the process of computing the expected value (weighting all possible net present values by their probabilities and adding).

Several other criteria closely related to NPV are sometimes used in CBA. One involves the internal rate of return, which is defined as the value of the discount rate r that makes the net present value equal to zero. Selecting alternatives on the basis of their internal rates of return has been recommended in some circumstances, usually as a method of allocating a fixed budget among a number of potential investment projects (e.g., McKean 1958). With this decision rule, projects are arrayed according to their internal rates of return, and the highest ones are selected until the budget is exhausted. Another decision rule, sometimes used when the investment level for each project is flexible, involves allocating resources in such a way as to equalize the marginal (strictly positive) rates of return among opportunities (Bohm 1974). In general, however, problems with the internal rate of return (e.g., nonunique solutions, difficulty in comparing projects with different time horizons) have led cost-benefit analysts to prefer the use of net present values.

In DA approaches, the objective is to select decision alternatives so as to maximize the expected value of utility. Thus, if U is the (cardinal) utility function representing preferences for the (uncertain) outcome to a decision, then the basic decision criterion for DA approaches may be expressed as:

$$\text{maximize } E(U)$$

where E, once again, denotes the process of taking expected values. Just whose preference judgments the utility function is meant to represent and the source of the preference judgments depend on the approach. With clinical DA approaches, the utility function represents the decision maker's personal preferences. With the supra DA approach, the utility function represents the (supra) decision maker's judgment of an appropriate way of aggregating the preferences of those who are affected by the decision. For both types of approaches, the probabilities for computing expected values and characterizing risks are the subjective probabilities held by the decision maker.

With SWT approaches, the criterion function also involves the maximization of utility. In this case however, the relevant utility function is a group utility function derived without reference to a single decision maker. This group utility function is a social welfare function which aggregates utility functions representing the individual preferences of members of the group. If preferences are established over the outcomes to the decision and uncertainty is explicitly represented with probabilities, the individual utility functions are von Neumann—Morgenstern utility functions. If the centralized approach is adopted for the assignment of probabilities, the criterion function may be expressed as

$$\text{maximize } E(W)$$

where W denotes the social welfare function derived by aggregating individual utilities for the uncertain decision outcomes. The probabilities in this case may be "objective" probabilities or some aggregation of the subjective probabilities held by members of the group. If uncertainty is not explicitly considered, the individual utility functions are devoid of risk and (as noted in the discussion of social choice theory above) may be refered to as value functions. In this case the criterion function is

$$\text{maximize } V$$

where V is a measure of social welfare derived by aggregating individual values that depend on the attributes of the selected alternative. The attributes for characterizing alternatives may be decision outcomes, as in the case of DA approaches, or other attributes that do not involve the consequences of the decision.

The criterion function used in social DA approaches attempts to integrate the views of SCT, DT, and CBT. Social DA essentially appeals to efficiency arguments to justify net present value as a reasonable measure for valuing the outcomes to a decision. It is, however, a value function rather than a von Neumann–Morgenstern utility function and as such is incapable of accounting for risk aversion (the generally held feeling that the worth of a risk is lower than its expected value). Social DA approaches therefore occasionally account for a presumed societal risk aversion by converting the net present value computed for alternatives to utilities, using some presumed form for the utility function (usually exponential) that is characterized by a coefficient termed a risk aversion coefficient. CBA approaches generally ignore risk aversion, although they occasionally incorporate people's presumed aversion to risk *ad hoc* by raising the number of lives lost in a single accident to some power (e.g., n^2) (Wilson 1975).[59]

Outcome Variables

To apply their decision criteria, decision-aiding approaches quantify attributes of the alternatives that affect preferences and establish a means for valuing those attributes. For most approaches, the relevant attributes are the outcomes produced by the decision. Thus most decision-aiding approaches define outcome variables to quantify and measure aspects of a decision's consequences that are relevant to determing the desirability of those consequences. To measure a decision's impact on health and safety, for example, natural outcome variables include the numbers of immediate and delayed fatalities, injuries of given severity levels, or health degradations of specific types or severities. Outcome variables often used for expressing environmental outcomes include reductions in the abundance or production of commercial or game fish populations, reduction in timber yield or agricultural production, and reductions in wildlife populations.

Decision analysis approaches generally define outcome variables through personal interviews and interactions with participants in the decision process. Several decision analysts have offered specific procedures for accomplishing this. For example, Keeney and Raiffa (1976) recommend that objectives first be established through questioning decision makers. These objectives are then organized into a hierarchy that distinguishes their relative importance and generality. The resulting objectives hierarchy defines general objectives (like health and safety) in terms of more specific, lower-level objectives (such as low incidences of specific types of injuries, sicknesses, and fatalities). The objectives hierarchy is recommended because its construction helps ensure that no "holes" (missing objectives) occur in the analysis and helps eliminate situations where double counting might result (because holes and redundancies are more easily identified from the hierarchy).

Once the objectives hierarchy has been agreed upon, the decision analyst works with decision makers to establish a quantitative variable for measuring the extent to which each of the lowest-level (most detailed) objectives is achieved. Specifying an outcome variable requires specifying the properties that it measures and the scale used. Outcome variables may serve as direct or indirect measures of objectives, and they may involve either natural or constructed scales (Keeney 1980b). A direct measure has a straightforward, well-defined physical relationship with the objective (e.g., measuring the safety of a nuclear waste repository in terms of the number of cases of cancer attributable to all radiation releases occurring over some time period). An indirect measure is an indicator whose relationship with the objective is less formal or precise. For example, an indirect measure for the above example might be the total amount of ionizing radiation released from the site. A natural scale is one that is in common usage and has a common interpretation (such as a scale of dollars to measure cost). A constructed scale is one developed specifically for the problem at hand. One way of developing a constructed scale is to develop several different descriptions of possible levels of impact and then to associate integer values, say from 0 to 5, with them. Outcome variables and their scales are constructed on a trial-and-error basis, with the analyst asking questions to assure that the variables are comprehensive, measurable, nonredundant, and minimal in number.

Cost-benefit analysis is less specific on how variables for quantifying benefits and costs should be defined, although a general rule is that benefits should consist of all effects producing an addition to real resources in the form of commodities (i.e., resources that even if not traded on the market, have some commercial value) (Layard 1974). For applications to social decisions regarding health, safety, and environment, reductions in undesirable environmental and other side effects of the production or consumption of goods are also obviously benefits. Social costs represent real resources actually consumed and therefore unavailable for private consumption, plus undesirable environmental and other effects that reduce the welfare of members of society (Gramlich 1981). In practice, cost-benefit analysts generally arrive at a specification of variables for measuring costs and benefits through examination of relevant literature and discussions within the analysis team and with the sponsors of their analyses.

Valuing Outcome Variables

The many valuation procedures used in decision-aiding approaches can be loosely grouped into two large and somewhat overlapping categories: market-based procedures and utility-based procedures. Market-based procedures, favored by most CBA approaches, assign a monetary value to a given level of an outcome variable through direct or indirect reference to the concept of a market for that outcome. Utility-based procedures derive values for outcome variables through the assessment and analysis of utility functions. The basic types of procedures within each of these categories are introduced below and summarized in Figure 6.

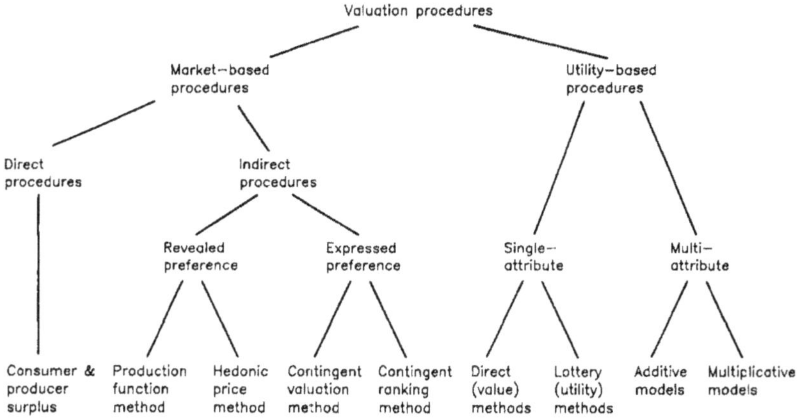

Fig. 6. Summary of major valuation procedures used in decision-aiding approaches.

Market-based procedures

Cost-benefit analysis and social decision analysis approaches almost always measure outcome values directly in monetary units established through reference to the marketplace. The link with the market may be direct, deriving values from existing market prices, or indirect. Indirect procedures may be based on revealed preferences, where monetary values are derived from people's market actions, or on expressed preferences, where monetary values are obtained from the stated preferences of individuals over hypothetical choices among alternatives. At the extreme, expressed preference procedures become utility-based procedures. The expressed preference procedures discussed in this section retain a link with the market in the sense that they involve the construction of a hypothetical market within which individuals are asked to value changes as if this market actually existed (Smith 1986).

The use of market-based procedures is significant for CBA and similar approaches that regard aggregate preferences as the appropriate basis for valuing decision consequences. Ideally, linking valuation to a market implies that a marketlike adjustment process implicitly takes care of the aggregation of individual preferences. Just as equilibrium prices for goods traded on the real market are assumed to balance aggregate demand with aggregate supply, surrogate prices for nonmarket goods are assumed to reflect aggregate equilibrium values for these goods produced in the hypothesized, implicit market adjustment mechanism.

The simplest market-based valuation procedure, the direct use of existing prices, may be applied if the outcome variable is expressed in units of some private good that is sold on the market (e.g., damage to an agricultural crop, property, or material, or losses of a commercially harvested fish or animal species). Thus, for example, crop losses due to air pollution could be valued at the price of the crop, assuming that the alternative being evaluated will not significantly alter the price of that crop. For many environmental, health, and safety decisions, of course, actions may be expected to produce substantial changes in the market, and these impacts must be taken into account when using market prices. Thus, valuing outcomes traded on the market typically requires a more complex analysis involving concepts of consumers' and producers' surplus.

To compute the values associated with a marketed good whose price is expected to change, the demand curve or supply curve for the good must be used. The demand curve (relationship between the quantity of a good purchased as a function of its per-unit price) shows the marginal utility (willingness to pay) of each item purchased. Because individuals generally obtain smaller and smaller amounts of satisfaction or utility from each additional unit of a good consumed, the demand curve is normally an inverse one; that is, willingness to pay decreases as quantity demanded increases. People are assumed to receive a "surplus" benefit from a purchased good because they need only pay the market price, which represents their willingness to pay for the last item purchased. Thus, the summed differences between the willingness to pay for a good and the cost paid for it measures the total benefit derived from that good by the consumer. It is for this reason that the area under the demand curve provides a measure of "consumers' surplus" — the sum total of benefit derived by all consumers from the purchase of the good. An analogous surplus arises for producers as a result of the difference between the amount producers receive for a good and their cost of producing it. By calculating changes in consumers' and producers' surplus, cost-benefit analysts have a means for computing dollar values associated with marketed commodities whose prices and quantities change as a result of government or private sector actions.

Implementing the logic of consumers' and producers' surplus for government health, safety, and environmental decisions can become rather complex. Normal demand curves, for instance, represent a situation where individuals' incomes are held constant. These curves (Marshallian demand curves) are inappropriate when the actions under consideration are likely to change substantially consumers' incomes. In this case, more complicated (e.g., Hicksian) demand curves are needed. Freeman (1979b) and Just, Hueth, and Schmitz (1982), among others, describe the concepts involved. Adjustments are also needed if it is clear that market values do not reflect the minimum price that would be necessary to attract the resources in an undistorted market (this minimum price is referred to in economic literature as the supply price or shadow price). Hurter *et al.* (1982), discuss the logic that demands adjustments for distortions caused by government-imposed excise taxes:

> . . . suppose that an environmental project requires the purchase of cement, which is subject to a sales tax. The supply price of cement, which is the social cost or opportunity cost to society of the cement, is the net-of-tax price of the cement. The tax receipts resulting from purchase of the cement for the project are part of project benefits transferred to the tax collector.

The procedures for adjusting prices to account for taxes and other market distortions typically involve judgment. Gramlich (1981), for example, notes that the proper way to handle taxes depends on whether or not the taxes are viewed as representing social external costs associated with the product.

Regardless of the logic followed, the demand curves used in calculations of consumers' and producers' surplus are rarely known with any accuracy. Lack of data for empirically deriving a demand curve generally forces the analyst to make simplifying assumptions regarding its shape. Usually, a simple form such as a linear or constant-elasticity demand curve is assumed, and one or two price/quantity measurements are used to estimate its parameters.

Although procedures based on consumers' and producers' surplus are designed to measure values associated with private goods that are sold on the market, they may also be used to value public goods, provided that some relationship between public and private goods can be established. The simplest procedures estimate a "production value" for the public good; that is, they value a change in the level of a public good, such as clear air, in terms of the changes it produces in the prices of goods individuals purchase or in their incomes. The procedures are applicable when the outcome measure for the public good may be regarded as an input factor in the production of a marketed product, and they are most appropriate for public goods such as forms of pollution where compensating actions to reduce the adverse impacts of the pollution are unavailable or not terribly effective. To value the benefits associated with improved water quality, for example, a CBA approach might use data on the relationship between biological productivity and water quality to determine the increase in net physical yield from commercial fisheries. The increased market value of the yield is interpreted as a measure of the benefit attributable to decreased water pollution.

A similar logic occurs in the opposite extreme where an environmental outcome measure affecting the production or consumption of goods may be entirely compensated for by increasing the purchases of some other (substitute) good. The increased spending required is a "direct cost" or "defensive expenditure" that may be used to value the environmental outcome measure. As an example, a CBA might value the benefits to households of reduced air pollution in terms of decreased costs incurred for cleaning or repairing damaged materials. Thus, some nonmarketed effects may be approximately valued by estimating the costs incurred by individuals who take steps to avoid having to experience them.

In general, the existence of complementary and substitute marketable goods for nonmarketed environmental, health, and safety outcomes often provides a relationship that permits values for the nonmarketed good to be inferred from the actions revealed by individuals in the market place. One popular procedure of this type is called the hedonic price method (e.g., Rosen 1974). The procedure involves identifying a "bundled" good that is traded on the market and whose price varies by whether or not it includes the nonmarketed feature. The price of real estate is a commonly used bundled good. Thus, for example, although peace and quiet may not have a market price, similar houses inside and outside airport flight paths do. Data on property values might be collected as a function of property location with respect to flight paths and a function of other characteristics affecting property values, such as the square footage of living space. Regression analysis may then be used to isolate the impact of the nonmarketed feature on prices.

Another popular procedure makes use of the complementarity between travel costs and recreational opportunities. Estimating the benefits associated with recreational activities is frequently an important part of an analysis of decisions whose outcomes affect pollution levels. Cost-benefit analysts have estimated, for example, that as much as one-half the total benefits from controlling water pollution can be attributed to improvements in water-based recreational opportunities (Freeman 1982). The travel-cost method assumes that individuals take trips to recreational sites to the point where their marginal trip costs equal the marginal benefits derived from the site. To conduct the procedure, the analyst generally uses a questionnaire to collect sufficient data from site visitors to estimate their trip costs, such as the length of the trip, nights stayed in motels, travel paths, meals at restaurants, plus their frequency of visits to the site in question and to substitute recreational sites. From the results, demand curves for the site may be developed and measures of consumer surplus derived (Vaughan and Russell 1982). To evaluate changes to characteristics of the site (e.g., reduced visibility due to air pollution) the respondent would be asked to indicate whether the given environmental change would affect his visitation rate and, if so, what alternative sites would be visited as a result of the change. The additional travel costs resulting from visits to the substitute sites are then interpreted as a rough approximation of the disbenefits produced by the site-specific environmental change (Cummings, Cox, and Freeman 1986).

A limitation of revealed preference procedures, such as those described above, is that at best they only measure "use benefits" — the values derived by individuals as they make direct use of the environment. Thus the travel cost method assumes that the marginal value of water quality in a lake is zero for those who do not use the lake for recreation. Obviously, such methods neglect "existence values" that are unrelated to the present or future use of the environment. To measure existence values, and the values of so-called intangibles, methods based on the derivation of values from "objective" market data are not very effective.

Cost-benefit analysis and related approaches deal with intangibles in one of two ways. One strategy, which is used by the Paretian approach to CBA, is not to attempt the monetization of such effects but to give a separate account of them in their respective units. The other strategy is to ask people to state in one form or another their willingness to pay to obtain a specified level or change in the level of the outcome of interest. If the preferences of very large numbers of people are sought, a survey might be conducted using questionnaires that ask respondents questions such as, "Are you willing to pay $100 more for a new automobile in order to have cleaner air?" Typically, however, a sequence of questions is posed so as to obtain more precise quantitative results.

The contingent valuation procedure is so named because it postulates a contingent market for the outcome measure whose valuation is sought. Specifically, a change in the level of the outcome measure is specified and the subject (a survey respondent) is asked to indicate how much that change is worth. For example, if an increase in a desired public good is being evaluated (e.g., cleaner air), then the subject would be asked to indicate the maximum amount he would be willing to pay to achieve that change. If the change is undesirable (e.g., more pollution), then the subject would be asked the minimum amount be would have to be paid to accept it or the maximum amount he would pay to prevent it. The appropriate form for the latter question depends on whether or not living without the undesired change is assumed to be a natural "right."

To obtain maximum willingness-to-pay (or minimum willingness-to-be-compensated) values, a "bidding game" is often used. With this procedure, the respondent is asked to indicate whether an amount proposed by the questioner is too high or too low for a specified change in the outcome measure. Photographs, maps, and other materials are often used to convey to the respondent the nature of the hypothesized change. The respondent might be told that the proposed amount will be added to his electric bill or established as a user fee for recreation. The questioning continues until the respondent indicates that the value is neither too high nor too low.

In addition to willingness-to-pay (or willingness-to-be-compensated) data, a contingent valuation survey collects basic demographic variables, such as the respondent's income, age, and sex, and possibly information concerning his attitudes toward certain fundamental issues. Regression techniques are then applied to identify the statistically significant explanatory variables. Aggregation across individuals with given characteristics for the explanatory variables provides a means for estimating total social benefit.

Contingent ranking procedures represent a variant to the contingent valuation method which requires subjects to provide rankings of preferences rather than direct willingness-to-pay estimates (Cummings, Cox, and Freeman 1986). The ranked order of preferences of subjects over specified combinations of levels of the public good and associated payments (or compensation) is used to determine parameter values for some assumed stochastic model of preferences (e.g., a utility model with some random "error" component). The resulting model is then used to define indifference curves and to estimate the change in the subject's income that would just offset utility gains (or losses) from a change in the postulated level of the outcome measure. This estimated change in income corresponds to the measure of compensating surplus which, in turn, is the subject's maximum willingness to pay for the public good. Schulze *et al.* (1981) used such a procedure to value visibility in national parks. In this application subjects were given a set of cards, each depicting a state of visibility and postulated park admission fee. Each subject was asked to order the cards in terms of preference. The rankings were then analyzed using a model in which individuals with similar characteristics were grouped under the assumption that the utility functions for individuals within each group were randomly distributed about some group mean utility function. The model's parameter weights on visibility level and price were set so as to maximize the likelihood of realizing the observed rank ordering. The resulting tradeoff between visibility and entry fee was then used to derive a measure of consumer surplus.

Because of the obvious impossibility of obtaining the preferences of every individual potentially affected by a decision, CBA approaches involving expressed preference methods must take care that respondents accurately represent their target population. Both political and statistical representativeness are important (Cox 1986). Political representativeness means that the sample of respondents adequately reflects the views of the various stakeholder groups whose interests and preferences are most significantly affected. Statistical representativeness means that the population included is large enough that the "luck of the draw" should result in a distribution of preferences similar to that obtainable if the whole population were sampled. Various stratified sampling procedures and rules regarding sample size provide some assistance for the selection of survey samples; however, the nonhomogeneity of the population, limited resources, and the self-selection of respondents associated with voluntary questionnaires often create difficulties in these areas.

Although the goal of a CBA is normally to estimate total costs and benefits without regard to their distribution, decision-maker CBA may involve altering market-based benefit estimates slightly to account for a subjective judgment that certain considerations are under- or over-valued by the market. A report prepared by the National Academy of Sciences (1977) summarizes a recommended procedure that provides room for a decision maker to apply weighting factors:

> In decisions that have a component that depends on human values, we propose the following: (i) The terms on both sides of the (cost-benefit balancing) equation should be given a monetary value based on the market place, public survey, or other appropriate means . . . (ii) Weighting factors should be applied to those terms which may be undervalued by market place economics . . . (iii) The values of the weighting factors have to be established by society in general, whether through the political process, public survey, or other means.

In practice, decision-maker CBA and related approaches have advanced many arguments to justify the use of weighting factors. Although there is no consensus for the use of weights, approaches sometimes weigh public fatalities more heavily than occupational fatalities and catastrophic outcomes more heavily than their aggregate outcomes would suggest. Selecting numerical values for weights generally requires creativity. For example, Gramlich (1981) describes an argument for assigning weights to account for equity concerns for the distribution of benefits between the rich and the poor:

> If we can rely on the ultimate rationality of the personal income tax as a measure of social judgment on the income distribution, . . . benefits to poor people from a program should be given a weight t_r/t_p times the weight given benefits to rich people, where t_r and t_p are the marginal effective tax rates paid by rich and poor people, respectively.

Utility-Based procedures

Social welfare theory and most decision analysis approaches measure values of outcomes in a common unit (utiles) which may or may not, depending on formulation, have a direct relationship with dollars. Both value functions (which measure strength of preference for multiattribute consequences known with certainty) and von Neumann–Morgenstern utility functions (which measure preferences for risks) are used.

Decision analysis approaches tend to use exacting and often time-consuming procedures for assessing utility and value functions. Because these approaches generally involve obtaining the preferences of a single decision maker or a small group of individuals, the application of detailed assessment procedures is feasible. Social welfare theory approaches may use simpler assessment procedures to enable the independent preferences of a broader range of individuals to be assessed. Some SWT approaches use a statistical approach for the assessment of utility functions that is similar to the contingent valuation and ranking methods of CBA approaches. With the statistical approach, respondents are presented with carefully designed sets of stimuli (e.g., cue cards) and asked to express preferences among them (e.g., by ranking or numerically rating them). The responses are then analyzed through regression-based methods to produce a utility function which is in some sense most consistent with the range of responses obtained. More often, however, a separate utility or value function is assessed for each individual whose preferences are to be considered. The emphasis for the discussion that follows is the available procedures for assessing value and utility functions from individuals and the procedures by which individual functions may be aggregated.

The most direct procedure for obtaining an individual's value function is to ask him to assign numbers to the possible decision outcomes according to some scale of desirability. For example, to apply the direct assessment procedure to a decision having several discrete outcomes, the individual is first asked to rank the possible outcomes from most to least preferred, thus establishing an ordinal preference scale. Next, the end-points for the scale are established by assigning arbitrary numerical values to the most- and least-preferred possibilities; for example, by assigning the least-preferred outcome a value of 0 and the most-preferred a value of 100. The numerical assignments serve to calibrate the value scale. The individual is then asked to assign numerical values to the other outcomes using the least-preferred and most-preferred outcomes as reference points. Thus, the subject is asked to assign to each outcome level z a number, denoted $v(z)$, reflecting that outcome's perceived "distance" from the two endpoints in terms of relative desirability.

A slightly more formal procedure involves direct value assignments based on preference differences rather than absolute preferences. To illustrate, suppose that there are four possible outcomes, z_1, z_2, z_3, and z_4. The subject would be asked to rank the various outcomes in terms of their desirability and to rank pairs of outcomes in terms of their preference differences; in other words, if the subject prefer z_2 to z_1 and z_4 to z_3, he would be asked whether he prefers z_4 to z_3 more than he prefers z_2 to z_1. Once the value scale has been fixed by determining its end points, then under some mild technical assumptions it is possible to determine a value $v(z)$ for each outcome. Again, the choice of the origin and scale are arbitrary, but once they have been made the rest of the values assigned to different outcomes may be uniquely determined.

If the outcome variable is a continuous variable that can take on any level within some range, the midpoint method (Chankong and Haimes 1983) is often used to derive a value function. The procedure involves successively identifying levels of the outcome variable whose values (desirabilities) are halfway between already established values. To initiate the procedure, arbitrary values are assigned to the least and most desirable outcome levels, denoted z_L and z_M. Various intermediate outcome levels are then selected until a point z' is found such that the subject is just indifferent between the prospect of exchanging z_L for z' and exchanging z' for z_M. The resulting outcome level is called the midpoint because the value function evaluated at this point is midway between the values of the other two outcome levels that were considered. The process is repeated to find other midpoints (e.g., the midpoint between z_L and z') until enough have been obtained to permit fitting a smooth curve. Kirkwood and Sarin (1980) propose checking whether the subject's preferences satisfy a "delta property" before applying the midpoint method. The delta property holds if for any quantity q, $z' + q$ is the midpoint between $z_1 + q$ and $z_2 + q$ whenever z' is the midpoint of z_1 and z_2. Under these conditions, the value function must have either a linear or exponential form.

A problem with the above assessment procedures is that they are conducted using arbitrary scales rather than a scale based on familiar units which would be more meaningful to the subject. For this reason, procedures are often used that involve exploring the tradeoffs that a subject is willing to make among pairs of outcome variables. Tradeoff (or marginal rates of substitution) methods are often useful when decision consequences involve several attributes, each of which is represented by a separate outcome variable. If there are two desirable attributes, for example, a DA approach might obtain a tradeoff function from a decision maker by asking by how much the variable measuring the first attribute would have to increase to offset a given decrease in the second. The resulting tradeoff function defines an indifference curve, which together with certain simplifying assumptions can be used to derive a value function.

Decision analysis approaches with an economic perspective, such as social DA, often use the tradeoff procedure to obtain a value function expressed in units of dollars. For example, a DA of a program to stimulate employment might obtain a value function for decision outcomes specified by the total costs of the program and the resulting level of employment. The analyst might then assume that the decrease in costs necessary to justify a given decrease in employment depends on the level of employment but not the level of costs. A dollar equivalent for different employment levels could then be obtained from the decision maker by an assessment procedure designed to elicit an indifference curve between costs and employment. For example, the dollar value of the lowest conceivable level of employment could arbitrarily be set at zero. The decision maker would then be asked to indicate what additional costs he would be willing to accrue to move from this level to various higher levels. After obtaining a few points, a smooth curve could be sketched in. Figure 7 shows the form of a curve that might result from such a process. The value (actually, tradeoff value) associated with any combination of program costs and employment levels would be simply found by subtracting program costs from the equivalent dollar value of employment obtained from the curve.

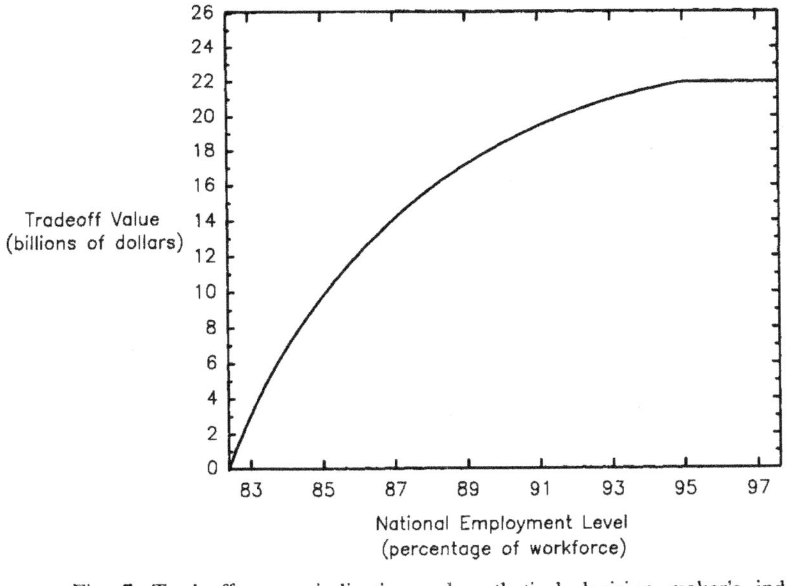

Fig. 7. Tradeoff curve indicating a hypothetical decision maker's indifference between increased program costs and increased levels of employment.

In situations where there are three or more outcome variables, value functions obtained using the tradeoff method would, in theory, have to be determined by varying each outcome variable one at a time so as to find indifference functions or surfaces. Because this is usually unmanageable, an assumption of preferential independence is used to permit the value function to have an additive form. In this case, the multiattribute value function is represented as a weighted sum of value functions on single outcome variables: if $x_1, x_2, \ldots, x_n$ represent n outcome variables and preferential independence holds, then the value function V is of the form

$$V(x_1; \ldots, x_n) = k_1 v_1(x_1) + k_2 v_2(x_2) + \cdots + k_n v_n(x_n)$$

where the $v_1(x_1)$, $v_2(x_2)$ are single-variable value functions. Preferential independence essentially implies that preferences for specific values of an outcome variable do not depend on the values of the other variables. The analyst establishes the presence or absence of preferential independence by asking a series of questions involving paired comparisons. The procedures normally used for constructing value functions thus consist of (1) verifying preferential independence, (2) determining single-variable value functions for each outcome variable, and (3) determining appropriate weights to be used in the summation (for example, by setting the outcome variables at extreme values). Keeney and Raiffa (1976) provide detailed descriptions of the procedures that are used.

Clinical and supra DA approaches use von Neumann–Morgenstern utility functions rather than value functions to represent preferences, because they permit establishing preferences over uncertain consequences (required for the application of decision theory). As with value functions, the assessment of a utility function is conducted in an interview. To assess a utility function describing a decision maker's preferences for uncertain outcomes involving a single outcome variable, two steps are required. Step 1 is to calibrate the scale by setting the upper and lower utility points. The worst possible outcome is normally assigned a utility of 0 and the best possible outcome a utility of 1 (or 100). Sometimes, however, there are specific, easily conceptualized outcomes that while not the worst or best are more meaningful to the decision maker. In this case these outcomes can be used to establish the arbitrary reference points for calibrating utilities. In the second step the analyst asks the decision maker to express preferences between pairs of alternatives involving lotteries. Normally this will involve a choice of the form "receive outcome z_C for certain or receive a lottery offering outcome level z_1 with probability p and outcome level z_2 with probability $1-p$." The purpose of these questions is to identify pairs toward which the decision maker is indifferent. Each time such a pair is found, the equality between the utility of the certain outcome and the utility of the lottery permits an equation to be written and solved for one point on the utility curve. For example, if z_1 is selected to be the least-preferred outcome level (assigned a value of 0) and z_2 is selected to be the most-preferred outcome level (assigned a value of 1), then the value of p that makes the decision maker indifferent between this lottery and receiving any other outcome z_C for certain may be regarded as the utility of that outcome level. By varying the outcomes and probabilities of the lottery, any number of points on the decision maker's utility curve can be generated. These points may be plotted and represented by a curve. Figure

8 shows an example for a case where the outcomes are varying amounts of money.

Decision analysis approaches often interpret the utility curve as indicating an attitude toward risk taking. If the outcomes on which the utility function operates are expressed in dollar values, then the shape of the curve indicates whether preferences reflect an aversion to risk, a preference for risk, or indifference (risk neutrality). A concave curve (such as Figure 8) indicates an aversion to risk — the decision maker assigns less

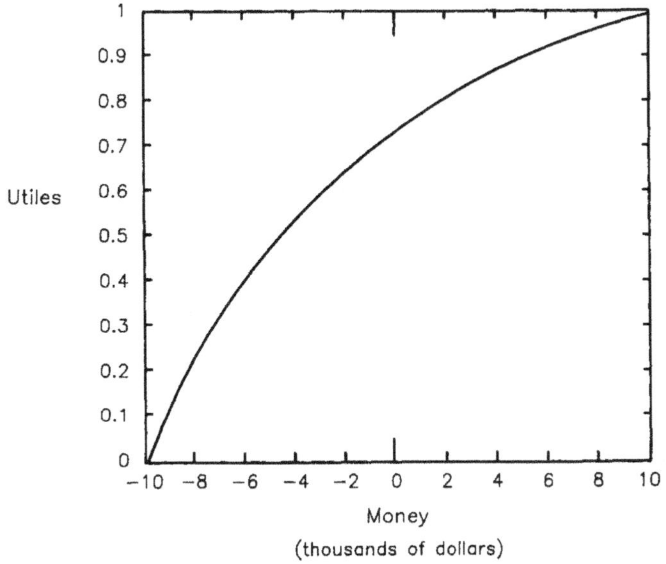

Fig. 8. Illustrative utility function over monetary outcomes.

utility to a 50:50 gamble for any dollar value (0.5 times the utility of that value) than to receiving one-half of that value for certain (the utility of the middle value). Decision analysts define the certainty equivalent of a lottery involving uncertain dollar payoffs as that dollar value which a decision maker is just willing to accept in lieu of the lottery. By the expected utility theorem of DT, the certainty equivalent has a utility equal to the expected value of the utilities associated with the possible outcomes. The difference between the expected value of a lottery (dollar values weighted by their probabilities) and its certainty equivalent is called the risk premium. The risk premium, in a sense, measures the amount the decision maker is willing to give up to avoid risk.

In many applications of utility, especially DA and applied SWT approaches, multiattribute utility functions are used which permit outcome variables such as those representing costs, health effects, and environmental degradation to be jointly considered in establishing utilities. Although any functional form for a multiattribute utility function is theoretically possible, applications almost always involve additive or multiplicative forms. The additive utility function U of n outcome variables, denoted $x_1, x_2, \ldots, x_n$, is

$$U(x_1, x_2, \ldots, x_n) = k_1 u_1(x_1) + k_2 u_2(x_2) + \cdots + k_n u_n(x_n)$$

where $u_1(x_1)$, $u_2(x_2), \ldots,$ are single-variable utility functions describing the decision maker's preferences over each of the outcome variables individually, and $k_1, k_2, \ldots,$ are scaling constants. The multiplicative form is

$$U(x_1, x_2, \ldots, x_n)$$
$$= (1/k)\{[1 + kk_1 u_1(x_1)]\,[1 + kk_2 u_2(x_2)] \ldots [1 + kk_n u_n(x_n)] - 1\}$$

where k is a constant satisfying the equation

$$1 + k = (1 + kk_1)(1 + kk_2) \ldots (1 + kk_n)$$

Actually the additive form is a special case of the multiplicative form where the scaling constants $k_1, k_2, \ldots, k_n$ sum to one. Just as in the case of additive value functions, the advantage of additive or multiplicative utility functions is that they permit a multiattribute utility assessment to be decomposed to assessments involving single attributes.

To justify the additive or multiplicative forms, the analyst uses an assumption of utility independence. Two outcome variables are utility independent if preferences for uncertain scenarios involving different levels of each variable are independent of the level of the other. Thus, the assessment process consists of (1) verifying that selected outcome variables permit an assumption of utility independence, (2) setting both the multiattribute and single-variable utility functions equal to 0 and 1 when their outcome variables are at their worst and best levels, respectively, (3) assessing single-variable utility functions for each outcome variable (using the procedures described above), (4) assessing the scaling constants $k_1, k_2, \ldots,$ and (5) using either the additive or multiplicative forms, depending on whether the assessed scaling constants approximately sum to one.

To check the validity of independence assumptions, a procedure derived by Keeney (1976) is often employed: if any one attribute, say z_1, can be verified to be utility-independent of the others, and all of the variables are preferentially independent, then the requirement of mutual utility independence is satisfied. The scaling constants essentially represent the value tradeoffs between the various pairs of attributes. Thus they are assessed by fixing all but two of the attributes (e.g., at their worst outcomes), and those two are varied until the decision maker is indifferent between them. This permits the development of $n-1$ of the n equations required to solve for the k's. To obtain an additional equation the analyst finds a lottery involving convenient values for outcome variables (e.g., all outcomes set at their best and worst levels) and asks the decision maker to specify the probability for the preferred outcome in this lottery that makes him indifferent between receiving that lottery and receiving some other convenient set of outcomes (such as the best value for one outcome variable and the worst for all others). This provides one more equation which permits calculating the k's.

Keeney (1980b), Bunn (1984), and Goicoechea *et al.* (1982) each give detailed examples of assessments of multiattribute utility functions. The individual component utility functions are usually assessed relatively rapidly by first asking questions designed to establish qualitative properties (such as concavity or linearity) and then assessing two or three points so as to approximate the shape of the curve. The curves are then adjusted and verified through a series of diagnostic questions (Keeney and Raiffa 1976). Despite this streamlining, the process is still time consuming if the number of outcome variables is very large. One of the applications described by Keeney was reported to have required roughly eight hours of interviewing time. Given the assessment demands for problems in which there are many relevant attributes, it is not surprising that most real-world applications involve further simplifications.

One common simplification is to group similar attributes so as to limit the number of outcome variables. Another is to make simplifying assumptions about the shape of the single-variable utility functions; for example, assuming an exponential form which allows a single assessment to specify each function. An exponential utility function embodies an assumption of constant risk aversion, which means that the risk premium (difference between the expected value and certainty equivalent) depends only on the magnitude of the differences between uncertain outcome values, not the absolute values of those outcomes. Another common simplification is to assume that the multiattribute utility function is simply a weighted average of performance measures:

$$U(z_1, z_2, \ldots, z_n) = b_1 z_1 + b_2 z_2 + \cdots + b_n z_n$$

In effect, this assumes that the outcome variables (interpreted as performance measures) are (mutually) utility independent and that each single-variable utility function is a linear function of the level of its outcome variable (which implies risk neutrality; i.e., no aversion to risk). In this case, specification of the utility function is simply a matter of selecting the weights describing the relative importance of each performance measure. Procedures for accomplishing this vary from simple ranking and rating to more elaborate methods involving such concepts as standard gambles, eigenvectors, or entropy (for discussions, see Hobbs 1979 or Hwang and Masud 1979).

Because it would be very difficult, if not impossible, to assess multiattribute utility or value functions for situations with three or more outcome variables without independence assumptions, successful application of utility methods requires the analyst to exercise skill in the definition of outcome variables. Practical experience suggests that it is usually possible to define the set of outcome variables so as to satisfy the appropriate independence assumptions, given sufficient thought. Procedures for assessing utility or value functions when independence cannot be verified, in fact, generally involve redefining outcome variables so that independence does hold (e.g., Keeney and Raiffa 1976).

Supra DA and SWT approaches use special forms for utility functions that are very similar to the additive and multiplicative functions described above, except that with supra DA and SWT the form of the utility function is meant to represent an aggregation of individual preferences rather than the personal preferences of a single decision maker for multiattribute outcomes. Supra DA and SWT approaches commonly compute the utility of a set of outcome variables, denoted x, using an additive function of the form

$$U(x) = w_1 u_1(x) + w_2 u_2(x) + \cdots + w_m u_m(x)$$

where $u_1(x)$, $u_2(x)$, . . . , are the utility (or value) functions representing the respective preferences of the m impacted individuals for the outcomes x. Occasionally used instead is a multiplicative form

$$U(x) = (1/w)\{[1 + ww_1 u_1(x)]\,[1 + ww_2 u_2(x)] \ldots [1 + ww_m u_m(x)] - 1\}$$

with w satisfying

$$1 + w = (1 + ww_1)(1 + ww_2) \ldots (1 + ww_m)$$

The constants w_1, w_2, . . . , may be loosely interpreted as the relative importance of each affected party's preferences.

With supra DA, the utility function $U(x)$ is meant to represent the (supra) decision maker's preferences for the impacts on the various group members. Keeney and Raiffa (1976) describe one procedure for obtaining a supradecision-maker utility function that involves assessing utility functions for all or selected representatives of the impacted parties and obtaining appropriate weighting functions from the decision maker. Harsanyi (1977b) suggests that rather than attempt to assess a utility function for a spokesperson from each group, the decision maker should use "imaginative empathy" — that is, imagine himself in each affected individual's shoes and hypothesize a utility function that might be expected to reflect reasonably the values of each group. Keeney (1980b) proposes that certain functional forms (such as the exponential) be postulated as utility functions for each identifiable group and that the parameters of these functions be set to specific values or ranges of values through inputs provided by individuals familiar with the values and viewpoints of each group (e.g., using the viewpoint of the mayor and a few council members to characterize the preferences of a local community). Kirkwood (1972) and Keeney (1980b) point out that the weights w_1, w_2, . . . , which must be obtained from the decision maker, should depend on both the decision maker's concern for the corresponding groups and on the implications of the range of possible values for the group. An assessment procedure analogous to that described for obtaining the k_i scaling constants may be used to obtain the weights. Keeney (1980b) suggests that simple rules — for example, that weights should be in proportion to the size of groups — may be used to reduce the assessment burden.

With SWT approaches, the utility function $U(x)$ is a group utility function. The methods described above for assessing value functions or von Neumann and Morgenstern utility functions (depending on whether risk is explicitly considered) are used in SWT approaches to obtain the utility functions for representing the preferences of each group member. Since there is no group decision maker for such approaches, the weights must be established by some group process or normative argument. Bodily (1979) describes a delegation process by which a group might iterate toward a consensus of weights that might be practical for small groups. Brock (1979) describes a procedure that essentially results in weights being assigned according to the relative need or intensity of desire of the group members.

Actual applications of SWT approaches, however, have tended to use weights that imply that all individuals are given equal influence, as members of groups tend to find it difficult to deviate from the norm of equality represented in majority-rule voting. Operationally, this may imply numerically different weights if scaling is needed to account for differences in the "goodness" or "badness" of outcomes to different individuals. For example, if a given option produces consequences that are twice as important or significant to one group member, that individual might be assigned twice the weight. Frequently, however, individual utilities are scaled in such a way that a given numerical utility may be assumed to indicate the same intensity of desire across different individuals. Cox (1986) describes the procedure by which this is normally achieved. First, extreme outcomes are defined as outcomes that everyone agrees are at least as good and at least as bad as any of the outcomes being considered. These are used to define common endpoints for the assessment of each individual's utility function. For example, to assess utilities of health outcomes, "normal healthy life" and "early and painful death" might be selected as common endpoints for scaling each individual's utilities. Second, it is assumed that the utility difference between these endpoints is the same for each individual. In other words, everyone prefers "normal healthy life" to "early and painful death" by the same amount. Such assumptions put the utilities for different individuals in the same units so that equality of influence is accomplished by assuming equal weights.

Valuing human life

Because most government programs dealing with risk are primarily designed to protect and improve human health, it is useful to give special consideration to the procedures by which various approaches place a value on human life and health. Zeckhauser (1975) and Linnerooth (1979) provide detailed summaries of some of the available procedures.

Assigning a dollar value to loss of life would appear to be especially difficult for CBA because life has no market value; under nearly all conceivable situations, no sum of money would induce individuals to give up their lives voluntarily. Some of the procedures that were first used in CBA approaches to rationalize finite values of life are now recognized by cost-benefit analysts as inappropriate. With the "human capital" calculation, for example, the dollar value assigned to the life of an individual is the present value of his or her projected future earnings. Because the human capital procedure is easy to implement, it is still occasionally used, even though the calculation views individuals as having worth only to the extent that they contribute economically to society. Another faulted procedure is to derive values of life from past decisions that have saved lives or from the damages that juries have awarded in compensation for death. A problem with each of these is that widely varying values are obtained, depending on the awards or past decisions considered (Graham and Vaupel 1981). Furthermore, such procedures are inconsistent with CBT logic — they fail to account for the value of the life to the individual who has died, and their implicit justification (that the democratic or legal process created them) runs counter to the CBT logic of deriving values from independent economic criteria.

Value-of-life calculations based on individuals' willingness to pay for some statistical reduction in the risk of death are now generally regarded as the appropriate basis for evaluating mortality risks in CBA. One accepted procedure is to derive values of life from wage differentials in riskier than normal occupations such as mining. This is an example of a hedonic price procedure as described above in the section on market-based procedures. Assuming there are large numbers of people facing choices among employment opportunities with varing wage and work hazards, over the population as a whole there will be some definite expected number of early deaths, which, in effect, have been traded off against increments to pecuniary social income. Studies have found that riskier jobs have annual wages that are between $4 and $70 higher for each 1 in 100,000 increase in the rate of mortality.[60] The hedonic wage procedure assumes, in effect, that workers know the risks to which they are exposed and have the freedom to bargain effectively with employers. Methods based on contingent valuation procedures are also used. For example, to establish a willingness to pay for safety, a questionnaire might ask the respondent to value a particular statistical reduction in mortality. In such applications, it is particularly important to distinguish what people are willing to pay to avoid the increased risk from how much they demand to be compensated when the risk is imposed upon them. The latter value tends to be much larger than the former. Because neither of the above procedures accounts for economic and psychic losses to survivors, additional procedures have been suggested that would augment individuals' willingness to pay with that of survivors .

Decision analysts and social welfare theorists have also devoted efforts to deriving procedures for valuing human life that are compatible with their approaches. Howard (1979) derives a decision theoretic model for estimating the value to an individual of increasing or decreasing probabilities of death and shows that a small-risk value of life exists.[61] Keeney (1981) describes a theory for evaluating mortality risks based on a social utility function describing group utility over probabilities of deaths experienced by group members and costs expended to reduce mortality risk.[62]

Despite the variety of procedures and perspectives for estimating a value of life, decision-aiding approaches that assign dollar values have tended to arrive at a relatively narrow range of conclusions. For this reason, the value assigned to life is frequently treated as a variable, and a sensitivity analysis is conducted to determine whether the results of the approach change depending on the value assumed. Typical values of life assigned by decision-aiding approaches range from about $200,000 to $4 million (Fischhoff *et al.* 1981b).

Valuing outcomes over time

Virtually all decision-aiding approaches discount values assigned to future outcomes. Two reasons are cited for doing this. First, resources received early are more valuable because they can be invested in productive areas to yield a return; that is, resources that are not consumed today can be invested to produce resources of greater value tomorrow. Second, irrespective of the existence of investment opportunities, individuals simply prefer to speed up the receipt of desirable outcomes and postpone undesirable ones; they demand compensation for reducing their consumption today in favor of consumption at a later date. Although there is little disagreement among approaches over the desirability of discounting, there are differences in the logic used for selecting discount rates.

Cost-benefit analysis approaches generally rely on market arguments for selecting a discount rate. To some extent, resources used for an environmental project in the public sector will force displacement of private investments. Following the logic of CBT, such investments will be economically efficient only if the rate of return per dollar outlay toward public goals exceeds the opportunities foregone per dollar in the private sector. Thus, the CBT perspective suggests that benefits and costs of public programs be discounted by the opportunity costs of shifting productive resources out of the private sector. In theory at least the various private institutions that supply investment and borrowing opportunities provide an indication of these opportunity costs because such institutions permit individuals to exchange commodities at one date for a larger or smaller volume at an earlier or later date.

The simplest CBA procedure for deriving a discount rate is to assume that each public dollar spent to achieve social benefits exactly displaces a dollar of private investment. Assuming that private and public investments should be equally profitable at the margin, this line of reasoning suggests that the before-tax rate of return on private investments is the appropriate rate of discount. Another procedure assumes that the dollars that make up costs in an application of CBA come partly from reduced private investment and partly from reduced consumption. The opportunity cost that governs the rate of consumption is argued to be the after-tax rate of return for savers. Thus, with this view, the appropriate discount rate is a weighted average of the before-tax return on private investment and the after-tax savings rates. Problems with implementing either of these procedures include uncertainty in where the money for public investment comes from and what, as a result, must be given up; the lack of a single before-tax rate of return on private investment or a single after-tax rate paid on private savings; and the many distortions to these market rates caused by the complexity of the tax system, imperfect competition, disequilibrium, and uncertainty.

Social weflare theory and decision analysis approaches typically use discount rates meant to reflect time preference rather than opportunity costs. Utility-based approaches arrive at this result by extending the logic of multiattribute utility theory to address outcomes in different time periods. For example, if value functions are being used, it might be assumed that outcomes in different time periods are (mutually) preferentially independent and that preference relationships among successive time periods are stationary (i.e., the preferential ordering established in the first period for time streams that differ only after the first period are the same as the preferential ordering starting from the second period). These assumptions can be shown to be sufficient to derive the discounting formula where the discount rate represents time preference (i.e., the scaling factors in the additive value model represent relative preference for receiving an outcome in an earlier rather than later time period) (Keeney and Raiffa 1976).

For clinical and supra DA approaches, the discount rate represents the decision maker's time preference. Thus, DA approaches often establish discount rates by relying on the utility assessment methods described in the previous sections. For SWT approaches, consistency with SCT would require aggregating the utilities obtained when each individual applies his discount rate to the time stream of outcomes. Sometimes, however, SWT and related approaches have relied on the concept of a social discount rate. This represents a time preference for society as a whole, which might be derived through some normative argument or through some explicit (perhaps political) process. For example, social DA views the social discount rate as a conscious value judgment for society, which, in effect, is a control variable that provides a means for deliberately apportioning costs and benefits among present and future time periods. Thus, the social decision analyst might interpret a low discount rate as a preference for pursuing projects that benefit lower income groups at the expense of those with high marginal tax rates. With this view, value judgments about the appropriate discount rate are ultimately tied to the question of the size of the government sector (Stokey and Zeckhauser 1978).

Decision-maker CBA approaches have also used for the discount rate a social time preference rather than a rate reflecting opportunity costs or the time preferences of a single individual, although analysts have recognized the inconsistencies with CBT (National Academy of Sciences 1977). Some CBA approaches, for example, have used discount rates derived from models designed to yield "socially optimal" discount rates. Economists have developed simple models for determining the rate of savings versus consumption that maximize per capita consumption across all generations. (If savings are 100 percent there is no consumption, and if consumption is 100 percent there is neither capital cumulation nor growth; thus, some intermediate value presumably provides a long-run maximum level of consumption.) Such models may be used to generate social discount rates that are consistent with assumptions regarding optimal growth.

Although relatively formal methods (such as those described above) may be used to derive time preference or opportunity-cost discount rates, it is rare for an approach to devote a great deal of effort to the selection of discount rates. Typically, the analyst will select a nominal or range of possible values in consultation with sponsors or decision makers. Selecting a range of values allows for a sensitivity analysis to determine if the recommended decision may be sensitive to the assumed discount rate. Values between 4 percent and 10 percent are most frequently used. Lower discount rates are sometimes selected for nonmonetizable benefits than for economic costs. It might be argued, for example, that a year of hearing loss at age 25 is no less important than a year of hearing loss at age 20, so that these health impacts should receive equal weight, despite the fact that one occurs five years later than the other. Alternatively, lower discount rates for health outcomes have been justified on the basis that such amenities are likely to become more valuable relative to other goods in society as time passes and the standard of living improves.

Models

Decision-aiding approaches use models to relate decision alternatives to the valuation measures that define its decision criteria. At one extreme, a decision model might be nothing more than a structured process by which participants subjectively rate each option according to each valuation measure. Applied SWT approaches often use such procedures. Likewise, the model used in some versions of DA is merely a simple process by which a decision maker works with the analyst to relate alternatives, uncertainties, and values to one another. Buede (1979) views this as the essence of the clinical art approach to DA and describes the valuation of alternatives as being based on a "simple structural model" that consists of a "structured platform for dialogue and debate among the decision maker and his associates." Behn and Vaupel (1983) describe a similar version of DA, referred to as "quick analysis," in which the model may be little more than a simple tree diagram that displays decision alternatives and their possible outcomes.

Increasingly, however, the models used in decision-aiding approaches are mathematical and logical representations of the processes relating alternatives to decision outcomes. Such models are most prominent in CBA and social DA approaches because these approaches strive for extrapersonal analysis. Regardless of the approach, models for relating alternatives to outcomes are used to improve objectivity and judgment by making experts' beliefs explicit in a format open to peer review, by reducing the need to combine large quantities of data in one's head, by facilitating the incorporation of new data or alternative assumptions (because the decision model can be recomputed at any time without having to rethink the whole problem), by creating a shared model of a complicated risk that can utilitze the knowledge of different disciplines (each expert addressing those components within his area of expertise), and by showing where judgment might be supplanted by research.[63]

Model building is a creative, iterative process in which the analyst explores the implications of alternative formulations and gradually delineates and refines a set of assumptions into a model that is compatible with the problem and information at hand. In general, the development of a model involves up to six steps:

1. *Structuring*: specifying problem variables and their relationships
2. *Estimating*: using empirical data obtained through measurements or subjective judgments to establish the free parameters of the model
3. *Analyzing*: exercising the model to estimate the behavior of the system under study
4. *Validating*: comparing the behavior predicted by the model with that of the aspect of the system the model is intended to approximate

5. *Diagnosing*: tracing differences between the model and the system for the purpose of determining how the model can be improved
6. *Refining*: modifying the model to improve its usefulness

Although all modeling procedures come from the general field of systems analysis, CBA and social DA tend to use different procedures. These differences may largely be traced to the different perspectives and roles adopted by the analysts favoring each approach type. For example, cost-benefit analyses, often conducted by members of the academic community, are frequently developed using "academic quality criteria":

> . . . models are designed to be nontrivial (permitting inferences not perceivable by direct observation), powerful (offering a large number of nontrivial inferences), elegant (using a minimum of carefully selected analytic tools to produce a model of great power), cost-effective (producing the answer with a minimal number of machine operations), valid (capable of reproducing results based on historical data), and so on (Quade 1980).

Decision analysts, because they view their role as one of aiding decision makers, tend to place greater emphasis on the model being relevant (inclusion of only those factors likely to influence the decision), acceptable (likely to reflect a view of the world acceptable to decision makers), and cost-effective (producing benefits that exceed the expense of their development).

Differences in the modeling procedures used in CBA and DA approaches also reflect the contrasting objective-subjective perspectives inherent in their underlying theories. An illustration is provided by modeling procedures involving directed graphs, which are diagrams consisting of nodes connected by arrows. Various forms of directed graphs are often used by different approaches to support the structuring steps of model building.

Cost-benefit analysis approaches sometimes use a form of directed graph called a digraph as a cause-effect diagram in which conditions or variables are represented by nodes (points) with connecting arrows to show which events or conditions cause other events or conditions (Chacko 1976). The symbols +, −, or × may be placed along the arrows to indicate whether a change in one variable or condition produces a change of the same sign, the opposite sign, or mixed effects of both the same and opposite signs. The diagrams may be further quantified by replacing each symbol with the signed derivative of the variable to which the arrow points with respect to the variable from which the arrow emanates. (Such diagrams are input-output models if the cause effects are interpreted as flows from one part of the system to another.) The diagrams may also be analyzed to identify the various system control loops and to classify them according to their ability to cancel system disturbance of a given size. Figure 9 shows an example of a digraph constructed to represent a simple

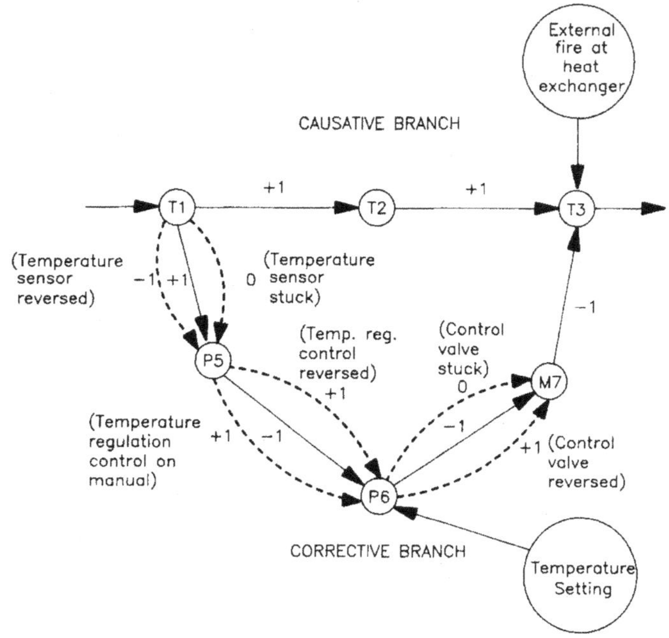

Fig. 9. Partial digraph for analyzing internal and external faults in a negative feedforward temperature control system, adapted from Philipson, Lambert, and Roland (1981). T1 = Temperature into sensor, T2 = Temperature out of sensor, T3 = Heat exchanger output, P5 = Air pressure at input to temperature control, P6 = Air pressure at output to temperature control, M7 = Water mass flow through the pneumatic valve to heat exchanger.

negative feedforward loop that controls the output temperature of a fluid cooled in a heat exchanger by mixing with water (Philipson, Lambert, and Roland 1981). The input temperature is sensed by a temperature regulation control which adds or subtracts air pressure to reset a pneumatic valve that controls the flow of cooling water. The variables are the temperature in and out of the sensor and the heat exchanger's output, the air pressure in and out of the temperature control, and the water mass flow through the pneumatic valve to the heat exchanger. The numbers adjacent to the directed arrows show that the gains between variables may change due to system faults. An external cause of system failure (fire at the heat exchanger) is also shown in the diagram.

Decision analysts use a similar directed graph called an influence diagram (Owen 1979; Howard and Matheson 1984). Influence diagrams are constructed as a joint process involving an analyst and experts in the substantive aspects of the problem. Like digraphs, the nodes in an influence diagram represent variables, but the arrows do not represent cause-effect. Instead, an arrow from a variable (represented by an ellipse or a rectangle) to an uncertain variable (represented by an ellipse) means that the probabilities subjectively assigned to possible values of the latter are "influenced" (in the opinion of the expert) by the value of the former. An arrow from a variable to a decision variable (represented by a rectangle) means that the value of the former variable is known at the time the decision is made.[64] The diagram indicates to the analyst probabilistic and informational dependencies that exist among the variables. Figure 10 shows an example of an influence diagram constructed to model the

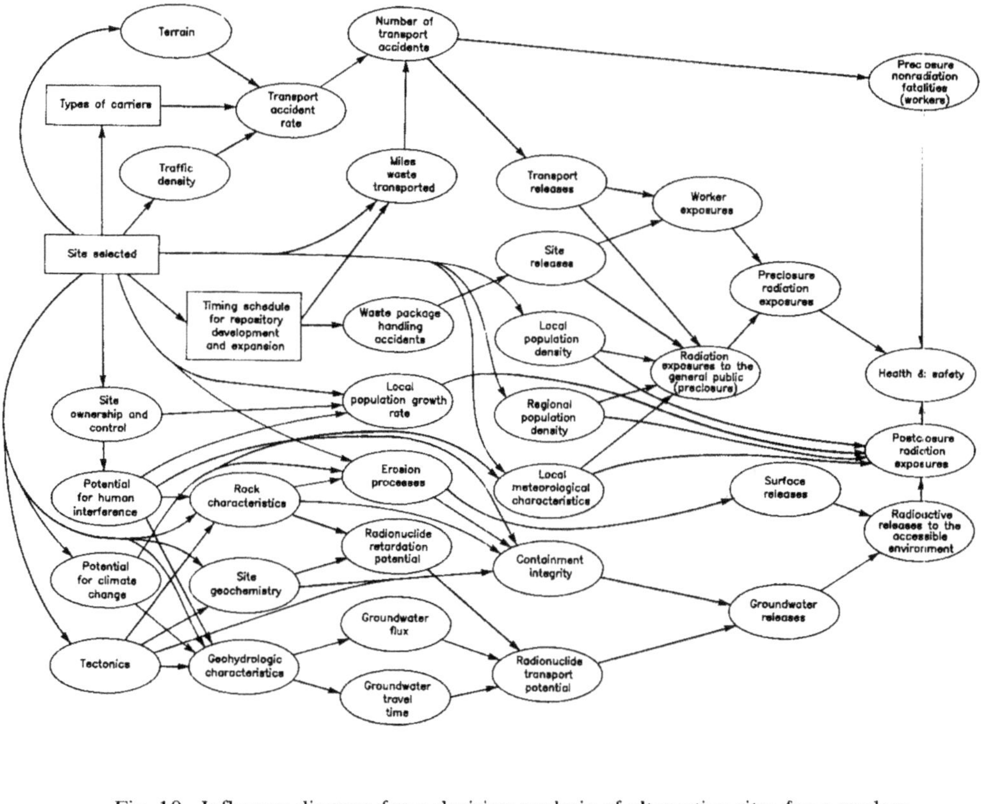

Fig. 10. Influence diagram for a decision analysis of alternative sites for a nuclear waste repository.

uncertain health effects of nuclear waste disposal siting decisions. The diagram indicates that the overall state of health and safety obtained from the selection of a site, development schedule, and mode of shipping radioactive waste depend on many factors that interact in a complicated way. Constructing the influence diagram helps the decision analyst identify the critical factors that must be accounted for in the evaluation of alternatives and suggests a logical framework for the development of more detailed models for estimating probabilities of possible decision outcomes.

In general, the modeling procedures used in CBA and DA approaches will differ in the extent to which the relationships assumed in the model and the model inputs are based on statistical inference from empirical data or on theoretical concepts and subjective judgments. Because CBA approaches attempt to build an objective or quasi-objective representation of the decision situation, they place greater reliance on statistical procedures. DA approaches, because of their acceptance of the inherent subjectivity of information, make greater use of judgmental models.

Most models are either analytical models (in the sense that predictions are generated from solving mathematical equations that describe the status of variables and the way in which they change and interact) or simulation models (in which predictions are generated by sequentially executing procedural steps designed to represent the dynamic relationships hypothesized to exist in the real situation). With very simple models the relationship between alternatives and outcomes may be direct; for example, CBA approaches have derived simple correlation models between production and measures of environmental quality through the statistical analysis of existing data. Most often, however, the models used in decision-aiding approaches are composed of submodels designed to represent the various components of the risk chain. Thus the model constructed for analyzing environmental, health, and safety decisions frequently consists of submodels for the source of risk, for exposure processes, and for effects processes. Since most decisions influence economic costs, cost models designed to estimate the economic impact of control alternatives are usually included. Some examples of these models are presented in the sections below.

Models for characterizing the source of risk

Models for the source of risk are often used in decision-aiding approaches to describe and quantify the characteristics of technologies, products, processes, or systems that have the potential for creating risk. These sources may contain substances that are inherently flammable, corrosive, explosive, toxic, or the like, or that possess high levels of energy. In such cases the model specifies the assumptions used in pinpointing the degree to which these risk agents are or might be released to human environments and thereby affect the level of danger posed. For example, if the concern is dam failure, then the risk-source model might represent the probability of various failure modes and the volume of water that would be released under each. In the case of automobile accidents, the concern is the release of violent kinetic energy into the passenger compartment. The model of the risk source might account for the likelihood of different types of collisions, the energy-dissipating characteristics of the automobile's body structure, and the integrity and dynamics of the passenger compartment under the various accident scenarios. If the concern is a pesticide used in agriculture and the hazard is regarded as agricultural products that contain pesticide residues, the risk-source model might be designed to account for the effect of the quantities of pesticides applied and the manner and timing of application on the quantities of residues present in the final product. Models for the source of risk are particularly important if the alternatives being analyzed are expected to reduce risk by altering the characteristics of the source rather than by altering exposure or effects processes.

The modeling methods employed to characterize a source of risk tend to be determined by the nature of the risk and by the type of decision-aiding approach used. For the case of continuous risks (activities that routinely or continously produce harmful pollutants), models are often used to estimate the characteristics, amounts, and locations at which polluting substances or undesirable forms of energy (such as noise) are emitted into the environment and how these emissions might change over time and location or with various controls. For example, to analyze a decision to reduce sulfur oxide emissions from electric power plants by taxing high-sulfur coal, a model might be developed to estimate the geographic distribution of emissions over time under different tax scenarios. A variety of procedures are available for constructing such models (for an overview of basic analytic modeling approaches, see Gass and Sisson 1975). A CBA might use an econometric model (a model based on economic and empirical data) to assess the effect of increased prices of high-sulfur coal on sulfur emissions resulting from a shift in use by the energy industry to oil and other low-sulfur energy sources (by comparing supply, demand, and price with interfuel competition, for example). A DA, on the other hand, might use engineering and process models based on judgment to estimate the effectiveness of changing fuels and improved technologies in reducing harmful emissions from representative plants.

For discrete risks, such as accidents, models of the risk source are generally based on a specification of the circumstances and sequence of events that must occur for the accident to take place. Several of the most important of the procedures used involve the development of graphic structures. For example, fault trees (Vesely *et al.* 1981; Fussel and Burdick 1977) are used to identify and quantify the sets of events leading to system failures. Used with methods of Boolean (binary) algebra, they provide a basis for computing system failure probabilities. Figure 11 illustrates the general structure of a fault tree for radioactivity release from a nuclear reactor.

A fault tree consists of events described by binary variables (the event either occurs or does not occur) which are related by logical functions, primarily OR and AND functions. When applied to two events, A and B,

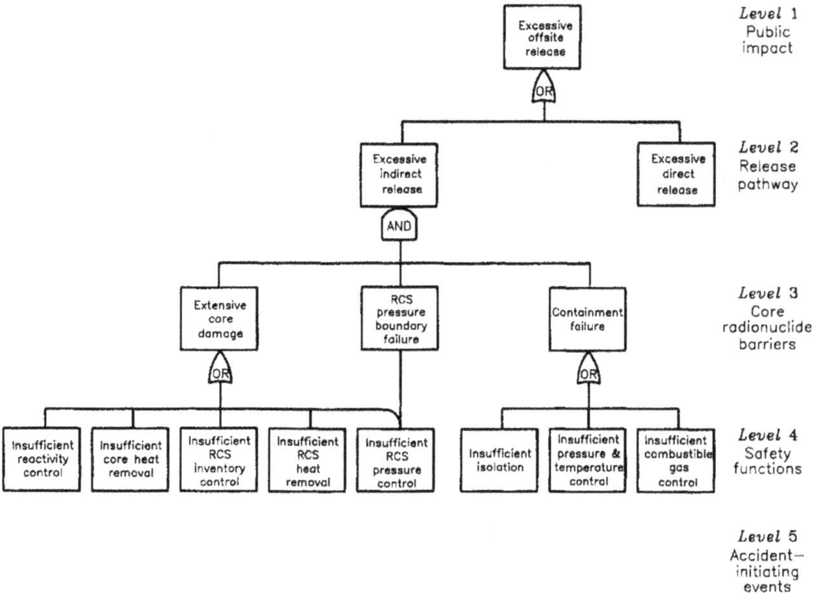

Fig. 11. Upper levels of a fault tree for radioactive release from a nuclear reactor (RCS = reactor cooling system), from American Nuclear Society and Institute of Electrical and Electronics Engineers (1983).

to define a third event, *Y*, for example, these functions operate as follows. With the OR function, event *Y* occurs if event *A* or event *B* occurs. With the AND function, event *Y* occurs if event *A* and event *B* occur. Graphically, these and other logical functions are represented by Boolean "gates." The output of a gate (depicted above it) is the corresponding Boolean function of its inputs (the events represented below it). Fault-tree diagrams may be interpreted to determine the logical relationships among the states of various system components. For example, Figure 11 indicates that the event of excessive indirect release of radioactivity will occur if, and only if, three other events occur: extensive damage to the core of the reactor, failure of the reactor cooling system boundary, and containment failure.

To construct a fault tree, the final, undesired (top-level) failure event (e.g., off-site release of radioactive material from a reactor) is defined. Using a functional description of the system, the events whose union or intersection could logically cause the failure are identified. These events in turn are related to lower-level events, and the diagram is expanded until a level of basic events is reached; for example, the failure of individual subsystems or components (such as pressure relief systems, pumps, and electrical components). Once constructed, the diagram implies a logical relationship between the top event and the set of basic events which in turn permits the probability of top event to be expressed as a function of the probabilities of the basic events.

The ability of a fault-tree model to relate the probability of failure or partial failure of a large system to the probabilities of basic events makes it well suited for CBA and related data-oriented approaches. In many cases operating experience will be insufficient to permit computing a system's failure probabilities directly from empirical data. The basic events appearing in the fault tree may, however, be sufficiently common that probabilities can be derived from empirical data (using methods that are discussed in the following subsection). Although many of the relationships expressed in a fault tree will necessarily be based on judgment, the modeling procedure's use of "objective" input data permits it to retain the appearance of a data-based procedure compatible with the perspective of CBT.

Another modeling procedure often used for discrete risks is the event tree, which identifies and associates probabilities with all possible sequences of events (scenarios) resulting from some undesired initiating event. To construct an event tree the analyst identifies an initiating event; for example, failure of a coolant pump in a nuclear reactor. He then constructs a tree structure displaying all possible sequences of events that together determine the possible responses to the initiating event. Each branch in the tree, with associated probabilities, represents a possible state (often simply success or failure) for each of the various subsystems that would be called upon as the accident progresses. Because the probability displayed for each event is conditional on the occurrence of events that precede it in the tree, the probability of the intersection of the events that constitute a given scenario may be found by multiplication.

Only rarely can the conditional probabilities required for an event tree be derived directly from available data. Therefore, event trees generally require the use of subjectively generated probabilities, making the modeling procedure better suited for DA and related approaches that adopt the Bayesian view of probabilities. Although reactor safety represents the most prominent application area, event trees have been applied to model sources of risk in many different applications. Figure 12 shows an event tree constructed as part of the decision analysis of an incinerator system for a possible space mission intended to return to earth a sample of material collected from the surface of Mars (Merkhofer 1977). The purpose of the proposed incinerator system was to reduce the probability that the mission might accidentally result in the contamination of earth with Martian organisms.

Models for exposure processes

Decision-aiding approaches use exposure models to estimate characteristics of human and other contacts with risk agents that influence the level of risk experienced. Exposure models are designed to estimate such factors as the intensity, frequency, and duration of human or other

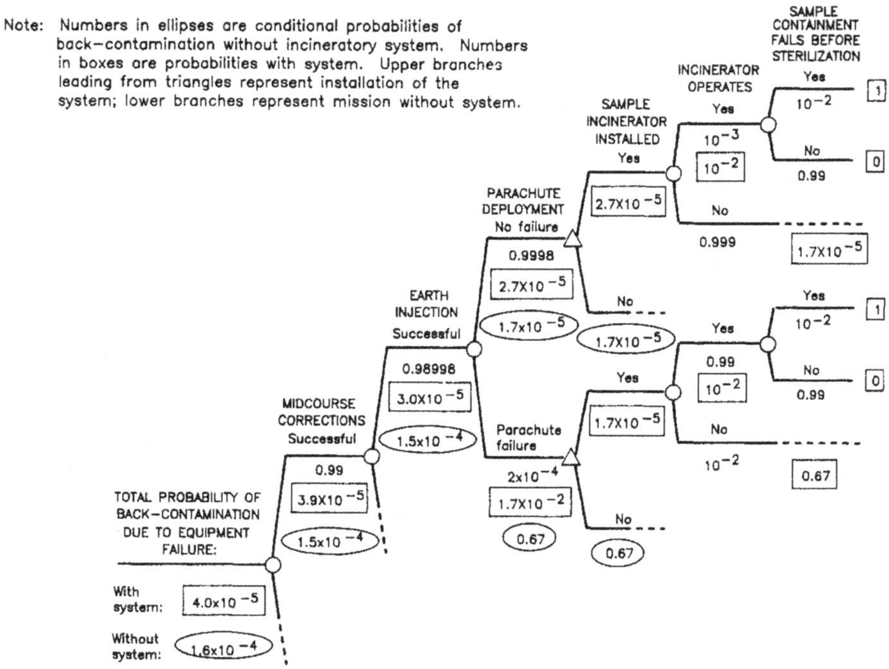

Fig. 12. Event tree showing effect on back-contamination probability of a system for incinerating a returned sample of Martian soil in the event of parachute failure, from Merkhofer (1977).

exposures to the hazardous substances or energies (risk agents) produced by a source of risk.

In situations where exposures can be conveniently represented as a series of discrete events, fault trees and event trees are useful procedures for modeling exposures. Event trees are especially versatile. For example, if the risk source is a defective brake light on an automobile during rush hour, an event tree might be used to estimate the probabilities and number of cars involved in chain reaction accidents.

If exposure processes fundamentally involve physically or temporally continuous processes, then event and fault trees are less useful modeling procedures. In this case, models consisting of relationships designed to represent physical cause-effect processes are often used. For example, exposure models based on cause-effect relationships have been developed to represent the transmission of ground motion from the source of an earthquake to a given site, taking into account the magnitude of the earthquake, local soil conditions, and the distance from the epicenter to urban areas or the facility at risk (e.g., a dam) (Patwardhan *et al.* 1980). Causal models have also been developed to learn about the dynamics and biomechanics of exposures to mechanical energies that occur during accidents involving consumer products or specific technologies. Causal models have been developed, for example, to study the accelerations and points of impact that occur during automobile-pedestrian collisions (Schneider and Beier 1974).

Within the general class of exposure models, the largest category of well-developed exposure models consists of specialized models for describing the environmental transport and transformation of pollutants released into the environment — so-called pollutant transport and fate models. Causal models used to describe the environmental transport and conversion of hazardous materials include models of the chemistry, transport, and deposition of acid rain; models of the photochemical processes associated with materials disposed of at sea; models of wind directions and velocities and precipitation patterns that might affect the transport of pollutants from normal or accidental industrial or power plant emissions; and models for the leaching and runoff of chemicals into streams. The various models differ from one another in terms of their approximations and the assumptions used in characterizing the source (e.g., point source or area source, instantaneous or continuous release), the media of transport (e.g., air, surface water, groundwater), and the manner of spreading and entrainment considered.

In the case of exposures produced through the movement of materials through air, transport modeling has reached a relatively high degree of sophistication. One procedure sometimes used in cost-benefit analyses of air quality decisions is to develop regression or time series models based on least-squares fitting to ambient concentrations obtained from monitoring data. Another approach, which has been used in DA in addition to CBA approaches, is the Gaussian plume model. This model assumes that the plume from an emission source spreads laterally and vertically in accordance with a Gaussian statistical distribution (Goodin 1981). Another, more computationally demanding, modeling procedure for air emissions is based on computing trajectories that released material might follow. The trajectories are typically composed of segments estimated for fixed intervals of time and based on historical wind data.

In addition to transport, many atmospheric models account for transformations of pollutants (e.g., photochemical transformations of sulfur dioxide emissions to various sulfate aerosols) and are useful in situations where transformation products may exhibit more or less bioactivity than the parent pollutant. Chemical transformation processes are often approximated by first-order rate reactions. Frequently, air transport models are used to estimate long-term average concentrations, while short-term concentrations are assumed to follow some specified frequency distribution (usually log-normal).

Aquatic transport models, although not as highly developed as atmospheric models, are also frequently used in exposure modeling. In the case of surface water, such models attempt to account for the major source of contaminants either as point sources (domestic wastes, industrial wastes, heated discharges from cooling technologies) or as storm water or normal rainfall runoff processes to water bodies from urban and agricultural areas. Surface water transport may be an important factor in situations where a principal exposure pathway is through drinking water or where local aquatic food is ingested by humans. Steady-state models are most frequently employed to account for the transport of pollutants in surface water. For instance, a river may be represented as a series of completely mixed reaches in which steady-state contaminant concentrations are estimated based on dilution and physical/chemical removal of contaminants from volatization, net absorption, net settling, photolysis, microbial degradation, and so forth (Travis *et al.* 1983). To represent spills, such as oil spills from tankers, dynamic models are available.

Groundwater transport is important if the risk source is a toxic waste disposal site. More generally, toxic materials can enter groundwater by deposition from the atmosphere onto the ground with subsequent infiltration into the water table, or through diffusion from surface water. In modeling groundwater flow it is important to account for the interaction of the contaminants with soil; in many cases, a large proportion of contaminants in groundwater will be adsorbed on soil particles or absorbed into the soil, and there their movement relative to that of groundwater will greatly decrease. Very simple models tend to be used to represent these processes; for example, a linear absorption model is often used to predict the relative concentrations of a compound in the soil.

Food-chain models are sometimes used in situations where there is a concern that contaminants may enter food chains and pose health risks to man. Aquatic food chains, for example, can sometimes concentrate harmful compounds. To estimate risks to man, intake rates for contaminants in the various food categories must be calculated based on the consumption rate of food. Three elements are generally required for such calculations: (1) the concentration of the component in question in each item of the diet, (2) the amount consumed of each item that contains the component, and (3) the frequency with which each item is consumed. Simple procedures are generally used to represent such effects. Market basket analysis, for example, looks at contaminants in prepared foods and typically produces an average or representative intake. Per capita disappearance is based on dividing the sum of annual production plus imports of a given food item by the number of people in the country. Since these procedures fail to reveal special groups with high intake rates, reliance is sometimes placed on dietary surveys in which consumers are asked to recall what foods they ate with what frequency and in what amounts (portion size) over a defined period of time (Food Safety Council 1980).

Exposure models also must account for the population at risk from a given pathway of exposure. This may entail determining who works or lives at various sites, and who breathes air, drinks water, eats food, or uses products coming from a particular source. This inference is generally painstaking but straightforward. Because the health consequences to exposed individuals are often known to depend on age, sex, health status, and other characteristics, it is usually necessary to disaggregate exposures to interface properly with dose-response models. Demographic models developed to support this task may include models for the population distribution within a geographic area surrounding a source of potential risk (e.g., dams, power plants, aircraft flight paths), models of the age distribution of a given population, and models of mobility and activity patterns of sensitive population groups.

Models for effects processes

Effects models (also called consequence models) are designed to translate a given exposure to a risk agent to the health and other consequences to exposed populations. These consequences can include early fatalities and injuries, latent cancer fatalities, genetic effects, environmental degradation, and economic losses. Most of the effort devoted to developing models for effects processes has been directed at understanding human health rather than toward the more general ecological consequences of exposures to hazardous substances. The discussion, therefore, emphasizes modeling procedures related to human health.

The principal component of a health consequence model is a dose-response model. A dose-response model is a functional relationship between the dose of a risk agent or biological stimulus received by a subject or population group and the corresponding effects on that subject's or population's health status. An effects model may also include health-endpoint models that translate physiological effects estimated by a dose-response relationship (such as reduced pulmonary function) to the end health effects that concern people most (e.g., chest pain). Health effects models also sometimes contain submodels for relating the exposure levels estimated by an exposure model to the specific measure of dose needed for input to the dose-response model.

Models for estimating health outcomes are generally based on epidemiological or animal studies and may be derived directly through the statistical analysis of existing quantitative data or through theoretical models. Empirical dose-response relationships are obtained by comparing the dose of an agent or biological stimulus administered and the response of a test subject. The dosage scale depends on the risk agent and the route of intake and might be expressed in terms of a cumulative exposure or rate per unit of time. Similarly, the response scale depends on the nature of the response and might measure severity, the probability of a given effect occurring, or the time until the occurrence of some effect. Often dose-response models estimate "excess" risk — that is, the increase in the incidence of some effect above the normal background level. The model form used may or may not be based on plausible cause-effect mechanisms. Dose-response models are most prominent when the risk agent is a hazardous chemical, but such models have also been developed to estimate injuries and other effects from accidents or exposures to other forms of kinetic energy such as noise.

In general, dose-response models are based on some assumed functional form of the relationship between dose and response. Simple relationships that are used include linear, linear with a threshold, quadratic, or sigmoidal (S-shaped) curves. Threshold distribution models are based on the assumption that each individual in the population has a particular threshold tolerance for the risk agent. Specification of a functional form for the distribution of tolerances then determines the shapes of the dose-response curve. Because the proportion of responses in a toxicity test generally exhibits a sigmoid relationship when plotted against the logarithm of the dose level, the frequency distribution of tolerances within a population is often assumed to be log-normal, the so-called probit model.

Another commonly used tolerance distribution model is the log logistic, or logit function. Like the probit, it reflects an assumption that the proportion of responses increases with dose and exhibits a symmetric, sigmoid shape, but it approaches the extreme (0 percent and 100 percent response) levels more gradually than the probit. Both model forms are frequently used to estimate the incidence of cancer resulting from a population's exposure to a carcinogen. In such applications, the logit and probit models appear so similar in the normal range of the dose-response curve (the range of empirical data) that it is nearly impossible to distinguish them. In the range of low doses, however, the logit model may predict doses producing no observable effects that are more than an order of magnitude lower than those produced by the probit model.

Depending on the level of understanding available, dose-response models may be developed based on theoretical assumptions for the biological processes involved. For example, in the well-studied area of carcinogenic risks, several mechanistic models are often used. One class of such models is based on the premise that a positive response is the result of the random occurrence of one or more biological events. The one-hit model, also known as the linear model (even though it is concave), has often been used to estimate the production of tumors from exposures to radiation. The model assumes that only one interaction between the target site and the hazardous substance is sufficient to produce an adverse condition, and that the probability of an interaction is directly proportional to the degree of inital exposure to the substance. The multi-hit model (also called the gamma multi-hit model) assumes that two or more hits of a substance are required to induce a response, and that the number of hits over time follows a Poisson distribution. The dose-response curve predicted by the multi-hit model at low doses may be linear, concave, or convex, depending on the values selected for its parameters.

The multi-stage model, also called the Armitage-Doll function, is based on the assumption that an adverse health effect originates as a predisposed cell that must undergo a series of mutational stages. In addition, it also assumes that the timing of transitions between stages is expressed by a probability function that is approximately proportional to the dose rate. The multi-stage model, like the one-hit model, is considered a conservative extrapolation function because the curve approaches linearity at low dose levels.

More sophisticated health-effects models may consist of submodels for the biological processes that control the uptake or residence time of the hazardous substance, mechanisms of cellular or organ damage or dysfunction, mechanisms of repair, and so forth. For example, a submodel might be developed to relate the dose of exposure to the quantities of pollutants delivered to sensitive parts of the body. In the case of analyses involving atmospheric particulate matter it is well known that particle deposition in the human respiratory tract is a function of the aerodynamic diameter of the particles and the individual's breathing characteristics. The health consequence model for such an analysis might begin, therefore, with a submodel for translating dose exposures, measured in terms of atmospheric concentrations, to delivered doses, measured in terms of quantities per unit time delivered to sensitive organs. This submodel might account for variations in the individual's activities over time that influence the size and chemical composition of the ambient particles to which he is exposed (e.g., whether they originate at home, outdoors, or at work) and the individual's breathing characteristics (e.g., ventilation rate and whether an oral or nasal breathing mode is being used, which would depend on physical activity).

Pharmacokinetic dose-response models are based on the principle that biological effects are the result of biochemical interactions between foreign substances (or products of the body's efforts to metabolize these substances) and parts of the body. These models generally assume so-called Michaelis-Menten nonlinear kinetics, which, like many of the dose-response models, produce estimated effects that are linear when extrapolated to low doses. Like the multi-hit and multi-stage models, pharmacokinetic models can describe convex or concave curvature in the dose-response relationship; but unlike the former models, they do not necessarily assume that nonlinear behavior is similar at both high and low dose levels.

For analyses involving risks of injury, dose-response models often relate accelerative forces or physical impacts to kinematic actions of the body and resulting physical damage. For example, simple two-dimensional mathematical models have been developed to estimate forces and impacts sustained by passengers and pedestrians in automobile accidents (Wismans, Maltha, and Van Wiljk 1982). Such models might treat the body as connected system of parts (head, torso, legs, arms) of various lengths, centers of gravity, and moments of inertia (Stainaker 1983). Similarly, if the concern is head injuries, a model that represents the human skull as a spherical shell under dynamic local loading might be used to predict the incidence of skull fracture (Perrone 1976).

Regardless of the type or form of the dose-response function, its parameters are usually selected by "fitting" the model to available toxicological, epidemiological, or other data. For example, a dose-response function relating driver survival probability to automobile collision velocity might be based on fatality data from actual crashes or from simulated accidents using anthropometric dummies with specified survival criteria. Regression analysis of one form or another is the typical means for model fitting. When animal-to-human extrapolation is required, it is often based on standardized dosage scales for making interspecies comparisons, such as milligrams per kilogram body weight per day. When data are lacking, medical experts are sometimes questioned directly to produce subjective, best-estimate, dose-response relationships.

Cost models

Cost models are of most significance to CBA and social DA approaches given their economic orientations, but they can be an important element of any decision-aiding approach. However, estimating costs can be difficult, because the most significant costs associated with health, safety, and environmental regulations are usually not direct government expenditures but increased costs incurred by the private sector in the production of goods and services. Thus cost estimation requires "second guessing" how the private sector will react to government actions (Smith 1986). To do this effectively, the analyst must work with individuals with sufficient technology and process expertise to identify least-cost methods of achieving the target levels of control. Even in the case of technology-only regulations, costs will depend on a producer's manufacturing process, technologies employed, plant size, and so on. Frequently the analyst will simply ask industry to supply estimated control costs. Those in industry, however, are generally a poor source of cost information, since they consider operating data to be highly proprietary and may have a natural incentive to alter their responses. Thus cost models are often necessary to estimate the economic costs associated with regulatory alternatives.

Cost are typically classified into capital and operating costs, with separate models used to estimate each. CBA approaches tend to use either econometric or engineering cost models, while DA approaches mainly use only the latter. As an example of an econometric model, statistical methods might be used to identify a "production function" relating the various inputs and outputs (including pollution) for an industry in question. Relationships between cost and pollution might then be derived from the production function. Econometric methods, however, have limited applicability for cost estimation because the estimates needed are typically too specific or detailed to be derivable from the data bases driving econometric models.

Engineering cost models must make a great many simplifying assumptions. The analyst might assume that a polluter uses some "typical" production technology, determine the control technology likely to be used, and estimate the cost of that control technology. This process might be conducted for "representative" plants from various categories. In the case of cost estimates associated with air quality regulations, plants might be classified by generating capacity, fuel characteristics (such as sulfur and ash content), and heating rates. The costs estimated for the representative plants would then be scaled in accordance with the numbers of plants within each category.

Procedures for Quantifying Uncertainty

A principal difference among approaches relates to their treatment of uncertainty. Although all approaches recognize probabilities as being the preferred way to quantify uncertainty, the interpretation given to probability, the emphasis placed on developing probability distributions, and the methods and models used all differ depending on the approach. CBA approaches tend to provide only a limited treatment of uncertainty and employ procedures that adopt an objective view of probability. DA approaches generally devote a great deal of attention to the quantification of uncertainty and interpret probabilities as subjective. Applied SWT approaches may or may not develop explicit quantifications of uncertainty, depending on whether decision outcomes and their uncertainties are explicity represented as attributes affecting individual preferences. When uncertainties in outcomes are explicitly quantified in applied SWT, either an objective or subjective interpretation of probabilities might be developed. A more detailed discussion of the procedures available for quantifying uncertainties is presented below.

Procedures adopting an objective view of probability

Most cost-benefit analysts adopt the view of classical statistics that risk is an objective property of the physical world. As such they assume that only repetitive phenomena can be analyzed with probabilities, and they equate probability distributions with frequency distributions derived from sample data. As a result, although a CBA might develop probability distributions to account for randomness demonstrated in nature, it is rare for a CBA to attempt a comprehensive accounting of the uncertainty in the outcomes of a decision.

The problem, frequently lamented by cost-benefit analysts, is the absence of sufficient empirical data. "Unfortunately, far from all project outcomes can be characterized by a well-defined probability distribution" (Abelson 1979). Lacking frequency distributions or symmetry arguments for estimating probabilities, Paretian CBA and other approaches adopting a strong objective perspective limit their quantification of uncertainty to a specification of a range within which the value of the uncertain quantity is judged "likely to fall." Some cost-benefit analysts have attempted to justify their limited consideration of uncertainty by arguing that accounting for uncertainty may not be that essential:

> Provided that the government can be regarded as having a time-horizon of several years during which it undertakes a large number of investment projects in many separate fields, these can be treated as essentially independent of one another. Hence we can anticipate that some projects will turn out worse and some better than expected. In fact, . . . we can expect that the negative and positive "surprises" would tend to cancel (Abelson 1979).

To the extent that a probabilistic analysis is undertaken in CBA, probability distributions are generally calculated statistically from empirical data. Cost-benefit analysts quantify uncertainty about such variables as weather conditions, failure rates, or health effects by treating them as random variables with probability distributions that are assigned on the basic of generic data. They regard the data as providing a sample from which "true," or "population," probability distributions can be derived. These distributions are "named" probability distributions (such as the binomial or normal distributions), and they are specified by one or more parameters (e.g., a mean and variance). The "population" probability functions are developed through the estimation of their associated parameters.

The simplest case is an event that is assumed to occur with constant probability per opportunity, independent of whether or not it occurred at any other opportunity. Examples might be a pump failing during successive operations at a chemical plant or a fatality occurring during an automobile crash. In this case, the number or frequency of occurrences of the event n for any given number of opportunities N is represented by a binomial probability distribution. The probability of occurrence per opportunity p is the parameter that together with N defines this distribution; the empirically measured fraction of times that the event occurred n/N, provides a point (single number) estimate for p. Analogous methods apply if the concern is the point in time (or distance) at which an uncertain event occurs, provided that the frequencies of the event may be assumed to be constant in time. The exponential probability distribution is commonly used to represent uncertainty in events that occur continuously in time. If the frequencies of occurrences are not constant but vary over time, several alternative procedures may be used -- for example, assuming a parametric function of time and estimating the values of the parameters from past data using similar statistical estimation techniques.

If the objective is to quantify uncertainty in the level of the consequences of an event, not merely its rate or time until occurrence, then the above methods may be used to estimate the frequency of all events having a given range of outcome values. By choosing successively larger outcome values and calculating frequencies of events having outcomes greater than these values, a complementary cumulative distribution may be obtained.

Similar statistical estimation methods apply if the outcomes of the random variable are continuous; that is, if they can take on any value within some range. For example, short-term ambient air pollutant concentrations are often assumed to be distributed as log-normal distributions. Assuming that the individual observations are independent, the data may be used to compute point estimates of the mean and variance of the underlying log-normal curve.

Dependencies among uncertain variables can be handled using classical methods, provided that sufficient data exist to estimate the nature of the dependencies. Two uncertain variables are dependent (or correlated) in the classical sense if a higher than average value for one variable tends to be associated with either a higher or lower value for another. Both multivariate correlations (situations where a particular variable is dependent on several other variables) and time-series patters of correlation (where the value of a variable in one time period has a relationship with values of that variable in previous time periods) may be important. Because of the data demands of multivariate and time-series analysis, simplifying assumptions are standard for approaches that use these procedures. For example, variables are usually assumed to be independent or totally dependent (to have perfect positive or negative correlation). Since total dependence technically requires that the variables have the same marginal probability distributions, an approximate notion of total dependence is needed for cases where the marginal probability distributions of the variables are different. One procedure is to assume that if one of the dependent variables takes on a value in a given percentile (e.g., a 90 percent value on a cumulative distribution), then the other variable would take on that same percentile value.

Error measures of classical statistics provide an approximate way for CBA approaches to quantify uncertainty due to limited data. The "standard error of estimate," for example, indicates the variability in the values that would be computed for the parameter if the process generating the data could somehow be repeated many times. Classical confidence limits (often interpreted as the likelihood that a given range of values about a point estimate contains the "true" value) similarly provide a measure of the ability of the data to yield an accurate estimate of the parameter. The probabilistic range of variation in each of the parameters of a distribution implies a corresponding range of variation in estimated probability curves. Thus confidence bounds for risk profiles can be generated using procedures from classical statistics. This sort of "error analysis" is, however, rarely conducted in actual application of CBA.

Because statistical methods for inferring probabilities require a data base, various extrapolation methods are needed to support applications in areas where data are limited. If the frequency of severe events is desired (such as the frequencies of extremely high wind velocities), estimates might be obtained by smoothly extrapolating from more common, less severe events (the frequencies of winds with lesser velocities). Extreme value theory (Vesely 1984) is sometimes applied if there are a large number of processes whose maximum or minimum states are involved in producing the risk and if there is not strong dependency among the processes. The procedure involves fitting a frequency versus consequence curve to past data and using the fitted curve to predict the frequencies of events with large consequences that have not yet occurred.

Models that decompose the uncertain variable into other variables are often used when direct data do not exist and extrapolation procedures are not applicable. Fault trees and event trees (described above), which allow the probability of an event to be computed from more basic events, are most commonly used for this purpose. Constructed data bases and surrogate data are also often used when data on the quantity of interest are unavailable. For example, although data may not be available for liquefied-natural-gas tanker accidents, oil tanker accident rates might be used as a surrogate data base. Similarly, nationwide per-mile accident statistics might be used in place of areawide accident statistics.

Some decision-aiding approaches employ Bayesian rather than classical methods for uncertainty analysis but are unwilling to abandon the objective view of probability. The advantage of this course is that Bayes theorem can be used to effectively integrate background knowledge with specific data. Bayes theorem specifies how a "prior" probability distribution (based on an initial state of knowledge) should be updated to produce a "posterior" probability distribution that accounts for additional data. The key to updating is a "likelihood function," which is a probability distribution describing the likelihood of the specific additional data conditional on given values for the uncertain quantity. Specifically, Bayes theorem states the following: if $P(H)$ is the prior probability that hypothesis H is true, $P(D)$ is the probability that datum D will be observed, and $P(D \mid H)$ is the conditional probability that D will be observed given that H is true, then the posterior probability that H is true given that D has been observed is[65]

$$P(H \mid D) = \frac{P(D \mid H)\,P(H)}{P(D)}$$

When used in CBA and related approaches, the Bayesian procedure is generally conducted by fitting a named probability distribution to data representing some prior state of information. If there is little or no prior information, "noninformative prior distributions" may be employed. These are distributions that, in one sense or another, minimize the importance of the prior, relative to new data, on the posterior estimate. In applications to nuclear power plants, generic data derived from a variety of plants and conditions are often used as a basis for obtaining prior distributions, while plant-specific information provides the basis for updating to provide a posterior distribution that "specializes" the prior to the specific plant under study.

Although Bayes theorem implies a subjective, degree-of-belief view of probability, its application to physically derived measures of variability provides an empirical basis for quantifying probabilities. From the point of view of analysts employing approaches that settle for a quasi-objective analysis (such as decision-maker CBA), such methods enjoy a degree of acceptance sufficient to qualify the probabilities produced as "objective."

Conjugate prior distributions are often used in "objective" Bayesian analysis. Conjugate priors are distributions that, for a given likelihood function, produce posterior distributions that are members of the same family. In this case, updating a probability distribution is simply a matter of updating the parameters of the distribution, and the process is computationally simple. Examples of conjugate priors are beta distributions (when a binomial likelihood function is used) and gamma distributions (when a Poisson likelihood function is used).

Figure 13 shows an example of Bayesian analysis based on empirical data taken from Apostolakis (1981). The prior curve describes the frequency with which diesel generators fail to start (expressed as a failure rate per demand). The curve is a log-normal distribution fitted to generic data representing different manufacturers and operating and maintenance conditions. The means and variance of the prior are 0.04 and 0.001. The posterior distribution was obtained using Bayes theorem and assuming specific data for the plant under study that indicate a total of 5 diesel generator failures in 227 tests (a plant failure rate of 5/227 = 0.022). As illustrated by the posterior curve (which has a mean of 0.025 and variance of 0.00008), the effect of the new evidence is to shift the distribution of the failure rate toward lower values and to reduce its dispersion.

Procedures adopting a subjective view of probability

Decision analysis approaches place a much greater emphasis on the explicit and comprehensive representation of uncertainties than is typical of cost-benefit analysis approaches. The subjectivist view inherent in DA argues that probabilities represent an individual's degree of belief about the world, not a property of the world. Hence, the decision analyst feels as comfortable eliciting probabilities for unique events (e.g., a new drug being

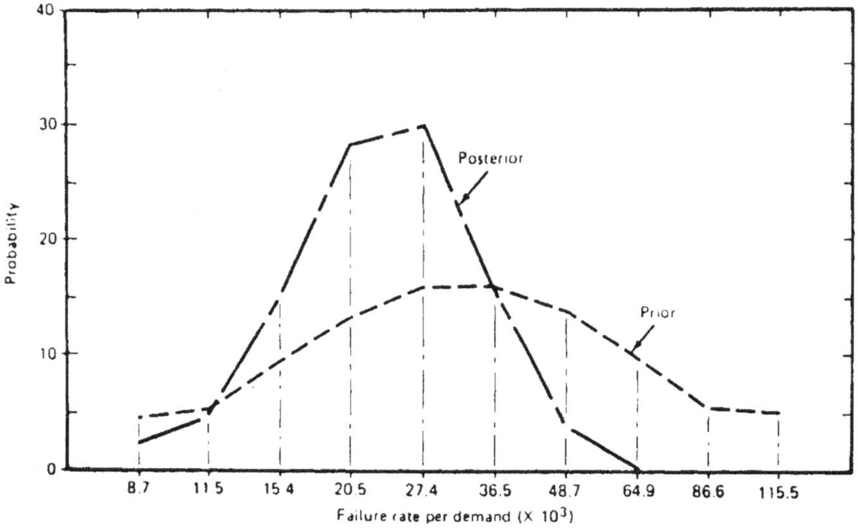

Fig. 13. Prior and posterior probability histograms for diesel-generator failure to start, from Apostolakis (1981).

carcinogenic) as for events for which frequency information is available (e.g., a valve failing to open during the course of 10,000 operations).[66] The arguments supporting the use of subjective probabilities in formal analysis derive from theoretical foundations that show judgments about uncertain events can be expressed as probabilities and also by practical assessment procedures. Probability assessment is normally conducted as a joint undertaking by an analyst and a subject. The analyst's expertise is in probability and methods for minimizing cognitive biases in the expression of judgment. The subject is typically an expert in what is known and unknown about the uncertain quantity of interest. The assessment process consists of several phases in addition to the actual quantification of the subject's judgment in probabilistic terms. These additional steps are designed to help the analyst identify and reduce the effect of cognitive and motivational biases held by the subject ("conceptual biases") and biases that might be introduced by the assessment method ("task biases").

Spetzler and Staël von Holstein (1975) describe an assessment process consisting of five phases: motivating, structuring, conditioning, encoding, and verifying. As part of this process the analyst describes to the subject how biases can arise so that the subject can try to prevent them or compensate for them. The analyst also actively tries to reduce biases. For example, probing is used to help identify hidden assumptions on which the assessment might be implicitly conditioned. This may suggest better disaggregations in which the probability of these conditions may be considered explicitly in a separate assessment. The process also attempts to use known cognitive biases to counteract one another. An extended description of this assessment process is described in a "probability encoding manual" prepared by the Decision Analysis Department of SRI International (Staël von Holstein and Matheson 1979).

Decision analysis approaches can draw from a wide range of elicitation and measurement procedures to conduct the encoding phase of a probability assessment process (Savage 1971; Hampton, Moore, and Thomas 1973; Brown *et al.* 1974). The basic types of encoding procedures are probability methods, which require the subject to respond by specifying points on a probability scale corresponding to fixed values of the uncertain variable; value methods, which require the subject to respond by specifying points on the value scale while the probabilities remain fixed; and probability/value methods, which ask questions that must be answered on both scales simultaneously (the subject essentially describes points on a probability distribution). Each encoding method may be presented either in a direct or indirect response mode.

In the direct response mode, the subject is asked questions that require numbers for answers, such as "What is the probability that the unknown quantity is less than X?" Probabilities may be expressed directly or in terms of odds (1:3 for a probability of 0.25).[67] With the indirect method the subject is asked to choose between two or more imaginary bets. One of the bets involves the uncertain quantity in question; the other is a "reference lottery" consisting of some physical device or conceptual situation where the outcomes are easily understood and computed. For subjects who are not experts in the assessment of probabilities, indirect methods are preferred to direct methods. Furthermore, fixed-value methods are generally preferred to fixed-probability methods, because they tend to produce more diffuse probability distributions and are therefore less likely to reflect expert overconfidence (Morgan *et al.* 1979).

A widely used encoding procedure in DA applications is an indirect, fixed-value method that employs a "probability wheel" as a reference lottery. Figure 14 shows the probability wheel developed for use by the Decision Analysis Group of SRI International. The wheel consists of a disc constructed so that two complementary orange and blue sectors can be varied in angle to occupy any fraction of the disc's area. The disc may be spun in relation to a fixed pointer and will come to rest with the pointer falling in one sector or the other at random.

To use a probability wheel, the analyst asks the subject whether he prefers a bet in which a prize is received if the spinner lands in a target color area or a bet in which the same prize is received if some event defined by the uncertainty occurs. To define the event based on the uncertainty, the analyst selects for the variable a value that the subject thinks is not too extreme (but not the most likely or central value). For

Fig. 14. The SRI International probability encoding wheel.

example, if the uncertain variable is the Dow Jones Industrial closing average for the end of the calendar year, a value X within a few hundred points of the current value might be chosen. The subject would be asked, "Would you rather bet that the Dow Jones average at the end of the year will be less than X, or that when I spin this wheel the pointer lands in the blue?" The relative area of the target color (blue region) is adjusted and the questions repeated until a setting is found for which the subject is indifferent between the two bets; the subjective probability estimate of the occurrence of the event is then assumed to be equivalent to the proportion of area covered by the target color. A scale on the back of the wheel gives the probability of the event. After a number of such points are encoded and plotted on probability group paper, a curve is drawn through the points and the result is assumed to represent the subject's subjective probability distribution. Figure 15 illustrates the typical form of the results. The curve shows a judgmental probability distribution for the average value of the effective porosity of a subterranean rock mass considered as a possible location for a nuclear waste repository. The curve was encoded from a geohydrologist and shows substantial uncertainty due to the lack of empirical data.

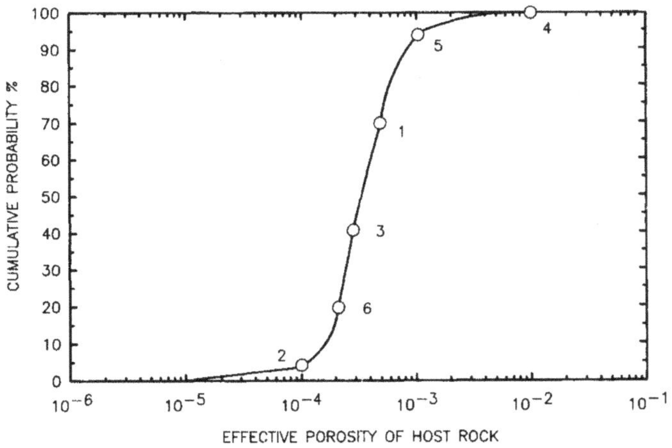

Fig. 15. Sample results from a probability encoding exercise.

The same basic encoding process may also be used with other reference lotteries. For instance, an urn containing a specified proportion of red and blue identically shaped balls has been used as a reference lottery for probability encoding. In this case the mix of red and blue balls is adjusted until the subject is indifferent between the two bets. As the reference lottery serves as an external scale for the subject, it should be selected so that its probabilities are as transparent to the subject as possible. Phillips and Thomas (1973) report investigations suggesting that assessed probabilities using the urn device may be 2 to 7 percent higher than probabilities obtained using the probability wheel.

Another popular probability encoding procedure is the so-called interval technique, an indirect, fixed-probability method. In the interval technique, the subject must specify values for the uncertain variable that serve as the boundaries for intervals over the range of possible values. The values are adjusted until the intervals are such that the subject thinks it equally likely for the actual value to lie in each. Typically the median value is determined first by dividing the range of possible values into two equally likely regions. Then, values for the 25 percent and 75 percent points on the cumulative distribution are found by subdividing each of those regions. This process is repeated as many times as necessary to obtain sufficient points that, together with the values judged by the subject as the lowest and highest possible, permit the analyst to draw in a reasonably smooth curve.

Probabilistic dependencies can be handled in probability encoding by assessing joint probability distributions describing the simultaneous values of the variables or by assessing a (marginal) probability distribution on one of the variables and then assessing the second variable conditional on each possible value of the first. The second approach, which is generally simpler and, therefore, more common, amounts to decomposing the joint uncertain event using the expansion rule of probability theory.[68] If conditioning is provided by continuous random variables, then a discrete approximation (e.g., high, medium, and low values) is obtained for each conditioning variable prior to assessing the probability distribution for the dependent variable. Influence diagrams (discussed in the section on models) are frequently used by decision analysts to identify probabilistically dependent variables. Because encoding probability distributions for dependent variables can be very time consuming, simplifying assumptions are often used. One such procedure is to assess marginal distributions for each dependent variable and then to elicit from the subject a judgment for the median value of one variable given some specific value (not equal to the median) for the other. Certain simplifying assumptions then permit a set of conditional distributions and a correlational measure of dependence between the variables to be derived (Hull 1977).

For social DA and similar approaches that strive to represent social rather than individual knowledge, group assessment procedures may be used. The Delphi method has been used as a procedure for promoting consensus when probability distributions are assessed from more than one individual (Runchal *et al.* 1984). With the Delphi method, probability curves and associated rationales are obtained independently from each member of a panel of experts. The results are summarized and conveyed to each panel member and the encoding process repeated. After several "rounds" of this procedure, it is hoped that most of the information held by the participants will be exchanged and the assumptions standardized, so that the probability curves obtained will converge. The structure of the process is designed to reduce the pressures associated with interpersonal interactions typical of most group meetings. In the supra DA approach and certain applied SWT approaches, group probability distributions are sometimes obtained as a weighted average of individual assessments. Winkler (1970) suggests four possible methods for selecting the weights to be used with this procedure: equal weights, self-weighting by the experts, subjective weighting by a third party (a supra decision maker), or weighting based on some measure of past experience.

To permit the results of probability encoding to be used meaningfully in an analysis the results must meet certain minimal properties required of a probability measure. For example, the probabilities assigned to all possible outcomes to an event must sum to one. For complex events with many possible outcomes, procedures are available for ensuring that the numbers assigned in probability encoding meet these requirements. In the modified Churchman-Ackoff method the subject is asked to define and rank in terms of their likelihood the possible outcomes to an event. To refine this initial ranking each possible outcome is compared with each other possible outcome for the purpose of estimating which is more likely. The outcomes are again ranked, this time by the frequency of being the more likely of each pair. After the subject has resolved any differences in the rankings, he is asked to provide likelihood ratios (odds) to the outcome pairs; for example, "outcome A is one-half as likely as outcome D." Given a consistent set of likelihood ratios for all outcome pairs, probabilities can be obtained by arbitrarily assigning a value to one and then scaling so that the results sum to one. To obtain a consistent set of likelihood ratios, the normalized geometric mean method is sometimes used. With this procedure the assessed likelihood ratios corresponding to all pair-wise comparisons of outcomes are displayed in a matrix such that the i, j element of the matrix represents the ratio of the likelihood of the ith outcome to the likelihood of the jth outcome (the diagonal elements are unity, and the j, i element is the inverse of the i, j element). The likelihood ratios will not be consistent unless the entries in the matrix can be multiplied so that every entry (i, j) times every entry (j, k) gives every (i, k) entry. An inconsistent $n \times n$ matrix can be converted to a consistent one that is in a specified sense "closest" to it by determining the normalized geometric mean of each row — that is, by multiplying the entries in each row, taking the nth root, and dividing by the sum of all n of such roots (Williams and Crawford 1980).

Decision analysis approaches, like cost-benefit analysis approaches, make use of probabilistic models to infer probabilities for the uncertain behavior of more complicated systems. Bayes theorem is used extensively to permit decomposing a complex uncertainty into simpler uncertainties whose combined probabilities can then be calculated. Probability trees (similar to event trees) and Markov models (e.g., Howard 1971) are common probabilistic model forms used in DA approaches.

Procedures for Developing Probability Distributions of Decision Outcomes

In accordance with the decomposition strategy inherent in all decision-aiding approaches, measures of uncertainty in decision outcomes (i.e., measures of risk) are obtained by identifying and combining individual sources of uncertainty. The model used to estimate decision outcomes provides a means for categorizing uncertainty sources. As noted earlier, irregardless of whether decision outcomes are estimated subjectively using a simple structural model (e.g., a tree structure identifying outcome scenarios) or through a complex simulation involving submodels for the risk-source, exposure processes, and effects processes, the basis in all cases is some assumed functional relationship (model) among the variables involved. Depending on the situation and the nature of the model applied to it, two possibilities exist:

1. The processes involved are sufficiently well understood to justify the functional relationships represented in the model. In this case, uncertainty is limited to uncertainty regarding the values of some of the variables and/or parameters in the model.
2. The physics, chemistry, biology, or other science and engineering aspects of the problem are so poorly understood that the functional relationships required by the model cannot be presumed to be known. Thus, in addition to uncertainties over the values of variables and parameters in the model, there is uncertainty over whether a selected model is an adequate representation of the real processes involved.

Morgan (1982) and Morgan *et al.* (1984) summarize the available means for dealing with each of these cases. In the first situation, the functional relationships among the various important variables are presumed known, so that the problem of dealing with uncertainty is one of characterizing and propagating the uncertainty in the parameters and variables through the model. The second situation is more difficult to handle. Typically, decision-aiding approaches address uncertainty in the form of the appropriate model in one of three ways:

1. By performing the analysis using the model form judged on the basis of current information to be most likely or the best of the available options (uncertainty in the value of the model's parameters is used as crude way of capturing uncertainty in the model form);
2. By performing an order-of-magnitude bounding analysis that is designed to determine the extent to which changes in model form affect the assessment results so as to establish bounds on the range of possible answers;
3. By performing a separate analysis using a variety of competing model forms, assigning probabilities to these alternative forms, and then combining the results probabilitically (this essentially converts a condition of Type 2 above to one of Type 1).

In each of these approaches, the principal procedure for estimating uncertainty in decision outcomes consists of quantifying uncertainties in the inputs or parameters of a decision model and then propagating those uncertainties through the model.

As described in the previous section, a variety of means might be used by a decision-aiding approach to quantify uncertainty. A simple range of possible values might be given, frequency distributions might be developed to represent statistical variability, standard error measures of classical statistics might be used to measure uncertainty due to limited data, or subjective probability distributions might be developed. The procedures used for propagating uncertainty depend on which of these procedures are used to quantify uncertainty and on the nature of the decision model.

If uncertainty is simply represented as a range of possible values, as might be the case in a Paretian CBA, then sensitivity analysis provides a means for propagating uncertainties. In a CBA sensitivity analysis, the calculations of costs and benefits are repeated, using alternative values selected for uncertain variables from the range of possibilities. The range of outcomes produced provides a simple measure of risks and indicates whether or not the conclusions from the analysis are robust.

If probability distributions are used to represent uncertainty, a variety of methods are available for converting model input uncertainties to the uncertainties they induce on the model's outputs. For complex decision models, such as those encountered in CBA and social DA approaches, probabilistic simulation procedures are often used. With these methods, probability distributions of the relevant input variables are fed into a computer, and a simulation program selects values for each variable on the basis of its relative likelihood of occurrence. If there are dependencies among the input variables, then these must be taken into account through sampling from the appropriate joint or conditional probability distributions. Through a sample of simulation runs of the decision model, the program estimates the distribution of possible outcomes. In the Monte Carlo simulation, input values are sampled at random from their specified distributions, and each scenario is given equal weight. In other methods, values of particular importance (e.g., the extreme tails of probability distributions corresponding to quantities that might cause a catastrophe) are sampled more often and given reduced weight, so as to obtain improved resolution for important parts of the distributions. In Latin hypercube sampling, sample values are constrained to cover the distribution uniformly, one form each equal probability interval. This can increase the stability of the output probability distribution if the uncertainty is dominated by only a few sources.

Discrete enumeration methods are also commonly used for the propagation of uncertainties. To use the method, continuous uncertain variables must first be represented by discrete approximations; that is, each input uncertainty must be characterized by a finite number of discrete values and associated probabilities. The procedure consists simply of evaluating the decision model for all possible combinations of input values. The probability associated with each computed output value is the product of the probabilities assocated with the corresponding input values. Several procedures are available for obtaining the necessary discretizations of input values and for smoothing output distributions (see, e.g., Miller and Rice 1983).

Decision analysts are particularly fond of the use of decision trees as a discrete enumeration method because they permit the dynamic nature of events and probabilistic dependencies (characteristics to which decision analysts attribute considerable importance) to be more easily addressed. Figure 16 shows an example of a portion of a decision tree constructed for a social DA of alternative options for a national synthetic fuels commercialization program (Tani 1978). As illustrated, a decision tree is a tree structure composed of decision nodes and chance nodes. The order of the nodes in the tree corresponds to the sequence in which decisions must be made and the order in which uncertain information is anticipated to be revealed. Branches emanating from decision nodes represent the alternatives available at points of decision, and branches from chance nodes represent the possible outcomes to uncertain variables. Probabilities of each outcome value for each uncertainty are displayed under the branches from chance nodes and are conditional on the decisions and outcomes leading to that node.

To conduct an analysis using a decision tree, the analyst uses the decision model to specify decision outcomes corresponding to the sequence of decisions and events represented by each path through the tree. If the model is a simple heuristic or structured process, these outcomes may be generated subjectively. If it is a model of the processes involved, the nodes in the tree will correspond to the model's input variables, so each path through the tree will require a separate run of the model. Figure 17 shows the structure of the decision model used in the synthetic fuels analysis. Values or utilities are assigned to each outcome (using the methods described in the discussion on Valuing Outcome Variables above), and these are displayed at the tree's end points. If the probabilities of all values (or utilities) less than each possible value (utility) are computed, by multiplying the probabilities along each path through the tree, and summed, a cumulative probability distribution may be obtained. A decision tree may also be "solved" to identify the decision strategy maximizing expected value (or utility). This is accomplished by the "roll-back" procedure: the decision strategy maximizing expected value (utility) is found by starting with the values (utilities) assigned to the end points of the tree and successively computing expected values (utilities) at each chance node and selecting the highest expected value (utility) at each decision node.

Cost-benefit and social decision analysis approaches that tend to use complicated models can be costly because simulation procedures and discrete enumeration methods require many evaluations of the decision model. To overcome this problem, response-surface methods are sometimes used in these approaches. The idea is to replace the complicated decision model with a simplified approximation called a response surface. Typically the approximation is a linear combination of certain simple functions whose coefficients are determined by least squares fitting. Because the response surface is much simpler to evaluate than the original

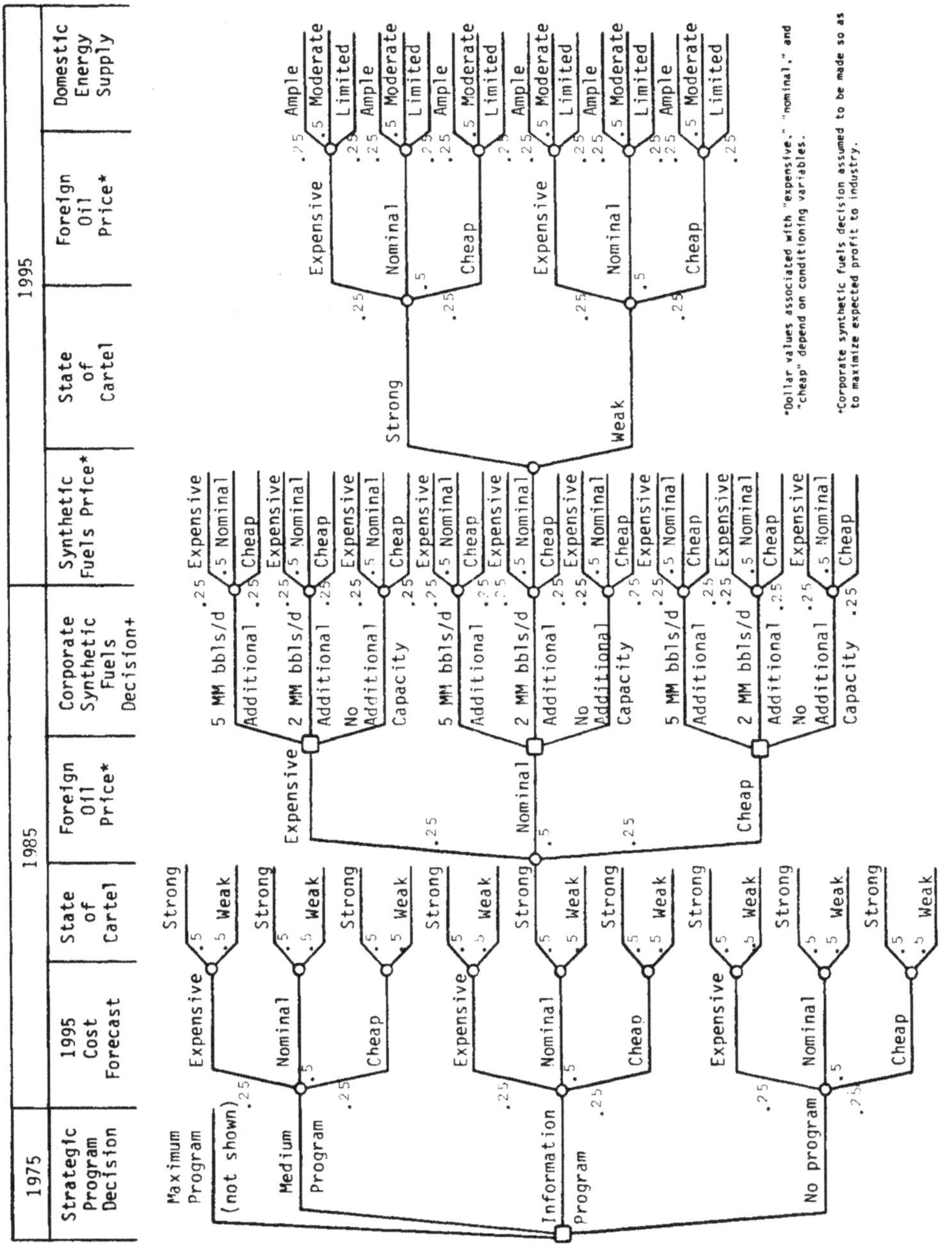

Fig. 16. Decision tree used in the analysis of a synthetic fuels program.

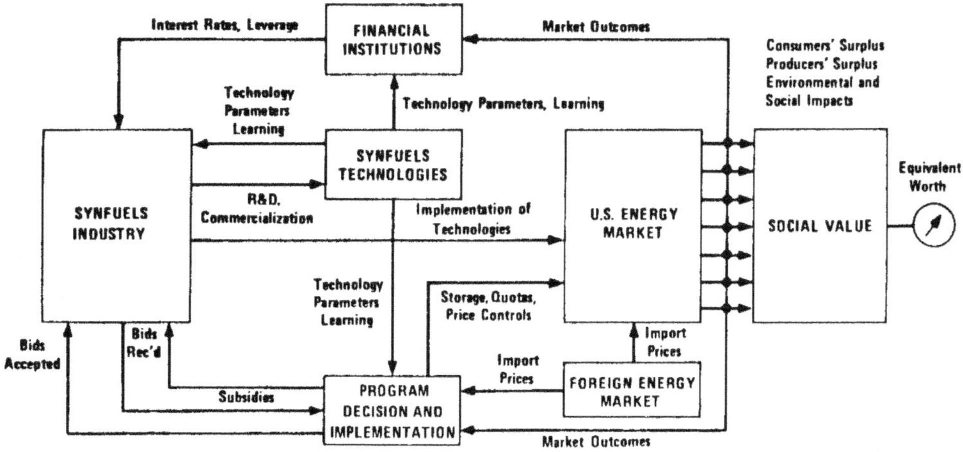

Fig. 17. Decision model used in the analysis of the synthetic fuels program.

model, it is less expensive to implement. If the decision model is simple enough (including situations where response-surface methods are used), outcome variables can sometimes be expressed as explicit analytic functions of the input variables. In this case, analytic methods (such as Taylor series expansions) can be used to express summary statistics (e.g., means and standard deviations) in terms of the parameters that define the distributions of the input variables. Thus analytic methods provide another means for propagatng uncertainty.

Because approaches that adopt the classical definition of probability distinguish uncertainties due to variabilities of nature from those due to lack of knowledge, this distinction may be retained when propagating uncertainties. Thus a decision-maker CBA might propagate frequency distributions generated from empirical data to obtain a frequency distribution describing decision outcomes and then superimpose judgmental bounds to reflect additional uncertainties related to lack of knowledge. Alternatively, the sampling distributions for estimated parameters may be computed and these distributions can be propagated (using the methods described above) to obtain standard errors or confidence intervals for computed frequency distributions. Figure 18 shows how the concept of a "cut curve" can then be used to make explicit the degree of belief behind various frequency curves. The figure shows a series of complementary cumulative probability distributions, each generated according to an alternative set of assumptions with associated judgmental probabilities. If a vertical line is drawn at any outcome level X_0, then the intersection of this line with the family leads to a cumulative probability distribution for the

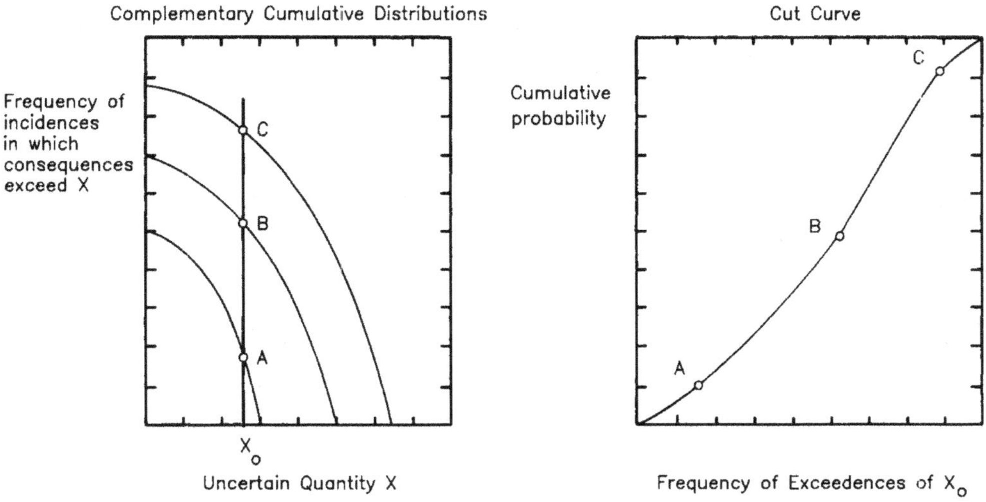

Fig. 18. Using families of risk curves to represent both statistical uncertainty and uncertainty due to lack of knowledge.

probability of the actual frequency of occurrences of outcomes with levels less than X_0.

Characterizing and Distinguishing Decision-Aiding Approaches

The preceding discussion has demonstrated that considerable flexibility is available for the selection of procedures that make up a decision-aiding approach. Some of the choices to be made in defining procedures are summarized in the analytic taxonomy of Table 7. While Table 7 contains only a small fraction of the choices regarding procedures, it serves to illustrate the wide range of options available to the analyst.

Because each of the decision-aiding theories implies the use of certain types of procedures (e.g., market arguments and frequency-based probabilities for CBT and the assessment of individual preferences and subjective probabilities for DT), logical inconsistencies can occur when the procedures associated with different theories are combined. Nevertheless, as noted previously, a typical decision-aiding approach will not rely entirely on the procedures associated with any single theory. Instead, while based primarily on one of the theories, an approach will most likely incorporate some concepts and procedures that are the logical results of the other theories. If the space of philosophical rationales were represented as a triangle whose vertices correspond to the distinct decision-aiding theories, then conceptually, as shown in Figure 19, the rationale for any specific approach could be represented by a point located somewhere

Table 7. Taxonomy of analytic choices

Analytic element	Consideration	Dimension	Distint cases
1. Alternatives	1.1 Scope	1.1.1 Decision variables	1.1.1.1 Account for all interrelated decisions
			1.1.1.2 Ignore some interrelated decisions
		1.1.2 Feasible values	1.1.2.1 Consider all available options
			1.1.2.2 Restrict set of options
	1.2 Generation	1.2.1 Source	1.2.1.1 Decision maker or client provides options
			1.2.1.2 Analyst provides options
			1.2.1.3 Other source provides options
		1.2.2 Assessment procedure	1.2.2.1 Obtain options through interviewing
			1.2.2.2 Obtain options through literature search
			1.2.2.3 Obtain options through brainstorming
			1.2.2.4 Other procedure for eliciting options
	1.3 Analytic representation	1.3.1 Characterization and specificity	1.3.1.1 Characterize actual options (e.g., minimum ambient air quality standards)
			1.3.1.2 Characterize effects of actual options (e.g., ambient concentrations not exceeding specified standards
			1.3.1.3 Make options more narrow than actual (e.g., analyze options for national energy policy by comparing nuclear plant A with coal plant B)
			1.3.1.4 Make options more general than actual (e.g., analyze choice among routes for transporting hazardous waste by comparing safety of rail v. trucking)
		1.3.2 Dynamics	1.3.2.1 Model decision as "one-shot" (e.g., no chance for subsequent modifications)
			1.3.2.2 Model subsequent decisions as uncertain acts
			1.3.2.3 Model subsequent decisions as contingent acts
2. Decision criterion	2.1 Efficiency	2.1.1 Cost-benefit theory	2.1.1.1 Maximize net present value of aggregate benefits minus aggregate costs
			2.1.1.2 Use other definition of economic efficiency
	2.2 Rationality	2.2.1 Decision theory	2.2.1.1 Maximize decision maker's expected subjective utility
			2.2.1.2 Use other definition of rationality
	2.3 Social choice	2.3.1 Social choice theory	2.3.1.1 Maximize social welfare function
			2.3.1.2 Deduce implication of "fair" decision process
	2.4 Mixed rationale	2.4.1 Cost-benefit/decision theory	2.4.1.1 Maximize extra-personal expected social value
			2.4.1.2 Use other definition of rational efficiency
		2.4.2 Decision theory/social choice theory	2.4.2.1 Maximize decision maker's subjective social utility
			2.4.2.2 Use other definition of "benevolent dictator"
		2.4.3 Social choice theory/cost-benefit theory	2.4.3.1 Maximize decision maker's definition of equitable, equivalent economic value
			2.4.3.2 Use other definition of equitable efficiency

Table 7 (*Continued*)

Analytic element	Consideration	Dimension	Distinct cases
3. Valuation	3.1 Outcome variables	3.1.1 Scope	3.1.1.1 Consider all relevant dimensions of value
			3.1.1.2 Consider only subset of relevant dimensions of value
		3.1.2 Source of outcome measures	3.1.2.1 Decision maker or client provides outcome measures
			3.1.2.2 Independent experts provide outcome measures
			3.1.2.3 Analyst provides outcome measures
			3.1.2.4 Other source provides outcome measures
		3.1.3 Identification procedure	3.1.3.1 Develop objectives hierarchy
			3.1.3.2 Identify outcome variables as all resources with economic value that will be produced or consumed
			3.1.3.3 Use other identification procedure
		3.1.4 Measurement units	3.1.4.1 Use a natural scale for outcome measurement (e.g., lives lost)
			3.1.4.2 Use a constructed scale for outcome measurement (e.g., quality of life on a scale of 1 to 10)
	3.2 Value assessment	3.2.1 Reference scale	3.2.1.1 Use an absolute scale for valuing outcomes (i.e., fixed zero scale)
			3.2.1.2 Use a relative scale for valuing outcomes (e.g., measure all outcomes relative to those produced under some possibly hypothetical alternative)
		3.2.2 Valuation units	3.2.2.1 Value outcomes in dollars
			3.2.2.2 Value outcomes in utiles
			3.2.2.3 Use other valuation units
		3.2.3 Assessment procedure	3.2.3.1 Value according to market prices
			3.2.3.2 Use surrogate values (e.g., what people pay to avoid undesirable outcomes)
			3.2.3.3 Value outcomes based on analysis of bundled goods
			3.2.3.4 Conduct survey of willingness to pay
			3.2.3.5 Encode value (utility) function
			3.2.3.6 Use other method for assessing values
		3.2.4 Procedure for valuing human life	3.2.4.1 Value life based on human capital
			3.2.4.2 Value life based on jury awards
			3.2.4.3 Value life based on previous government decisions
			3.2.4.4 Value life based on wage differentials for dangerous work
			3.2.4.5 Value life based on willingness to pay for statistical changes in probability of death
			3.2.4.6 Use other procedure for valuing life
		3.2.5 Procedure for valuing outcomes over time	3.2.5.1 Discount future outcomes using market-based discount rate
			3.2.5.2 Discount future outcomes using discount rate based on time preference
			3.2.5.3 Do not discount future outcome values

Table 7 (*Continued*)

Analytic element	Consideration	Dimension	Distinct cases
		3.2.6 Procedure for accounting for risk attitude	3.2.6.1 Assess utility function over value measure
			3.2.6.2 Use power function to account for risk
			3.2.6.3 Adjust discount rate to account for risk
			3.2.6.4 Use other procedure for representing risk preference or risk aversion
			3.2.6.5 Assume risk neutrality
	3.3 Value aggregation	3.3.1 Means for achieving comparability	3.3.1.1 Combine all value criteria into a single value measure
			3.3.1.2 Treat subsidiary value criteria by making tradeoff comparisons to the main criteria
			3.3.1.3 Use other method to obtain single valuation index
		3.3.2 Function for expressing multi-attribute value	3.3.2.1 Use linear multiattribute value (utility)
			3.3.2.2. Use multiplicative multiattribute value (utility) function
			3.3.2.3 Use other multiattribute functional form
4. Predictive model	4.1 General characteristics	4.1.1 Size	4.1.1.1 Model contains few system variables
			4.1.1.2 Model contains many system variables
		4.1.2 Complexity	4.1.2.1 Model contains simple relationships and algorithms
			4.1.2.2 Model contains complicated relationships and algorithms
		4.1.3 Level of aggregation	4.1.3.1 Model contains relatively large number of highly specific but sparsely interconnected system variables
			4.1.3.2 Model contains relatively small number of aggregated but highly interdependent system variables
	4.2 Structure	4.2.1 Source	4.2.1.1 Decision maker specifies elements of model
			4.2.1.2 Independent experts specify elements of model
			4.2.1.3 Analyst specifies elements of model
		4.2.2 Procedure for assessing model structure	4.2.2.1 Rely primarily on top-down modeling methods (e.g., influence diagrams, function graphs)
			4.2.2.2 Rely primarily on bottom-up modeling methods (e.g., scenario generation)
			4.2.2.3 Use other modeling procedures
	4.3 Analytic relationship	4.3.1 Source	4.3.1.1 Derive relationships among model variables through statistical inference (e.g., regression analysis)
			4.3.1.2 Base relationships among model variables on theory and judgment
		4.3.2 Deterministic v. probabilistic	4.3.2.1 Model inputs uniquely determine model outputs
			4.3.2.2 Model contains some probabilistic relationships among its system variables
		4.3.3 Static v. dynamic	4.3.3.1 Model relationships vary as a function of time
			4.3.3.2 Model relationships are not a function of time

Table 7 (*Continued*)

Analytic element	Consideration	Dimension	Distinct cases
		4.3.4 Continuous v. discrete	4.3.4.1 Model contains continuous or discrete variables but continuous transformations only
			4.3.4.2 Model contains discontinuous or discrete transformations
		4.3.5 Constrainted v. unconstrained	4.3.5.1 All system variables are unconstrained
			4.3.5.2 Some variables are constrained to lie within specified intervals
		4.3.6 Time horizon	4.3.6.1 Model operates over short time horizon
			4.3.6.2 Model operates over long time horizon
		4.3.7 Special model forms	4.3.7.1 Model contains special model forms selected for their analytic tractability (e.g., Markov processes)
			4.3.7.2 No special model forms are employed
	4.4 Inputs	4.4.1 Source	4.4.1.1 Decision maker or client provides model inputs
			4.4.1.2 Experts provide model inputs
			4.4.1.3 Analyst provides model inputs
	4.5 Validation	4.5.1 Source	4.5.1.1 Model validated against historical data
			4.5.1.2 Model validated against expert judgments
5. Representation of uncertainty	5.1 General characteristics	5.1.1 Extent	5.1.1.1 Include limited consideration of uncertainty based on credible intervals, sensitivity analysis, etc.
			5.1.1.2 Include explicit development of probability distributions
		5.1.2 Precision	5.1.2.1 High precision adopted for representing uncertainties
			5.1.2.2 Low precision adopted for representing uncertainties (e.g., binary groupings)
		5.1.3 Aggregation	5.1.3.1 Aggregation of uncertainties is high
			5.1.3.2 Aggregation of uncertainties is low
	5.2 Analytic representation	5.2.1 Logical basis	5.2.1.1 Interpret probability as frequency
			5.2.1.2 Interpret probability as degree of belief
		5.2.2 Values	5.2.2.1 Represent uncertainty as continuous value random variable
			5.2.2.2 Represent uncertainty as discrete value random variable
		5.2.3 Time	5.2.3.1 Represent uncertainty as discrete event (i.e., value of random variable X at time T)
			5.2.3.2 Represent uncertainty as random process (i.e., scenario $x(T)$)
		5.2.4 Dependencies	5.2.4.1 Assume probabilistic independence
			5.2.4.2 Assume some probabilistic dependencies

Table 7 (*Continued*)

Analytic element	Consideration	Dimension	Distinct cases	
	5.3 Assessment procedure	5.3.1 Statistical inference	5.3.1.1	Derive probabilities through statistical inference of "population" probabilities
			5.3.1.2	Derive probabilities through statistical inference of named probability distribution parameters (beta, binomial, normal, etc.)
		5.3.2 Probability encoding	5.3.2.1	Encode probabilities using indirect fixed value method (e.g., probability wheel)
			5.3.2.2	Encode probabilities using indirect fixed value method (e.g., interval technique)
			5.3.2.3	Encode probabilities using direct method (e.g., elicitation of odds)
6. Processing of probabilities, solution, and representation of results	6.1 Computational procedures	6.1.1 Analytic methods	6.1.1.1	Employ Taylor series expansions, response surfaces, etc.
			6.1.1.2	Do not employ analytic methods
		6.1.2 Simulation methods	6.1.2.1	Use Monte Carlo analysis to derive uncertainty over outcome variables
			6.1.2.2	Use tree methods to derive uncertainty over outcome variables (e.g., event trees, decision trees)
		6.1.3 Solution methods	6.1.3.1	Use dynamic programming to solve for optimal decision
			6.1.3.2	Use search methods to solve for optimal decision
			6.1.3.3	Use other solution methods
	6.2 Representation of risk	6.2.1 Units for expressing potential consequences	6.2.1.1	Use vivid measurement scales which may evoke emotional response (e.g., cumulative deaths attributable to hazard)
			6.2.1.2	Use unexpressive measurement scales which may hinder interpretation (e.g., fatalities per 10^6 lifetimes per year)
		6.2.2 Display of uncertainty	6.2.2.1	Display range of uncertainty
			6.2.2.2	Display cumulative probability plots
			6.2.2.3	Display probability density plots
			6.2.2.4	Use other displays
		6.2.3 Summary measures of risk	6.2.3.1	Calculate expected values
			6.2.3.2	Calculate probability of exceeding specified consequence levels
			6.2.3.3	Calculate other summary statistics

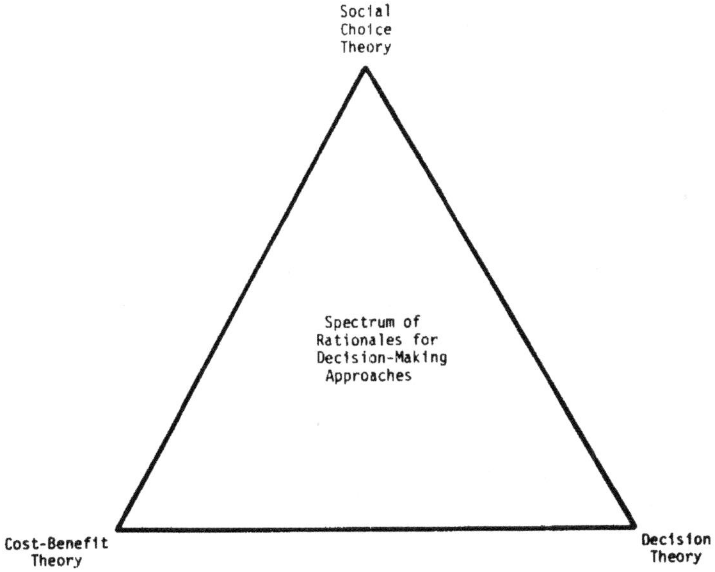

Fig. 19. The space of implicit rationales for decision-aiding approaches defined by mutually exclusive, internally consistent theories.

within the triangle (not necessarily at one of the extreme corners). What is gained in moving away from a single well-defined theory is an approach whose procedures (ideally) are tailormade to the situational needs of the specific application. What is lost is some consistency in underlying logic.

As described earlier, variations of CBA, DA, and applied SWT provide a link between procedures and theory. These approach types rely on procedures to implement the decision-making theories; and although much variation occurs from application to application, specific procedures may be generally associated with each type of approach. Figure 20 shows the various approach types discussed in this section as they might be represented in the space of philosophical rationales, taking into account the procedures typically used by authors who describe their approaches using the indicated labels. Because of the association of procedures with theories, and the internal inconsistencies that result when an approach relies strongly on procedures consistent with different theories, the nature of the internal inconsistencies and conflict that are present is indicated by the location of the approach within the triangle.

The triangle of Figure 19 may be used to help distinguish approaches: if the procedures applied or specified by an approach are first identified and used to locate the approach within the triangle, the philosophical rationale and internal consistency of the approach may then be inferred. Locating an approach within the triangle requires that the approach be assessed

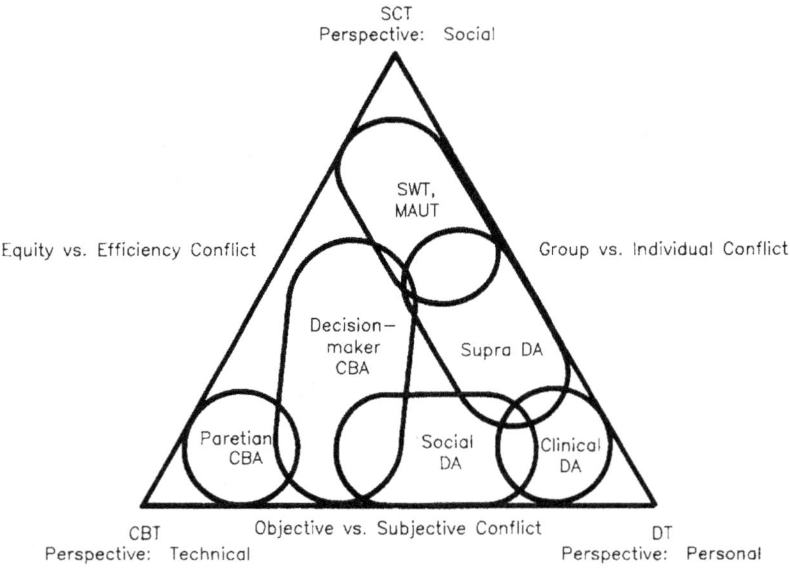

Fig. 20. Implicit rationales and perspective conflicts inherent in various types of decision-aiding approaches described in the literature.

along three dimensions: objective versus subjective, individual versus group, and efficiency versus equity. Objective versus subjective refers to the extent to which quantitative estimates are objectively or subjectively obtained in the approach.[69] Individual versus group reflects the extent to which information and values used within the approach are those of a specific decision maker or represent society as a whole. Efficiency versus equity refers to the degree to which the decision criterion used in the approach represents aggregate (summed) value versus accounts for distributional considerations.

Although applications literature generally provides a detailed account of the procedures applied in a given approach, the underlying rationale for the approach is rarely made explicit. Hence, a careful consideration along the lines described above may be necessary to identify the underlying rationale. Determining the theoretical rationale for an approach helps distinguish problems of poor application from inherent limitations. Furthermore, understanding the philosophical justification for an approach is important for approach evaluation because, as shown in the next chapter, higher-order criticisms (i.e., those relating to philosophical, institutional, or ethical issues as opposed to operational issues) are directed primarily at the foundations of an approach. Criticisms leveled at an approach may be inappropriate if they misinterpret the underlying rationale.

The discussion in this chapter has been designed to highlight the differences among approaches. By focusing on distinctions, our intent has been to facilitate the comparison of approaches and to emphasize the wide range of analytic procedures and perspectives available to the analyst. It should be recognized, however, that there is considerable consistency and complementarity and among the different types of decision-aiding approaches. This may be demonstrated by noting that the decision criteria associated with the various approaches can be logically derived from each of the theories by making the appropriate assumptions. For example, SWT approaches may be regarded as special cases of DT in which the decision maker adopts a utility function that is an aggregation of the utility functions of the affected individuals (i.e., the decision maker feels that the criterion for the choice should be the preferences of the individuals in society). Similarly, CBA may be regarded as a special case of SCT with (1) the efficiency criterion used to provide the principle for aggregating individual preferences and (2) an additive form for each individual's utility function which decomposes individual utility into costs and benefits, and where (3) a market value is imputed for all benefits and costs, (4) the assumption is made that each individual values benefits and costs at their market values, and (5) the value of each additional dollar is constant and equal among all individuals (Dyer and Sarin 1978a). The existence of this logical equivalence among approaches demonstrates that the inconsistencies introduced through the "borrowing of procedures" are inconsistencies of perspective more than inconsistencies of logic.

Chapter 3

Criticisms and Limitations of Decision-Aiding Approaches

This chapter explores the basic criticisms that have been directed at formal decision-aiding approaches. The purpose of presenting these criticisms is not to suggest that formal approaches have little value, but to provide the reader with an improved awareness of the limitations of formal analysis. Such information can help indicate when certain approaches can be appropriately applied, how much precision and reliability can be expected in the results, what the nature is of any existing controversy surrounding the use of an approach, and what questions should be asked about the methodology in examining the results of any application. This knowledge is not only essential to making informed judgments about the comparative value of alternative approaches, it is also essential to making informed decisions based on the application of a formal approach.

Critics have attacked decision-aiding approaches at a variety of levels. Discussed first are what might be called fundamental criticisms: limitations of existing theory with respect to characteristics inherent in social decision making that make available approaches incapable of identifying an alternative that is in any absolute sense optimal. Second, operational criticisms are addressed; these are practical considerations that limit or subvert an approach's value in actual applications. The third category discussed is institutional incompatibilities — characteristics of approaches that are at odds with existing institutions and processes. Finally, criticisms raised on ethical grounds are presented.

As the reader will see, many of the most serious criticisms of formal decision-aiding approaches not only apply to all approaches but will be present in any social decision-making process. Consequently, when judging the value of an approach in light of a criticism, the reader is encouraged to consider whether traditional decision-making methods that do not involve formal analysis are more or less subject to that same criticism.

Fundamental Criticisms

Fundamental criticisms are alleged errors in logic or reasoning that stem from weaknesses in the theoretical foundations of decision-aiding approaches. The premise of such criticisms is that formal approaches cannot identify an optimal decision either because of fallacies in the underlying assumptions or because of limitations of their philosophical rationales. The basic criticisms in this category are discussed in the sections below.

Impossibility of Finding a Socially Optimal Decision Rule

As noted in the section above on decision theroy, through postulating a set of reasonable axioms of rational behavior, decision theorists have successfully derived a logical rule for individual decision making (maximization of subjective expected utility). Social choice theorists have sought an analogous rule for social decision making based on aggregating individual preferences without offending one or more of a similar set of axioms of reasonableness. Arrow's impossibility theorem says in effect that this may not be possible. So long as group decisions are made on the basis of the rank order preferences of individuals without directly or indirectly using indications of intensity of preference, no consistent decision-making rule exists that is not either dictatorial or completely arbitrary.

Thus, the social decision-making problem does not have a solution in the same sense that individual decision making does, and this reduces the applicability of decision theory to social decisions. According to Howard (1980), "Decision analysis is a paradigm suitable only for the individual decision maker .... I can see no way that the [public policy decision] problem can be 'solved' using decision analysis in the sense that would be correct for an individual decision maker."

Most approaches avoid the impossibility theorem by incorporating measures of strengths of preferences in the form of a social utility function. Howard objects to this on ethical grounds relating to implications on individual rights, but if the social utility function is meant to represent the preferences of individual members of society, then fundamental elicitation problems also exist. Asking people what they want may not produce good answers because what people say does not always agree with what they do. Furthermore, as soon as intensity of preference is introduced into the decision-making process, the procedure is subject to subversion: individuals have an incentive to misrepresent the strengths of their preferences so as to produce decisions that they personally prefer. Dyer and Miles (1976) report that such misrepresentation of preferences was observed in their application of social welfare theory (see Appendix C). The same problem would seem likely to occur in cost-benefit analyses that rely on polling to determine willingness to pay:[70] "Taking into account their likely tax burden and other respondents' likely responses, [people] may understate or overstate their real willingness to pay and the resulting tax burden in ways helpful to them" (Wilson 1975).

Even if strengths of individual preferences are accurately obtained, shortcomings in underlying theories may still prevent an approach from identifying an optimal social action. In the case of approaches based on CBT, for example, the decision rule of potential Pareto improvement will fail to yield an optimal alternative if there happens to be no alternative that is a potential Pareto improvement with respect to all others. This possibility is the subject of a situation known as the Scitovsky paradox (Thompson 1982). Suppose that the social choice is between alternatives A and B. Typically, those individuals preferring A will be willing to pay less to obtain A instead of B than they would consider fair compensation to willingly accept B in place of A. Thus, it is entirely possible that A advocates will not have a total willingness to pay sufficient to provide the compensation desired by B advocates, while at the same time B advocates could not adequately compensate those favoring A. The CBT decision rule fails in this case to identify a preferred action.

Approaches based on SCT are by no means immune to theoretical shortcomings. In expressing his preferences over some proposed government action, an individual will likely consider his uncertainty as to the actions that others may take that affect the ultimate consequences to him. This can create problems when individuals try to second-guess each other. For example, suppose that each member of society can, in principle, identify a most-preferred sequence of social actions for addressing some issue, and that each such preferred sequence happens to begin with the same first action. Suppose, however, that this action is also the first step for another sequence of actions that represents a very undesirable policy. Because each individual is uncertain about what future acts might be implemented, the nonzero probability that each assigns to the otherwise preferred initial act leading to the undesired sequence may cause that first act to be rejected. Thus, for example, each member of society might be willing to agree to invest more for future generations if that member could only be assured that intermediate generations would not reap the increased savings in increased consumption for themselves.

Destruction through Decomposition

As noted earlier, divide and conquer is the essence of formal approaches. If a problem is too complex, it is simplified by dividing it into subsystems. If the subsystems are still too complicated to comprehend, they are reduced further. Unfortunately, almost no research has been conducted on the learnability, applicability, or effectiveness of decomposition (Fischhoff 1979). Linstone (1981) suggests that von Foerster's first law aptly characterizes the process: "The more complex the problem which is being ignored, the greater are the chances for fame and success" (von Foerster 1972).

Basic to the decomposition principle in decision-aiding approaches is the separation of information and preference. Philosophically, however, no neat distinction can be made between facts and values. All alleged facts reflect to a considerable extent the perspective of the investigator toward reality. "Facts do not simply present themselves; interpretation and analysis are always necessary" (Callahan 1980). Similarly, facts shape values: "Values are acquired by rote, by imitation, and by experience. The world we observe tells us what issues are worth caring about . . . . Insofar as that world is revealed to us through the prism of science, the facts it creates shape our world outlook" (Fischhoff *et al.* 1978b).

Tribe (1972) is concerned that trying to separate facts and values eliminates context. He offers the following illustrations of the importance of considering context when assessing outcome value:

> Convicting an innocent person should be deemed a worse outcome when the jury feels unsure of the person's guilt than when the jury feels fully confident of guilt (but simply happens to be wrong) . . . . Similarly, destroying a species of wildlife should probably be regarded as a worse outcome when it results from disregard of a high known risk than when it results from the materialization of a highly unlikely contingency. The tradition in many legal systems of distinguishing among acts in terms of the mental state accompanying them . . . rests on this sort of proposition. Yet the objectivist's fact-value dichotomy, leading to an insistence on separating assessments of probability from the valuation of outcomes, tends to exclude this important dimension of human choice.

Howard (1980) replies that there is nothing to prevent assessing preference on the combination of decision (or probability assessment) and outcome, a fact that Tribe recognizes as well: "a sufficiently careful policy analysis could define the 'outcomes' in question so as to include information about the associated probability assessments." Both authors agree, however, that these criticisms point out the difficulty of using decomposition effectively in practice.

A more general problem associated with emphasizing systematic decomposition may be the diversion of attention from more creative approaches to problem solving. It is well known that the most original and creative solutions often come from viewing a problem in unconventional terms. Thus, there is a danger that too much attention given to analyzing a particular decision model might prove to be myopic in this wider sense. Similarly, decomposition may oversimplify the problem. For example, it has been suggested that the contrived questions often used in utility assessment may put a decision maker in a less serious frame of mind than that which he would like to face the real problem (Thompson 1982).

Drawing on the work of Gestalt psychologists, critics argue that people think most naturally and adequately by analogy with past experiences (Polanyi 1962). They conclude, therefore, that any attempt to evaluate separately the attributes of a particular event or to designate the importance of attributes in the abstract is likely to produce spurious results.

"In addition to destroying the respondent's natural understanding, decomposition procedures impose a foreign response mode that does not allow people to articulate their understanding of (holistic) value issues" (Fischhoff *et al.* 1978b).

Psychological studies show that individuals who must decide between complex alternatives usually do not naturally decompose their thinking into separate components and formalize each component task separately (Raiffa and Zeckhauser 1981). Critics charge that because decomposition deviates from the "normal" way in which judgments are made, the respondent's "feel" for the issues at hand is likely to be eroded. Furthermore, decompositions are not unique, and, therefore, different decompositions may lead to different judgments of the same issue. Some theories of choice, for example, predict shifts in the perceived attractiveness of simple gambles as a function of their decomposition (Aschenbrenner 1978).

Howard (1980) notes that "those who criticize the reductionist viewpoint in analysis are more concerned by what may be lost in the process than by what is gained." Yet, if the error in unaided decision making is caused by biases resulting from the counter-intuitive nature of probabilities, rather than by complexity, there may be little reason to expect component judgments to be less prone to bias.

Tribe (1972) argues that decomposition itself carries a message that some may find offensive: "It stresses ends over means. It proclaims the superiority of the elicitor's overall perspective (and the overall social importance of analysis and its purveyors). It conveys a message of analyzability or solvability where that may be inappropriate."

Inherent Incompleteness

The model of a decision situation can never be an absolutely complete representation of reality. This is true because the resources available for analysis are finite, but the possible interactions and implications of a decision are infinite. Any complex real-world problem will always have more dimensions than a model can capture. Therefore, the results derived from an approach based on a necessarily incomplete model will not necessarily apply to the real world that the model is meant to approximate.[71]

Because of "unknown uncertainties" (e.g., the possibility that undiscovered pathways to disaster exist), risk estimates based on analytical models may overlook significant risks. An example, cited by Page (1978), is provided by a DA of alternative standards for the flame-retardant Tris that was conducted for "the major purpose of making sure that no important aspects of the problem were neglected." The principal tradeoff considered was the balance between the greater cost of a garment with increased amounts of the retardant and the increased likelihood that the standard would be subverted by parents making their own less expensive, nonretardant sleepware for their children. At the time of the analysis, no investigation had been undertaken of toxicity for Tris or the other chemicals proposed as retardants; five years later, however, the National Cancer Institute found the chemical to be one of the most potent carcinogens ever tested. As Green (1975) notes,

It is possible to identify respects in which an activity is or may be hazardous. This can be done only by reference to the existing body of scientific knowledge and existing experience with the activity or related activities. It may be possible to conclude on these bases that the activity involves risks, but it is rarely possible to reach a reliable conclusion that there is no risk. The fact that existing knowledge provides absolutely no basis for concern does not establish that there is zero risk, since future experience or scientific knowledge may demonstrate that a present activity, either by itself or synergistically, does indeed involve hazards.

Inability to Account for the Costs of Irreversibility

Some environmental risk decisions may produce consequences that for all practical purposes are irreversible. Such consequences have the effect of limiting future options and thereby reducing our ability to react to new information. For example, the extinction of a species would be an irreversible consequence. From our current perspective, the eradication of some species, such as the mosquito, might appear to have no significant social, environmental, or economic costs. At some future date, however, it might be discovered that the mosquito played some previously unrecognized role in the ecosystem, or alternatively could be used to provide some valuable function in some new process if it still existed. Regardless of how great this new value might be, the extinct mosquito could not be brought back. Many potential consequences are only partially reversible. For example, the decision to use certain pesticides has temporarily irreversible consequences because of the pervasiveness of pesticide residues in the food chain. If at some future date it should be discovered that a currently used pesticide produces an increase in the rate of cancer, the use of that pesticide can be discontinued; but the cancer rate will most likely remain elevated for a very long period of time. If irreversible consequences are avoided, the possibility of reversing earlier decisions remains open and decision makers can adjust to new information. The greater the uncertainty that the future is expected to resolve, the more costly is the foreclosure of choice through irreversibility. Conversely, avoiding irreversible decisions ensures the maximum potential for utilizing future knowledge.

Accurately accounting for the costs of irreversibility or, conversely, the value of flexibility requires that all future contingencies be identified and represented in the analysis (Merkhofer and Saade 1978). As noted above, however, every analysis is necessarily limited: it can account for only a subset of all possibilities. Therefore, formal decision-aiding approaches will tend to undervalue the costs of potentially irreversible decisions to the extent that important future possibilities and options are not factored into the analysis.

A report prepared by the National Planning Association for the EPA expresses the irreversibility problem from the specific perspective of CBA:

> A serious, but not fatal, deficiency of the [cost-benefit analysis] Pareto criterion for evaluating social decisions involving low probability, high consequent occurrences is that it changes as knowledge and expectations change [because expectation changes may change market prices]. Thus, at any particular point in time, an evaluation of a social decision with uncertainty might conclude that a decision is correct. Yet, with additional information, obtained only with passage of time, the conclusion might be that the decision is incorrect. If decisions are reversible, then this limitation of the Pareto criterion is not overwhelming. However, if decisions are irreversible, it may be that any original efficient equilibrium would impose substantial welfare losses over time (National Planning Association 1978).

Critics charge that because formal approaches tend to undervalue the cost of irreversibility, these approaches may be an inappropriate means for society to deal with risk.

> Cost benefit analysis . . . [and related techniques] . . . are typical of the search for an optimal solution. It usually comes as a shock to those nurtured in this perspective that complex living systems have not organized themselves in accordance with this principle . . . ecological systems sacrifice efficiency for resilience . . . . Evolution shows that their safe-fail strategy is eminently suited to a world which is inherently unpredictable at certain times (Holling 1977).

Extending this thought, Wildavsky (1980) warns,

> In the past we relied on resilience. We anticipated the worst and we relied on a general capacity to be flexible to respond. By attempting a politics of anticipation, we shall not only destroy any potential we have for growth and for group accommodation, but we will make people poorer and sicker.

Resiliency and flexibility can, of course, be accommodated and explored in a formal approach through sensitivity analysis. Critics appear to be concerned that institutionalization of quantitative approaches may result in the fallibilities of the analyst being forgotten.

Operational Criticisms

Even if decision-aiding approaches could in theory identify an optimal social alternative, practical problems may prevent this capability from being realized in real-world applications. Operational criticisms, described below, are practical limitations that critics claim reduce or subvert the value of formal approaches.

Omissions and Inaccuracies

The principal reason for omissions and inaccuracies in the application of decision-aiding approaches is that scientific and technical knowledge is deficient. Lack of knowledge, coupled with the complexity of risks, severely taxes the insight of the analyst who must correctly identify the many aspects of a problem that bear on a decision. As indicated in Chapter 1, difficulties exist in each component of the risk chain:

1. Characterizing the source of risks is complicated because of lack of data that permit the frequency and consequences of infrequent accidents to be measured, the difficulty of distinguishing the causes of pervasive chronic health effects (due to masking by naturally occurring background levels of the effects and the typical time-lag between exposures and harms), the existence of confounding or intervening causal factors.
2. Predicting exposure is difficult because of inherent limitations of available environmental pollutant transport and fate models.
3. Estimating dose-response relationships is difficult because of the shortage or nonexistence of epidemiological data (most populations are exposed to multiple hazards, and isolating effects is difficult) and the problems of collecting human clinical data.
4. Extrapolating the effects on animal species to humans is highly uncertain, but testing new potential hazards by intentionally exposing large numbers of people is neither feasible nor morally acceptable.
5. Estimating consequences is complicated by the high variability of the human population and the consequent existence of sensitive individuals.

In the face of such complexity and armed only with imperfect modeling procedures and assumptions, it is not surprising that analysts make errors. Some of the common sources of error that may be significant include:

1. Biases in collected data (e.g., in the case of exposure or accident data, events often go unreported and those that are reported tend to fall into one class of interest more than others).
2. Data reporting and processing errors (often, such errors are not random but are more likely to occur across special classes of data values).
3. Stationarity assumptions upon which extrapolations are based (systems may change over time, so inferences about the future may not be valid).
4. Fitting models to sparse data (e.g., different named probability distributions may fit available data equally well but differ considerably in the "tails" of the distribution, which may be most significant).

5. The use of data aggregation or surrogate data (if sparse data is overcome by aggregation or using data from "similar" situations, the methods may produce errors to the extent that problem specifics are missed).
6. Discretizing continuous decision variables or event outcomes to obtain mathematical tractability.
7. Errors of judgment in subjective inputs.
8. Errors in models of physical processes (resulting from the lack of, or incompleteness of, available theories and data).
9. Incompleteness of models (e.g., omitted accident scenarios in an event tree).

Incompleteness resulting from omissions of important considerations is likely to be responsible for the largest errors in applications of decision-aiding approaches. Fischhoff (1977) identifies four sources of omissions that have historically haunted analyses:

1. The first is overlooking possibilities for human error. Hazard characterizations and exposure assessments generally assume that people behave wisely and follow established procedures; people are assumed not to expose themselves deliberately or irrationally to harmful substances and dangerous situations. Thus, the possiblity of operator or participant error is often igored in assessments of the likelihood and consequences of accidents. As an example, the multi-million-dollar "Reactor Safety Study" (Nuclear Regulatory Commission 1975) was an unusually thorough analysis of the probability of a core melt in a nuclear reactor. The detail considered in some analyses reportedly produced a level of complexity that exceeded the capabilities of the computer program designed for the study (Weathermax 1975). Nevertheless, the most significant accident, at Three Mile Island, was largely due to human error and represented an accident sequence not specifically considered by the study.[72]
2. A second source of omission is failure to anticipate human response to safety measures. For example, dams are constructed for the purpose of preventing flooding. The partial protection offered by dams and levees, however, gives people a greater sense of security and promotes the development of the flood plain. When a rare flood does exceed the capacity of the dam, the damage may be considerably greater than if the flood plain had been left unprotected (Slovic *et al.* 1979a). Similarly, driver training programs, designed to create safe drivers, have actually increased the number of high-risk persons entering the ranks of drivers and the number of younger people involved in automobile accidents (Robertson and Zaldor 1977).
3. A third source of omission is the failure to consider unanticipated changes in the environment. Analyses are always predicated on some

assumed constancies in the external environment, but these may prove to be erroneous. For example, an analysis of the decision to construct the Alaska pipeline failed to foresee the threat now posed by the retreat of the Columbia Glacier. As the glacier retreats, it discharges large icebergs in the direction of shipping lanes for tankers loaded with North Slope oil (Fischhoff 1977). As another example, risk analyses of nuclear power plant designs normally assume the availability of back-up electrical power — an assumption that must be questioned in view of the Northeastern power blackout of 1965 (Linnerooth 1979).

4. A fourth source of omission results from failure to appreciate large-scale interactions and dependencies. Most applications of decision-aiding approaches to regulatory decisions measure impacts relative to the "status quo," assuming, in effect, that things will remain as they are. In fact, regulatory inaction may lead to increased industry self-regulation, more or less court involvement (as parties initially turn to the courts for redress and then industry adjusts its behavior to avoid liability), reduced business activity (due to the rigidity of the regulatory system), or increased business activity (in a regulatory environment that treats new products predictably). Ashford *et al.* (1979) note that analyses of the benefits of regulation generally fail to consider the impact of regulation on the rate of technological change.[73] Similarly, though the respiratory risk of emissions from fossil-fuel plants has been recognized for several years, the related effects of acid precipitation on ecosystems have only come to be appreciated much more recently (Slovic *et al.* 1979a).

Fischhoff and Cox (1986) conclude that omissions are most likely to occur for novel risk situations which lack an established protocol identifying elements that experience has shown to be relevant, and for multidisciplinary analyses for which no one has a comprehensive mental model of the entire system. Whether or not important omissions will occur in these circumstances is partly a function of the diligence and imagination of the analyst and partly a function of the specific approach used. Certain approaches are simply ill-equipped to deal with certain considerations. For example, the tendency for CBA approaches to rely on frequency distributions for quantifying uncertainty and market values for establishing preferences tends to discourage cost-benefit analysts from including explicit subjective information and nonmarket outcome attributes. This can be a problem in situations where the state of information is changing rapidly or where the only sources of information are the judgments of technical experts. Similarly, an event tree, which reduces behavior to discrete, connected actions, is a difficult representation for modeling human behavior as a factor in the assessment of nuclear reactor safety risks. Approaches such as DA that adopt a subjective view of probability often have difficulty incorporating large data bases such as actuarial data.

Regardless of the approach, identified factors may be excluded because they are too costly or complex to assess or measure (e.g., the impacts of proposed urban housing policies on family structure) or because of the tenuous quality of the causal connection between a planned action and the possible outcome, as in the case of higher-order effects, In general, there is a danger that decision-aiding approaches will concentrate on aspects that are easy to quantify and omit or place less emphasis on considerations that are difficult to represent quantitatively. Even if the attributes have been identified, included in the analysis, and properly characterized, they may be inadequately measured or perhaps are unmeasurable.[74]

In addition to omissions, another important source of error results from the treatment of probabilistic dependencies, which can complicate analyses tremendously. As a result, most probabilistic modeling is based on independence assumptions which frequently go unquestioned (Howard 1980). In practice, however, dependencies often occur; for example, the discovery that a set of pipes in a nuclear power plant were all made from the same batch of defective steel would significantly increase the likelihood of a simultaneous failure resulting in a major disaster (Slovic *et al.* 1982). Because the electrical cables controlling the multiple safety systems of the Brown's Ferry reactor were not spatially separated, all five emergency cooling systems were incapacitated by a single fire. Influence diagrams, used in some decision analyses, are a useful procedure for identifying probabilistic dependencies.

Another criticism relating to omissions and inaccuracies is the charge that their inevitability makes formal approaches useless in situations in which the events of importance are extremely rare. In criticizing risk assessments of nuclear power, Holdren (1976) states:

> I am among those who believe it to be impossible in principle to support numbers as small as these with convincing theoretical arguments (that is, in the absence of operating experience in the range of 10,000 reactor years or more) .... The reason I hold this view is straightforward: nuclear power systems are so complex that the probability the safety analysis contains serious errors ... is so big as to render meaningless the tiny computed probability of accident.

Difficulty of Measuring Benefits and Costs

As described in Chapter 2, decision-aiding approaches incorporate a variety of means and perspectives for establishing quantitative measures for the benefits and costs associated with social decisions. All of the standard procedures have been criticized, however, for various logical deficiencies and practical problems of implementation. Most of the approaches, for example, make one or more of the following questionable assumptions: (1) informed rationality of individual choices, (2) harmony between individual and collective rationality (e.g., no externalities), and (3) truthful revelation of preferences (e.g., no strategic misrepresentation of preferences for the purpose of manipulating social decisions). Criticisms related to these and other assumptions as they apply to specific approaches and procedures are described below.

Cost-benefit analysis and related approaches rely heavily on the concept of revealed preferences; that is, they attempt to infer and measure the values individuals place on outcomes by the choices they make in the marketplace. As indicated earlier, the procedures for implementing this logic can get rather complex because (1) market values are often distorted by government interventions, private monopolies, externalities, and so forth; (2) marginal per-unit values derived from market prices will not reflect the per-unit value of large impacts on resources or product supplies; (3) changes in prices in one market can create price changes in related markets (so-called secondary gains and losses of consumer and producer surplus), and (4) many of the environmental outcomes for risk decisions are not traded on the market. Since the procedures used to address these considerations are imperfect, the analysis will be inaccurate to the extent that it produces poor estimates of the values of concern.

At the simplest level, using market prices for valuation can be criticized for its unrealistic assumption that displaced resources from one activity will be reallocated to other activities. Unemployed Detroit auto workers may find it very difficult to fulfill excess demand for mass transit workers in California. Also, price is often not the only factor that rations scarce resources. If congestion or capacity constraints effectively limit access, then those with maximum willingness to pay may not always be consumers. Furthermore, the use of aggregate demand curves to provide estimates of aggregate benefits presupposes that demand curves adequately represent aggregate preferences. If the distribution of income is believed unfair, then market demand curves may not be regarded as acceptable guides for social decisions.

Kelman (1982) criticizes the hedonic price procedure often used by CBA approaches to compute an economic value for a nonmarket good as the price increment to bundled goods that include the nonmarketed good as a feature (e.g., using property values to infer a price for clean air). Kelman notes that people vary in their preferences and the constraints to which they are subject. Therefore, people with an unusually weak aversion to the undesirable feature or with unusually strong constraints on choice will be willing to take the bundled good in question at less of a discount than the average person. Thus, to use the property value discount of homes near airports as a measure of people's willingness to pay for quiet, means accepting as a proxy for the rest of the population the behavior of those least sensitive to noise. Similarly, the use of wage premiums associated with hazardous work as a measure of the value of life, requires accepting as proxies for the rest of us the decisions of people who do not have many choices or who are exceptionally risk seeking. One of the lowest value-of-life estimates obtained using this procedure, in fact, involved workers in very high risk jobs having an annual risk of fatality of approximately 1 in 1,000 (Thaler and Rosen 1976).

Cummings, Cox, and Freeman (1986) note that the hedonic price method may fail to account for the full value of the public good. Clean air, for example, affects people at work as well as at home; therefore, wage differentials along with property value differentials may have to be considered to get a full picture of the value of cleaner air. As a practical problem, using regression methods to impute a value for one attribute of a bundled good is not easy. In addition to problems of choosing an appropriate functional form and explanatory variables, estimation of a price function is often hindered by severe colinearity, a pitfall which reduces the power of regression techniques.

Even if revealed preference methods do reflect true preferences, it might not be correct to assume that the values measured in one situation are appropriate for another. One frequent error is to assume that preferences revealed by individuals who have a choice of what risks or disbenefits to accept (e.g., workers choosing among occupations, or house buyers among locations) apply to situations where similar risks or disbenefits are imposed involuntarily (such as pollution effects on homeowners who are not in the market for new houses). The value of a reduction in a voluntarily assumed risk may be very different from the value of a comparable reduction in a risk that must be accepted as part of an externality produced by the actions of others.

The CBA procedure of measuring an adverse effect in terms of defensive expenditures taken by individuals to avoid those effects is also flawed. Freeman (1986) describes the income-effect problem. Suppose that a disbenefit (such as pollution) and some market good (such as paint, air-conditioning equipment, or cleaning materials) are perfect substitutes in the sense that a reduction in the disbenefit can be entirely compensated for by a reduction in the spending on the market good which would be required to keep the individual on his original indifference curve. In reality, however, a reduction in the disbenefit will not cause the individual to reduce his spending so as to stay on the original indifference curve. Because of the decreased spending, the individual will reallocate expenditures among all goods (including the defensive purchase) so as to maximize the increase in total benefit. Therefore the observed decrease in spending on the defensive good will be less than the decrease required to maintain indifference.

At a more fundamental level, revealed preferences may themselves be criticized. It has been argued, for example, that preferences are not freely chosen but are created by the economic system:

> Not only do producers determine the range of market goods from which consumers must take their choice, they also seek continuously to persuade consumers to choose that which is being produced today and to 'unchoose' that which was produced yesterday. To continue to regard the market . . . as primarily a want-satisfying mechanism is to close one's eyes to the more important fact that it has become a want-creating mechanism (Mishan 1967).

Another fundamental criticism of the logic of revealed preferences is that consumers do not always know what is best for them; or even if they do, the means of achieving their wants. The large body of literature demonstrating inconsistency in individual decision making may lead one to question the accuracy of market-determined prices as true measures of individual preferences. Lack of information is also a problem. Abelson (1979), for example, argues that home prices do not accurately measure relative preferences for living near commercial airports, because home buyers typically underestimate aircraft noise when they purchase a house under flight paths and then are deferred from leaving by the moving costs.

When insufficient data are available to rely on revealed preferences, analysts turn to "expressed preferences" and use polls, surveys, or interviews with small groups of citizens to assess individuals' willingness to pay. Although some studies have found attitudes elicited in surveys that correlate highly with behavior (Liska 1975), the use of questionnaires has been criticized on the grounds that responses are strongly influenced by what people think the elicitor wants them to say (Rappaport 1974). People are especially poor at answering hypothetical questions about important events, because the mood and motive of actual choice are difficult to simulate. Questions regarding the value of life are particularly problematic: people simply do not like to think about the subject of death. For example, asking whether an individual would buy trip insurance at a particular price if it were available at the airport may not give the same result that the individual's actions would reveal (Holdren 1976). Rhoads (1978) notes that respondents frequently give identical willingness-to-pay answers for different changes in the probability of death, and also that the order in which questions are asked affects responses.

Usually, willingness-to-pay methods ignore dependencies among individual preferences. Mishan (1976), though, presents a methodology for estimating the value of life that adds the preferences of all concerned in order to determine a composite value for a health-improvement program. Thus, if some of the non-poor are willing to pay to help the poor achieve a minimum standard of safety and medical care, their willingness to pay should be added to that of the poor when deciding if a life-saving program is justified. Such attempts to account for plausible interdependencies among individual preferences (e.g., empathy, malevolence, jealousy, or tastes for interfering with others) run counter to the philosophical rationale of the efficiency criterion and create conceptual as well as measurement difficulties.

Another problem is distinguishing between what people are willing to pay to obtain some favorable change and how much they would insist upon being paid to assent to a worsening of their situation. For a marketed good, willingness to pay for an additional unit and demand to be compensated for the loss of a unit are assumed to be equal to each other and equal across all individuals: even individuals who do not enjoy the good would buy it at below-market price for the purposes of arbitrage. For nonmarketed goods, such as peace and quiet, there is no mechanism to drive everyone toward an equal buying and selling price. Most applications use buying prices (willingness-to-pay values) rather than selling prices. The correct interpretation, however, depends on the assignment of rights. For example, do people have a preexistent right to peace and quiet, or does the airport (perhaps by virtue of being there first) have a right to make noise? In effect, the choice can be regarded as a policy decision about the initial allocation of rights. The issue has practical significance because evidence suggests that most people would insist on being paid far more to assent to a worsening of their situation than they would be willing to pay to improve their situation. Such factors as habituation and psychological attachments may account for the difference (Kelman 1982).

A general problem with all market-based valuation procedures is that they impute social preferences solely from individual preferences in the context of private transactions. Kelman points out that by doing so, cost-benefit analysts slip an important and controversial value judgment into their analyses; namely, that there should be no difference between private, individual values and behavior, and the values and behavior displayed in public. An alternative view is that public decisions provide an opportunity to give certain things a higher valuation than we choose, for whatever reason, to give in private decisions. For example, people may occasionally gamble but be opposed to legalizing gambling.

Decision analysis and other approaches that rely on methods of encoding individual utility functions have their own problems. Thompson (1982) notes that the utility assignment process is not easy:

> Theoretically, there is exactly one probability p for which the decision maker is indifferent between the intermediate outcome for sure and the lottery on the best and worst outcomes. In practice, it is extremely hard for the decision actor to decide whether an intermediate outcome is more like an 0.68—0.32 lottery on the best and worst outcomes or more like an 0.63—0.37 lottery.

Von Winterfeldt and Edwards (1986) suggest that respondents can typically only identify indifference probabilities to within about ±5 percent.

Because utility assessment procedures are time consuming and require asking a great many questions, applications to multiattribute problems almost always use the least demanding assumption of the linear additive form, as discussed in Chapter 2. Oftentimes, however, the independence assumptions necessary to justify the additive form are difficult to achieve when the outcomes of concern involve health effects. For example, if reduced air pollution tends both to extend one's life and to improve the quality of life, these outcomes must be regarded as complementary rather than preferentially independent. Torrance, Boyle, and Horwood (1982) suggest that the multiplicative form is often a better approximation than the more commonly used additive form.

Fischhoff *et al.* (1980a) question the utility assessment procedure by arguing that values do not actually exist in a decision maker's mind; rather they may be formed and molded during the assessment procedure that seeks to elicit them. Different procedures may therefore affect these values differently and actually change the preferences of the subject. Tribe (1974) takes this a step further:

> . . . in environmental matters even more than others, most people lack clearly articulated ends and values at any given time and have only vague ideas about what they might regard as desirable or undesirable; such inchoate values are crystallized into distinct preferences or criteria of choice only through the concrete process of seeking means to attain them and gradually discovering what such means entail.

Another fundamental difficulty in measuring benefits, which applies to all procedures, is that values are not static but are constantly evolving. As a result, social values may change within the planning horizon for the decision. Decision analysis approaches can, in principle, ask people what they want today and what they expect to want in the future (when the consequences of today's decisions will be experienced), but such questions would be very difficult for a subject to answer. By inferring preferences from historical market data, CBA approaches assume unchanging values. Mishan and Page (1979) note:

> Inasmuch as the untoward consequences of consumer innovations tend to unfold slowly over time, their valuations at any point of time . . . as determined by market prices . . . may bear no relation whatever to the net utilities conferred over time. Indeed, the very pace of change today . . . is such that it is no longer possible for the buying public to learn from its own experience . . . . In consequence, society can have no confidence that the valuations of such goods have any ex post correspondence with people's subjective wants . . . .

Redmond (1978) has described five possibilities for dealing with changing values: (1) ignore them, (2) attempt to forecast value changes, (3) make plans flexible to permit adapting to value changes, (4) limit planning to the immediate future where value changes are insignificant, and (5) attempt to control future values in such a way as to make them mesh with present values. Redmond points out the error of (1), the difficulty of (2), the general infeasibility of (3) and (4), and the moral unattractiveness of (5).

Lovins (1977) cites cost overruns as evidence of the inability of analysts to estimate costs reliably. In fact, estimating costs can be as difficult as measuring benefits. Double counting is a common error. Zeckhauser and Shepard (1976) illustrate the difficulties:

> Consider a hypothetical cost analysis of a coal-based electric-generator technology. First, the dollar costs of producing a unit of electricity are computed to be $100. One input to this calculation is the miner's wage, which reflects his valuation of the risks he runs. Let us assume that in deciding what wage to accept he values his life implicitly at $300,000, at least where small probabilities are involved. If one miner's life is lost for each hundred thousand units of electricity produced, then $3 of the $100 unit cost of electricity is attributable to risks to miners' lives. Next, we total up the other costs, one of which is expected lives that are lost .... To attach an additional value ... would be double counting.

Hurter *et al.* (1982) describe three practical difficulties that arise in the estimation of costs incurred by industry as a result of pollution control regulations. First, control procedures tend to alter emissions of several pollutants simultaneously, and it is therefore difficult to allocate a specific cost to a particular pollutant. Second, industry cost estimates are generally derived on an industry-by-industry basis, not on a pollutant-by-pollutant basis. It is therefore difficult to translate the costs of cleaning up the "stone, clay and glass" industry to the costs of cleaning up sulfur dioxide emissons. Finally, there will typically be several alternative control strategies to deal with each source of pollution, and the least costly response for one firm will often depend on the control strategies adopted by other firms. Such interdependencies make it difficult to estimate overall program cost.

Engineering cost estimates can be exceptionally sensitive to the modeling assumptions used to define them. Changes in the detail of the description of processing options, for example, can lead to quite substantial differences in estimates of costs (Smith and Vaughan 1980). Because a regulatory action will have a continuing influence on industry costs, the appropriate baseline is not simply a single definition of nominal plant (or set of plants) but can involve a time sequence of plants as they would evolve in the absence of the action, as well as a specification of the form they would take with it. This level of detail, as essential as it may be, is extremely difficult to adopt. Engineering estimates, for example, routinely ignore the fact that retirement and scrappage decisions can be influenced by the policies under review.

Other difficulties come into play if cost estimates are based on forecasts of industry purchases of the control technologies. A major source of uncertainty in this case will be the number of plants that will have to install pollution controls. For example, EPA's cost-benefit analysis of effluent limitations guidelines for the iron the steel industry found that uncertainty in future demand for the product was a major source of cost uncertainty (General Accounting Office 1984). Contributing to uncertainty in demand were such difficult to predict factors as business cycle fluctuations, foreign competition, and reduction of steel content in automobiles. Other direct costs of environmental regulation which may not be easy to predict include government costs to issue permits, monitor performance, and enforce compliance. Likewise, companies can incur significant costs in applying for permits, including the cost of long delays before projects can be started and completed. The direct expense of regulation can also trigger indirect or secondary costs which are potentially significant but difficult to measure, such as worker unemployment, underuse of capital, and declining innovation and productivity. These effects are difficult to predict because information on future demand and the reaction of industry to new regulation is generally not available.

The distortion on market values produced by government activities is another problem in estimating costs. Howard (1980) asks,

> In computing the investment for a new energy development, should one use the cost of domestic steel or perhaps the lower cost of imported "dumped" steel? . . . If federal law requires that union wage rates be paid for construction, should these rates be used as the actual opportunity costs for labor in the calculation even when there are unemployed nonunion laborers of equal competence available? In fact, is cost not just as uncertain as benefit? Think about any cost-benefit analysis you have ever seen and determine if it still makes sense after such effects are included.

Assessment, Modeling, and Analysis Bias

The creative and judgmental nature of decision-aiding approaches ensures that they will be prone to distortion due to the judgmental limitations of the individuals who participate in their use. Furthermore, the specific analytic structure of a decision-aiding approach forces those who must supply the judgments to think in a somewhat unconventional way. Thus, even if the required knowledge is possessed, there is no assurance that the possessor will be able to express it in the form required by the decision-aiding approach. Experimental research in psychology has recently begun to clarify the cognitive limits of human judgment and their implications for assessment and decision making. A central result is the conclusion that biases are systematic and very strongly prevalent.

The most immediate implications of research on cognitive psychology relate to the estimation of subjective probabilities. Research findings suggest that in most cases individuals make judgments on the basis of simplified inferential rules, referred to as heuristics.[75] Although valid in some circumstances, heuristics can lead to large and persistent biases with important implications for probability assessment. Cognitive biases affecting the assessment of probabilities include:

1. *Adjustment and anchoring* (Tversky and Kahneman 1974; Slovic and Lichtenstein 1971). Individuals tend to produce estimates by starting with an initial value (suggested perhaps by the formulation of the problem) and then adjusting to yield a final answer. The adjustment is typically insufficient. As a result, initial ideas play too large a role in determining final assessments.
2. *Availability* (Tversky and Kahneman 1974, 1981). If it is easy to recall instances of an event's occurrence, that event will tend to be assigned a higher probability. Thus individuals tend to overestimate the probabilities of vivid or dramatic events or events that have just recently occurred.
3. *Base rates and the law of small numbers* (Makridakis and Wheelwright 1979; Bar-Hillel 1980; Tversky and Sattagh 1981). Statistical, abstract information (such as the normal rate at which an event would be expected to occur) is given insufficient weight relative to concrete, specific data. People place too much weight on predictions based on small samples of data and are insufficiently sensitive to the quality of such data.
4. *Coherence and conjunctive distortions* (Tversky and Kahneman 1974). The probability that a sequence of events will occur seems higher than it should, especially if the events fit into a plausible scenario. Thus, a good story makes events seem more likely.
5. *Representativeness* (Tversky and Kahneman 1974; Tversky 1973). Event likelihoods tend to be estimated on the basis to which they are representative of the larger population of possibilities. In considering tosses of a fair coin, for example, people incorrectly regard the sequence H-T-H-T-T-H to be more likely than the sequence H-H-H-T-T-T because the latter does not appear to be random.
6. *Overconfidence* (Kahneman and Tversky 1973; Lichtenstein and Fischhoff 1977; Fischhoff *et al.* 1977). People generally ascribe too much assurance to their estimates (and therefore underestimate confidence intervals). They will tend to see patterns in data where none exist, reinterpret data to increase consistency, and ignore evidence that does not fit their worldview.

A few studies, scattered in fields such as meteorology, economics, forecasting, and radiology, have attempted to compare subjective probability assessments with empirical outcomes. Results typically show that the quality of judgments is quite sensitive to the particular procedures used to elicit them and that predictable biases can often be ameliorated by the careful selection of method or the application of corrective procedures. Kahneman and Tversky (1977) have proposed steps to minimize biases in probability encoding, based on preliminary models for how people naturally encode, store, retrieve, and manipulate data in their minds; and decision analysts have integrated these into formal probability encoding processes (Staël von Holstein and Matheson 1979). The most innocuous debiasing step is to warn people of potential biases. Experiments indicate that this can be effective when individuals have their own biases directly demonstrated to them, but merely reporting the results of research conducted on others seems to have little effect (Lichtenstein *et al.* 1978). The persistent tendency is to think "I'm not like that." In general, the effectiveness of the debiasing procedures used in decision analysis approaches and the extent to which they introduce undesirable biases of their own can only be conjectured at this time.

Cost-benefit analysis and related approaches that rely on frequency data and models to estimate probabilities are by no means immune to biases. Not only do some events go unreported, these events may fall into one class of factors of interest more than another. If probabilities are derived from aggregating data arising from "similar" situations, biases may be introduced because of the failure to account for important specifics or because system changes have occurred over time. For risks associated with nonroutine system functioning (e.g., accidents) there may be few actual incidences of the risk-initiating event, and the predictive ability of the sample will be poor. If probability distributions are obtained by fitting mathematical forms to sparse data, several different forms are likely to fit the data with equal precision and yet have very different implications for the probabilities of events of greatest significance (such as rare, catastrophic events) because they lie outside the range of typical input data.

Logic tree methods (event trees and fault trees) are preferred procedures for estimating probabilities for rare events when the event can be decomposed into combinations of basic events (such as component failures) whose probabilities are well understood. In a study of 50 safety systems in the chemical, nuclear, and conventional power industries in Great Britain, for example, observed failure rates were within a factor of 4 of the predicted values in 96 percent of all cases (Rasmussen 1975). Systematic biases, however, frequently crop up in such estimates. It is common engineering practice to attempt to use conservative values whenever there are uncertainties in an analysis; therefore, individual probability estimates are generally likely to represent upper bounds on the probability of failure. In fact, in the British study the observed failure rate was on the average about 25 percent below the predicted rate. Another source of bias in fault tree and event tree analysis is the previously mentioned problem of common-mode failure. Because of the complexity of major engineering systems and the number of interactions possible, it is virtually impossible to eliminate totally all conceivable sources of common-mode failures. Such dependencies must be considered in probabilistic analysis in order to adjust probability factors appropriately. Any existing common-mode failure sequences not identified in the analysis tend to cause calculated probabilities to understate actual likelihoods. One other important source of bias in the probability estimates produced in CBA is, of course, the neglect of uncertainties associated with ignorance of structure and relationships. Thus even if CBA succeeds in properly reflecting uncertainties about random processes that are well understood, it often neglects those for which there is little or no experience.

Little in general can be said about the degree to which biases in probabilities produce distortions in the results of decision-aiding approaches. Von Winterfeldt and Edwards (1982) showed that a class of decisions with continuous options was not particularly sensitive to individual probability judgments. On the other hand, Lichtenstein *et al.* (1977) demonstrated that modest changes in probabilities can substantially alter computed expected utility with discrete decision alternatives. Experience in conducting sensitivity analyses suggests that in most real-world problems, conclusions are not highly sensitive to small changes in probability estimates.

Because of the apparent trend in the application of decision-aiding approaches to rely increasingly on sophisticated hazard, exposure, and dose-response models rather than on simple extrapolations from data or direct judgments of technical experts (Hammond *et al.* 1980), the question of bias in modeling is gaining importance. Very little is known about the effect of human judgment on formulation and modeling. Parkinson (1979) has pointed out that people have a tendency to allocate excessive time and attention to analytically tractable or familiar problems and insufficient time to intuitively important yet analytically difficult problems. This is consistent with the common criticism that analysts tend to concentrate on those aspects of the problem that are easier to treat, such as those with quantifiable, commensurable variables, and to ignore more subtle, difficult-to-quantify issues.

Because of overconfidence bias, experts can be expected to place too much faith in theoretical models derived from limited data. Chapman and Chapman (1969) observed that clinical psychologists frequently see the patterns they expect to find even in randomly generated data. Fischhoff *et al.* (1980b) describe the natural process by which a scientist might mistakenly identify a relationship as causal:

> . . . as a toxicologist, you are "certain" that exposure to Chemical X is bad for one's health, so you compare workers who do and do not work with it in a particular plant for bladder cancer, but obtain no effect. So you try intestinal cancer, emphysema, dizziness . . . , until you finally get a significant difference in skin cancer. Is that difference meaningful? Of course, the way to test these explanations or theories is by replication on new samples. That step, unfortunately, is seldom taken . . . .

Lovins (1977) argues that the distortions inherent in models are particularly significant for the analysis of decisions dealing with long time horizons:

> One can be quickly transported to the ethereal realm by using computer modeling to simulate the supposed interactions of large numbers of estimated variables over long periods of time. Such models are inflexible; they are obliged to assume constants in lieu of slow variables, and their structure, boundary values, and objective functions do not evolve like those of a real society.

Finally, the vast array of modeling procedures available and the tendency of many analysts to specialize create their own potential for bias. Strauch (1980) warns,

> Reliance on method-oriented approach and neglect of the focusing stage can restrict attention to the particular class of models called for by the approach. Give the problem to an analyst committed to a particular methodology, and you are almost certain to find it structured in a way which ignores those aspects of the problem not considered by the methodology. As the saying goes, "to a small boy with a hammer, everything looks like a nail."

Propensity Toward Interference by the Analyst

While interference by the analyst may seem to be of greatest concern for DA and related approaches that make explicit use of subjective information, significant subjective elements occur in the model structuring and data development procedures used in all formal approaches. The judgments made in approaches adopting an objective perspective are often so deeply imbedded in fundamental assumptions that they are taken for granted. For example, the scientific judgment to require a 95 percent level of confidence when deriving correlations from empirical data, rather than, say, 90 percent or 99 percent, could under certain conditions have major implications for the conclusions of an analysis. Although DA approaches make greater use of expert judgment, they also place the greatest emphasis on making such judgment explicit and using carefully applied elicitation procedures designed to minimize analyst interference. Despite such efforts, even the most carefully designed approach leaves considerable potential for subtle, intentional, or unintentional interference by the analyst in the development of probabilities, values, and model relationships.

Contrary to the supposition of DA and SWT approaches, people do not always have well-defined opinions regarding the desirability or likelihood of various events. Assessment procedures can therefore become major forces in shaping the values expressed, or apparently expressed. The procedures used to assess these "inputs" for an approach can introduce random error (e.g., by confusing the respondent), systematic error (e.g., by hinting at what a "correct" response might be), or unduly extreme judgments (by creating unreasonable perspectives).

Fischhoff *et al.* (1980a) list decisions made by an elicitor that are likely to affect a respondent's judgments. First, the way the problem and issues are defined establishes a framework for thinking which may have a significant impact on the expression or formulation of values and judgmental information. The elicitor's definitions determine whether or not a "problem" exists, the options and consequences to be considered relevant, how they are labeled and measured, and whether and to what extent the problem is decomposed.

Considerable experimental research has demonstrated the influence of problem definition on judgments. For example, Fischhoff *et al.* (1980b) explored the importance of labels. They presented people with two versions of mathematically equivalent gambles. In one a consequence was labeled a "sure loss"; in the other an "insurance premium." Most people were found to dislike the former and like the latter. Similarly, whether an outcome is measured in "lives saved" or "lives lost" creates a very different impression. Kahneman and Tversky (1979) were able to produce a variety of reversals in the preferences stated by individuals through the use of different forms of decomposition. Particularly effective was decomposing possible outcomes into a component that was known for certain and a remainder that was uncertain. Apparently, certain losses and gains tend to be weighted more heavily than uncertain outcomes when determining overall attractiveness. The importance of measurement scales is discussed by Fischhoff *et al.* (1980a) in reference to a study that attempted to establish a dollar value for aesthetic degradation of the environment:

> The researchers asked visitors to Lake Powell how much they would be willing to pay in increased users' fees in order not to have an ugly (coal-fired) power plant looming on the opposite shore. They asked, "Would you pay $1, $2, $3?" and so on, until the respondent answered "No" .... Rather different numerical values might have been obtained had the bidding procedure begun at $100 and decreased by steps of $10 .... Any respondents who were not sure what they wanted in dollars and cents might naturally and necessarily look to the range of options presented ... for cues as to what are reasonable and plausible responses.

A second means by which an elicitor may affect a respondent's judgments is through control of perspective. Perspective may be altered through affecting the salience or importance of issues, judiciously choosing the time of inquiry, or changing the confidence or apparent degree of coherence expressed. As an example of how perspective may alter responses, Parducci (1974) found that judged satisfaction with one's state in life depends highly on the range of states considered. According to Turner and Krauss (1978), the order of questions presented in a survey has had marked effects on people's evaluation of the health of the nation and its institutions. People are typically unaware of the extent of such shifts in their perspectives, and they often have no guidelines as to which perspective is the appropriate one (Slovic *et al.* 1979b).

The elicitor not only may alter the respondent's perspective but also may irreversibly shape it, for example, by destroying existing perspectives or creating an opinion where none previously existed (Kahneman and Tversky 1979). Thus an elicitor's control can be reversible or irreversible (Fischhoff *et al.* 1978b). Furthermore, some form of impact is unavoidable; questions must be posed in some manner, and that manner may have a large effect on the responses elicited.

Susceptibility to Manipulation

Regardless of the approach used, the opportunities for the analyst to manipulate outcomes are innumerable. Some forms of manipulation are relatively innocent, such as the tendency of analysts to have too much faith in their product and therefore to oversell it. Others are more devious, as when experts submit scientific evidence of low quality or play a "numbers game" (Fischhoff 1977). Ida Hoos (1972) has a particularly unflattering view: "Obedient consultants in think tanks and accommodating academics of all stripes constitute a reservoir of talent . . . . They produce 'data' to substantiate any position. They dutifully interpret 'facts.' "

Analysts who are advocates can deliberately influence the results of analysis in a variety of ways. Judicious selection of costs and benefits, to which values are carefully assigned from within a wider range of plausible values, can predetermine the outcome of a cost-benefit analysis. More generally, the selection of the valuation procedures used in an approach may have a significant impact on results by establishing whether benefits are defined in terms of what people like or what paternalistic others believe to be good for them; whether people have entitlements, the violation of which requires special dispensation; whether consideration is given to the benefits that some people derive from knowing that a natural resource has been preserved even though they themselves will never use it; and whether distribution effects are ignored (Fischhoff and Cox 1986). The cost-benefit balance of any capital-intensive project whose benefits are deferred but whose real costs are immediate is sensitive to the assumed discount rate. Therefore, results can often be manipulated through the selection of the discount rate and inflation rate. In decision analyses, results can be manipulated through the selection of the arbitrary values assigned to identified costs and benefits, strategic selection of experts, the exclusion of costs that would tilt the outcome against the preferred option, the use of self-serving assumptions about distributional fairness, and so forth. Very conservative assumptions can be used to drive estimated risks up, omissions of inconvenient pathways can drive in down, clear presentations of expert disagreements can decrease the confidence in the results, and so on.

Lathrop and Linnerooth (1983) compare three risk assessments of a proposal to locate a liquefied natural gas terminal at Oxnard, California. One study was funded by a company representing gas interests, another by the city of Oxnard, and the third was conducted by the Federal Power Commission (FPC). The Oxnard-sponsored study estimated expected annual fatalities contributed by the terminal that were 380 times larger than that estimated by the industry-sponsored study. Whereas the FPC and industry-sponsored reports described the risks as "extremely low" and "negligible," the Oxnard study concluded "it is not now possible to state confidently that the proposed facility possesses a low probability of a high consequence accident." Most significant, the Oxnard report presented plume maps that essentially displayed maximum credible accidents. By vividly portraying to local residents the specter of a deadly methane plume covering their own homes, the Oxnard report, in the words of one participant, "electrified opposition to the terminal" (Ahern 1980).

Peer review of analyses, which occur only infrequently, would be helpful in preventing the most blatant manipulations, but more subtle tactics are available which would be difficult for even other experienced analysts to detect. In assessing information or preferences the analyst might present information in ways that induce countervailing biases. For example, the vividness of events that seem in danger of being overlooked could be artificially enhanced. Reliance on market data rather than expert judgment in no way renders CBA immune to value manipulation; the essence of marketing is to manipulate people's values, altering their preferences and creating desires that did not previously exist (Fischhoff *et al.* 1981b).

Even if the analyst is well intentioned, unconscious motivational biases can affect results. Raiffa and Zeckhauser (1981) describe an experiment in which students were asked to play roles in a simulated merger exercise between a large company and a small company. The students were told that future profitabilities of the companies depend on the state of the industry in the subsequent three years. Before being assigned their roles, the students were given identical data about industry outlook. Each student was then assigned one of three roles — an analyst for the large company, an analyst for the small company, or a third-party intervenor — and then asked to generate a probability distribution over industry-wide sales on which the student would base further analyses. The resulting assessed probabilities turned out to be dramatically different, depending on the simulated role played by the student.

According to Wilson (1973),

Rarely does anyone deliberately fudge the results of a study to conform to pre-existing opinions. What is frequently done is to apply very different standards of evidence and method. [Agency studies] accept an agency's view about what it is doing and with what effect, adopt a time frame (long or short) that maximizes the probability of observing the desired effect, and minimize the search for other variables that might account for the effect observed.

Formal analysis is certainly not unique in its susceptibility to manipulation. The vulnerability of analysis is, however, discomforting given the apparent precision of the quantitative results reported.

Susceptibility to Misuse and Misinterpretation

Even if analysts do not manipulate their results, approaches may be misused by sponsors, and the results may be misinterpreted. The potential economic incentives to stakeholders for tailoring environmental regulations can be very high. Despite analysts' desires to be impartial in their evaluations, difficulties in communication (resulting, e.g., from the technical complexity of the analytic techniques involved) frequently prevent users from completely understanding the approximations and assumptions on which conclusions are based, if not the conclusions themselves.

There are a number of ways in which sponsors might misuse analysis. Because formal approaches make clear what is not known, they might be employed as an excuse for delay or inaction.[76] Howard (1980) observes that in many instances assessments seem to be commissioned not to aid impartial decision making, but rather to advocate a particular position:

I have recently been asking decision analysts how many of their studies were prompted by a belief in systematic analysis per se rather than by a desire for some party to the decision to advance either his own proposals or to defend them against attacks. A very large proportion of responses has been in favor of the self-interest hypothesis.

It is not surprising that many applications are motivated by a desire to support a particular point of view. As discussed in Chapter 1, our existing decision-making processes rely heavily on bargaining, voting, delegation, material incentives, and other procedures for which persuasion is a central element. Majone and Quade (1980) argue that although many analysts are likely to view persuasion as an illegitimate use of analysis (because this use violates central values of the scientific method — objectivity and rationality), there are "reasonably well-defined situations in which persuasion can be, and has been, used legitimately in support of rational analysis."

Another way in which an analysis can be misused is by selective omission of results or caveats. Fischhoff (1977) describes the aftermath of a National Academy of Sciences study on Supersonic Transport:

Shortly after the report was completed, the Department of Transportation produced a summary of the same data which completely neglected the conclusion that for each SST flying, we should expect three to four more cancer deaths per year worldwide, due to reduction in the earth's ozone shield. The Academy's scientists were outraged at this misinterpretation of their work. Since the details of the analysis were part of the public record, they were able to have the misinterpretation rectified.

The most frequently expressed concern involving misinterpretation of analysis involves the apparent precision of numerical estimates. "Numerical benefit-cost estimates can carry with them a spurious precision that no disclaimer can dispel" (National Academy of Sciences 1977). Freeman (1979a) observes that the qualifications and limitations associated with cost-benefit numbers tend to be ignored in public discourse. Application of a quantitative approach may thus create the false impression that more is known than is really known.[77]

Institutional and Political Incompatibilities

Some criticisms are focused not on whether formal decision-aiding approaches are capable of working as they are designed to, but rather on whether they are acceptable in view of established institutions, procedures, and social norms. Institutions can and do change, and might well do so in response to new technologies such as decision-aiding approaches. Thus incompatibilities with current practices do not negate an approach's value. However, whether or not an approach is compatible with existing institutions and processes will have a major bearing on its acceptability to participants in the decision process. Questions of the compatibility of formal approaches with democratic processes, existing constitutional and legal principles, organizational decision making, and cultural realities are discussed below.

Incompatibility with Principles of Democratic Decision Making

A general theme running through criticisms in this category is that whether a decision-making process conforms with democratic principles is more important than the logical "correctness" of the decisions it produces. Tribe (1974) states, "The 'way of acting' to which we commit ourselves must be a process valued for its intrinsic qualities rather than for its likely results." Thus, Rowe (1979) observes: "Acceptable risks may be defined not by their levels, but by how they are dervied . . . the process, its visibility, its fairness and credibility, and its opportunity for participation, may itself be the method to achieve acceptable risks under the operating definition."

The processes that have evolved for making environmental decisions are not always compatible with the assumptions underlying formal decision-aiding approaches. For example, agency regulatory hearings and judicial review both are adversarial processes that seek to resolve conflict rather than achieve objective truth:

> . . . the purpose of the adversary proceeding is not to discover truth but to settle disputes without violence. Truth is valued in the courts because it usually makes the judgment based on it more acceptable . . . . But other considerations are more important than truth . . . predictability of results, blindness to individual differences and respect for social values are examples. The adversary process has survived for centuries because it has developed rules and procedures that balance these elements, of which truth is merely one (Abrams and Berry 1977).

A formal approach that strives to produce an integrated, comprehensive, and balanced presentation of issues will not hold much appeal to an advocate, particularly if the analysis provides evidence supporting conclusions intermediate to the antagonists' positions. Both sides will tend to fear that the introduction of such evidence, however relevant and extensive, will reduce the strength of their arguments.

Critics of analysis have charged that formal approaches adversely affect the democratic decision-making process by concentrating power, limiting opportunities for participation, and damaging credibility. Decision analysis applied in its supradecision-maker mode, for example, has been criticized for delegating too much social decision-making power to individual officials (or small groups of them). Harsanyi (1977a) argues,

> To be sure, if these officials are persons of great ability, and if the decision-making tasks delegated to them do not give rise to unmanageable problems of collecting and processing information, then this arrangement may produce highly efficient and altogether satisfactory administrative decisions . . . . But, it is notoriously hard to ensure that officials entrusted with important decisions will always be persons of superior ability . . . most important of all, any system based on great concentration of power is sooner or later bound to give rise to intolerable political abuse.

Formal approaches may reduce opportunities for participation, it is claimed, because they restrict the representation of diverse values and viewpoints. Slovic *et al.* (1979b) note that by requiring a single, aggregated expression of social preferences, decision-aiding approaches assume that society is sufficiently cohesive and common-goaled that problems can be resolved by reason and without confrontation. Those who in fact hold nonestablished views will not be represented by the aggregated expression of social preferences. As a result, they might well resist formal analysis, reasoning that "avoiding disenfranchisement" is more important than determining an optimal way of reducing risk: "Sitting down to discuss a risk-benefit analysis would, in this view, itself constitute the surrender of important principles. Cooperation may even be seen as a scheme to submerge the opposition in paper work and abrogate its right to fight the outcome of an analysis not to its liking" (Slovic *et al.* 1979b).

Simply treating social values as variables (as in many appliciations of social DA) and investigating the implication of different opinions using sensitivity analysis does not eliminate the problem. As Fischhoff (1979) notes,

> It is . . . unclear how sensitivity analyses accommodate the variety of perceptions and misperceptions held by members of the public. Ignoring unusual views means abrogating their holders' right to an opinion; considering them may force the analyst to produce results that seem patently absurd. It is hard to choose which values to use and doubly hard when the decisions must be made publicly.

Risk acceptance, of course, is ultimately inseparable from the psychology of risk perception, since the political decision-making process must be responsible to people's perceptions. As noted in Chapter 1, the technical perspective of risk adopted by formal approaches does not account for many psychologically important factors. Perhaps most important in this regard is the extent to which people react to the possibility of severe outcomes irrespective of their estimated low probabilities.

Lovins (1977) observes that the use of formal analyses that leave out politically significant factors may create results that the public perceives as absurd, and can therefore adversely affect the credibility of the decision-making process. To illustrate, he cites a cost-effectiveness analysis of remote siting of nuclear reactors which considered only biological effects of potential radiation exposure and the costs of transmission lines. The study concluded that "remote siting of nuclear power facilities would not seem to be a cost-effective way to control potential radiation exposures." Lovins remarks, "anyone who tried on this basis to site reactors in cities would be in for a shock." Fischhoff (1977) cites an example of a National Academy of Sciences study of the effects of thermonuclear war. The Academy panel decided that the anticipated reduction of the earth's ozone shield would not imperil the food supply of survivors because many crops could survive the increased ultraviolet radiation. Fischhoff notes, however, that the study failed to point out that the increased ultraviolet radiation would make it virtually impossible to work in the fields to raise those crops. Although such oversights might be largely addressed by ensuring that the assumptions and resulting caveats for the conclusions of an analysis are clearly delineated, Lovins fears that the public cynicism created by politically naive studies may give rise to a "general public suspicion in all political matters" (Lovins 1977).

Incompatibility with Constitutional and Legal Principles

A frequently voiced constitutional concern is that basing decisions on formal analysis removes power from elected officials held accountable to the public through the democratic process and transfers that power to a "technical elite." Furthermore, because formal approaches produce redistribution of costs and benefits, they imply some concept of distributional equity reached without regard to representative processes.

The fundamental individual safeguards provided by the Constitution, particularly the evolving law on due process, equal protection and property rights; the enactment of policies promoting particular redistributions of costs and benefits by government action through representative processes; and the ability of individuals to alter distributive patterns by using the common law and administrative law, are of greater significance to solving the problem of distribution than the development of assumptions about fairness by economists and analysts. The distribution and redistribution of costs and benefits of government actions is essentially founded on these Constitutional and other value-based principles which lay no claim to rationality (Baram *et al.* 1977).

According to Lovins (1977) and Green (1979), the "language of ordinary political discourse" preserves "accountability through the ballot box and the courts," but letting "ultimate value judgments . . . be based to a substantial extent on the value judgments inherent in the risk-benefit calculus performed by the elite" would "seriously inhibit political accountability and judicial review."

Cost-benefit analysis approaches seem to be especially prone to such criticisms because they provide the least flexibility for the independent input of social values. Decision analysis is criticized for the arbitrary way in which social values for the analyses are typically established. Baram (1980) says, "It is irresponsible for such analysts to make arbitrary assumptions on issues which lie beyond their or anyone else's expertise. Decisions on the valuation of human life or environmental quality, are not properly dealt with by unaccountable analysts, but should be made by those vested with responsibility to make these subjective choices."

In theory at least, social decision analysis, with its concept of making preference inputs the responsibility of a specific body (such as Congress), is less susceptible to these sorts of objections. All approaches, however, rely to a greater or lesser degree on technical specialists for estimating decision consequences, and the inputs provided by these experts may be influenced by their own self-interests and values. Whether existing, unaided decision processes involving experts are less likely to yield excessive influence to these experts is debatable. Hammond and Adelman (1976) describe how the decision of choosing a new bullet by the Denver Police Department nearly passed into the hands of ballistics experts:

As a result of focusing on bullets and their technical ballistics characteristics, legislators and city councilmen never described the social policy that should control the use of force and injury in enforcing the law; . . . Instead, the ballistics experts assumed that function. When the legislators requested their judgment as to which bullet was "best," the ballistics experts implicitly indicated the social policy that should be employed."

Lovins (1977) apparently rejects the use of technical experts as well as formal approaches, believing instead that "democratic decisions should rest on common-sense judgments made by generalists" who, according to Green, "have greater affinity with public values leading to greater (and more deserved) confidence in the conclusions reached . . . Any other approach smacks of government by the elite, a concept that is at variance with democratic principles."

Another alleged incompatibility concerns the establishment of a fixed logic by approaches in which competing social preferences are traded off against one another. This runs contrary to the concept of social decision making as being largely a matter of the assignment of "rights and responsibilities" (Green 1979). Senator Stafford, for example, has opposed the use of cost-benefit analysis in environmental issues because monetizing costs and benefits converts an intangible right-to-health to a property right, which is then "involuntarily alienated." Such a system permits a polluter to injure others because it would cost him too much to avoid harming them. Such a system, he claims, would be skewed in favor of pollution because benefits are difficult to determine but costs are easily calculated (Stafford 1981). According to Baram (1980), "The varied and often conflicting needs and desires of many segments of our society must be weighted against fundamental individual rights in order to establish ultimate societal values and reach an optimal governmental choice. This process is subverted when cost-benefit analysis is the basis for decision making." Lovins (1977) adds, "To reduce the [decision-making] process to purely incremental, instrumental terms distorts the concept much as 'rights' are flattened by making them a contingent grant rather than an inherent part of a person."

Legal realities also pose compatibility questions. Lawyers often caution their clients against the disclosure of finite risks associated with company or agency decisions because of the possible increase in the organization's vulnerability to liability actions. Also along this line, Ashford (1978) has called attention to the legal mandates[78] of some regulatory agencies to protect the public against the worst possible outcomes and not of maximizing social utility.

Finally, Baram (1980) points out that, depending on how proposed regulatory reforms involving risk analysis are implemented, questions of the constitutional balance of power may be raised. In general, formal decision-aiding approaches imply a more autonomous and managerial role for regulatory agencies. To implement such approaches, especially DA and SWT approaches, the agency must in principle have both discretionary authority and responsibility to make value judgments balancing conflicting goals and, presumably, to select from the full range of regulatory and nonregulatory means for achieving the optimal accomplishment of its objectives. The use of formal approaches, especially CBA, will tend to increase agency accountability to those whose benefits and costs are most concentrated and measurable and decrease accountability to the legislature and to the other groups affected by the problems the agency was established to address. Proposals[79] that give the Office of the President authority to oversee and review applications of formal approaches necessarily produce a shift in power from Congress and the regulatory agencies to the Executive.

Incompatibility with Organizational Structure and Processes

The view of the decision-making process implicit in formal approaches implies someone or some collective mechanism actively selecting from among distinct alternatives in a stable decision environment so as to achieve a well-articulated set of goals and objectives. It is widely recognized that this model is not an accurate characterization of most public sector decision making.

Models of organizational decision making that more accurately reflect empirical evidence have been suggested by a number of authors. March (1978) describes organizational decision making as a process of limited rationality whereby individuals and groups simplify a large problem by addressing it as a series of smaller decisions to avoid the difficulty of simultaneously considering all information and alternatives. Allison (1971) describes an "organizational process model" in which government behavior is better understood as the output of a large organization functioning according to standard patterns of behavior than it is in terms of deliberate choices. In one version, which he calls a "bureaucratic politics model," action is the result not of maximizing some objective, but of political bargaining. Holling (1981) observes that random events (such as the destruction of a forest by an outbreak of the spruce budworm) play a critical role in organizational decisions, because of the ambiguity of many situations and the limited attention that can be given to any particular problem by interested parties unless the problem is perceived as being critical.

Several authors have suggested that these and other realities present incompatibilities for using formal approaches. For example, Boulding (1966) notes that information overload typically forces decision makers to limit the agenda for inquiry. He argues that the application of formal approaches can contribute to information overload and cause creative solutions to environmental problems that cannot be translated easily into mathematical equations to be overlooked in favor of those that are easily quantifiable. Lathrop and Linnerooth (1983), in their study of the unsuccessful attempt to site a liquefied natural gas (LNG) terminal in California, observed that the sequential nature of decision making tends to limit possibilities for comprehensive analysis and locks the process into a given course of action: "resolution of the question whether a site was needed necessarily preceded site slection . . . . During the seven-year course of the

California proceedings, the need for imported natural gas diminished greatly. Instead of reexamining this need, the process [became] locked into a commitment for an import facility."

Carter (1972) notes that incompatibilities between analysis procedures and established management procedures can prevent the successful implementation of formal analysis and suggests that a strongly decentralized organization may be able to tailor analysis to its needs more easily than a strongly centralized organization. Use of some decision-aiding approaches may, however, tend to promote centralization. Brown (in press) argues that the use of DA within a hierarchical organization may shift the locus of power in the direction of greater centralization because the approach makes it easier for higher-level authorities to intervene selectively and effectively into lower-level decisions.

Some critics have argued that the political nature of decision making prevents risk decisions from being usefully analyzed in abstract. "Whether we like it or not, the risks to which we have access are processed risks" (Thompson 1980b). "Risks do not constitute independent objective facts but are modeled and defined in the process of social communication; risks are neither 'right' or 'wrong' nor 'small' or 'large' but depend in their qualifications upon the consensus of those concerned . . ." (Conrad 1981). Noting that "the risk from any given action will tend to ramify, more or less strongly, through all other areas of human experience," these critics have concluded that risk problems cannot necessarily be isolated, evaluated, and solved through a technical fix as assumed by formal approaches (Ravetz 1980).

While the differences between the rational actor model and reality do not necessarily negate the value of formal approaches, they would appear to limit their applicability and utilization. Lathrop (1980) concludes from his investigation of the LNG siting process:

> The most serious problem in the use of a decision analytic siting evaluation model, such as the one proposed by Keeney [1980b], is that the decision structure assumed by the model does not match the institutional structure making the siting decision . . . . It is not clear where a decision analytic evaluation model would fit into such a process, whose trade-offs should be used to set the parameters, or even to whom the analysis should be delivered.

Despite this pessimistic picture, organizations do, of course, commission formal analyses of risk decisions. To obtain a clearer understanding of the decision environment existing in situations in which formal approaches have been applied, the author asked professional analysts to describe characteristics of the decision situation encountered in analyses of decisions for government organizations. Conclusions derived from this survey tend to confirm aspects of organizational and process models of decision making but suggest that a formal approach will frequently be perceived as playing an important role:

1. Procedural rules for decision making are prevalent and appear to be most highly developed in organizations that have evolved in stable environments.
2. Decisions beyond the routine application of procedural rules occur relatively infrequently and tend to relate either to changes in organizational structure or process or to one-of-a-kind decisions, typically of a strategic nature.
3. Changes in an organization's environment often provide the source and recognition of a need for decision by creating situations for which existing procedures do not apply or are inadequate.
4. Conflict is generated in such situations, and the search for compromise is a characteristic response.
5. The initiative for the application of a formal approach frequently comes from a particular interest group within the organization rather than from the highest level of decision responsibility, and promoters tend to be oriented analytically and otherwise predisposed to the logic of rational choice.
6. Organizational objectives and values are not well defined nor immutable but seem to be largely created in the minds of participants in the course of the analysis.
7. The requirement that statements of information and preference be made explicit is sometimes resisted, apparently for fear of ending in a position that may prove to be untenable.
8. Subjects occasionally admit to the existence of decision criteria that they wish to exclude deliberately from formal consideration. As a result, analyses may produce solutions that ignore undeclared preferences and are therefore unacceptable.
9. Institutionalized analysis often tends to become a form of ritual which has to be gone through to make a decision, something that the decision maker may use as a subsequent defense.
10. The crisis management atmosphere of agency regulation, promoted by congressional requirements with short deadlines and political forces, mitigates against thoughtful and timely analysis.

Incompatibility with Social Norms

Some critics argue that formal decision-aiding approaches run significantly counter to existing social attitudes, preferences, and practices, and consequently, regardless of their potential value, will not gain any real acceptance. Issues discussed in this category include the reluctance of society to quantify environmental and health amenities, and the unwillingness of decision makers to admit that the basis for decisions is largely subjective and inconclusive.

The most frequently made cultural incompatibility criticism is that the general public considers valuation, quantification, and monetization of environmental and health outcomes morally distasteful and irrelevant to decision making. This view is consistent with the prevailing attitude of doctors and lawyers who interpret their "professional ethics" as dictating that they maximize their clients' legal or patients' health interests without attempting to balance these interests against resulting economic burdens. Doniger (1980) states,

> There is simply no morally and socially acceptable way to trade human life and such things as the survival of other species for conventional economic products. Any decision calculus which, even if it had precise numbers, attempted to put a price on human life would be widely rejected as morally repugnant and as practice unbecoming a democratic government.

A less extreme view is that people recognize the need to place finite values on health and safety but prefer to find ways to avoid explicitly talking about it. MacGregor and Slovic (1984) found that respondents preferred that safety decisions be based on CBA rather than on less formalized methods such as looking at past decisions and employing standard industry practices, but they also found the CBA was judged less attractive than a comparable approach that avoided explicit value of life tradeoffs. Rhoads (1978) illustrates the reluctance people have to compare health and economics by citing a speech by Judge David Bazelon, who observed that his court rarely if ever encounters candid arguments against taking actions that will improve health and safety: "No one says that a program will save lives but is not worth the cost. Instead there is a tendency to hide behind scientific uncertainties ('There is no credible evidence at this time . . .') or to use some other subterfuge." Rhoads also observed a similar circumvention of economic arguments in debate surrounding the government's involvement with kidney dialysis: "A basic asymmetry was manifested throughout this complicated policy debate. The proponents of an expanded Federal government role were able to carry their case to the public in a variety of ways, while the opponents made their case in *sotto voce* fashion."

Cost-benefit analysis and other approaches using willingness-to-pay measures are most direct in their assignment of dollar equivalents to loss of life and other health effects. Decision analysts frequently avoid the problem of trying to establish a market value for human life but require instead that a "decision maker" reveal his own subjective judgment regarding the value of a human life. Because such statements are extremely controversial, it is not surprising that decision analysts must typically resort to the use of "illustrative" and "arbitrarily selected" values, perhaps derived from previous government decisions. Multiattribute utility theory approaches do not necessarily require health and environmental consequences to be assessed directly in dollars, but they do necessitate that relative preferences be made explicit. So long as "dollars spent" is a criterion, however, expression of relative weights in a multiattribute utility function is mathematically equivalent to assigning directly a "value of life." Furthermore, because personal criteria may be involved, such as the impact of a decision on the decision maker's career, "decision makers" are often reluctant to take an explicit public position on relative social values.

Decision analysts who incorporate a value of life in their analyses often take great pains to explain that the dollar figure does not reflect a judgment of "what a life is worth" but rather that it represents a tradeoff judgment established for consistently and efficiently allocating economic resources for life-saving programs:

> While some might see these procedures as callous, it is important to observe that we are not attempting to value a human life or injury in a moral sense, we are rather assigning values to life and other loss categories as a means of making social decisions. The values are not measurements but the results of a decision made by society concerning the balancing of social benefit and social risk . . . . Thus, the problem is to assign a value to human life (and to injury, and so forth) that reflects what we can afford to pay as a society to avoid a death from involuntary risk (Barrager *et al.* 1976).

Judging from criticisms, opponents either fail to grasp this distinction or do not view it as significant. In a harsh review of the SRI International study from which the above quotes were taken, Cochran (1976) asks, "Do any of you really think this $60,000 study is worth the random incidence of six additional serious genetic effects ($10,000 for each according to SRI) . . . ?"

MacLean (1979) argues that the use of a value of life to permit a more efficient reduction of risk is itself wrong. Believing in a moral responsibility "to create the best world possible," he argues that there are circumstances in which societies ought to take actions that cannot be justified in terms of lower overall risk or reduced costs:

> . . . the argument assigning a social value to human life is wrong because it fails to treat human life as sacred. The argument does not deny that at some point we should stop paying to reduce our risks and spend our money more wisely; nor does it deny that, once we have made our decisions we can calculate the costs and estimate how much we are willing to spend to save lives . . . . Life-saving policies are also symbolic; they are public decisions undertaken to show that human life has a special, intrinsic value.

MacLean's view is that the values that determine acceptable risk are not systematic but are expressed instead by a more complicated set of responses, the appropriateness of which is determined by details of particular situations.

Andrews (1982) argues that the basic problem is the limited view of the role of government implied by formal approaches based on optimization, particularly the economic efficiency role inherent in CBA logic. Government, he argues, is not simply an instrument for correcting economic markets but an arena in which members of society choose and legitimize, however imperfectly in practice, their collective values. President Lincoln did not sign the Emancipation Proclamation on the basis of its economic efficiency and could not have justified it on that basis had he wished to, nor does society condone murder, theft, perjury, or many other forms of behavior even if their overall economic benefits exceed their costs (U.S. Congress 1979). Although like economic optimization the underlying philosophy is utilitarian, it may require absolute standards as constraints on unacceptable behavior; and its decision methods rationally rely on constitutional concepts and procedures, such as representative government, due process, and equal protection, rather than on economic theory and expert balancing, as a source of decision criteria.

Hapgood (1978) argues that quantitative analysis is incompatible with social norms by describing a confrontation between a parks commissioner and a parent whose child was killed when he hit his head against a diving board at a municipal swimming pool. The parents insisted that the diving boards be removed from the city's pools for the protection of other children. "How many kids are going to have to die before you do the right thing?" they demanded. Hapgood asks: "who could imagine, or would want a commissioner to answer, 'Actually, I've got that right here, the answer is two. My analysts have established that two children would have to die each summer before our net psychosocial quality-of-life figures would tell us to take the boards down.'" Hapgood concludes:

> When most of us look at these analysts all we can or will see are mean-minded and heartless bureaucrats thinking like computers, and we will be demoralized at the very sight. The average person's sensibilities are too crude, too much hostage to the short-run and the tangible, to appreciate the subtle and abstract ways in which "thinking like computers" works to the benefit of everyone. In fact, the damage done to the ethical self-confidence of the citizenry by analysts by being open about risk-benefit would probably cost more than could be gained by using the technique.

According to Howard (1980), though, who quotes Smith (1958),

> There are those who believe that there is something "cold and inhuman" about rational analysis. I believe that to be human is to be reasoning as well as compassionate. My ideal here is Buddha: Perhaps the most striking thing about him, to use the words of J. B. Pratt, was his combination of a cool head and a warm heart, a blend which shielded him from sentimentality on the one hand and indifference on the other. He was undoubtedly one of the great rationalists of all times, resembling in this respect no one as much as Socrates. Every problem that came his way was automatically subjected to the cold, analytical glare of his intellect. First, it would be dissected into its component parts, after which these would be reassembled in logical, architectonic order with their main import laid bare.

Another source of discomfort is that formal approaches explicitly represent uncertainty in the outcomes to decisions, which runs counter to the social decision maker's desire for surety and objectivity. Slovic *et al.* (1975) suggest that many people make decisions by searching for or constructing a good justification, one that minimizes lingering doubts and can be defended no matter what outcome occurs. People accustomed to relying on such justifications may find the explicit recognition of uncertainty in outcomes unsatisfactory. Borch (1968) has observed that corporate managers are often annoyed with consultants who give them probabilities of possible events instead of telling them exactly what will happen. Senator Edmund Muskie has pleaded for "one-armed" scientists who do not respond "on the one hand, the evidence is so, but on the other hand" when asked about health effects of pollutants (David 1975). Just before a blue-ribbon panel of scientists reported that they were "95 percent sure" that cyclamates do not cause cancer, FDA Commissioner Alexander Schmidt said, "I'm looking for a clean bill of health, not a wishy-washy, iffy answer on cyclamates."[80] Rivlin (1971) partly defends administrators who shun uncertainty: "They are afraid to admit that they do not know. And they may be wise. The Office of Economic Opportunity may have told Congress: 'We don't know whether preschool programs will work, or what kind would be best, but we have designed a program to find out.' But, would they then have gotten the money?" Levine (1982) notes, however, that denying uncertainty can sometimes be a dangerous course for public authorities. In her book dealing with toxic waste contamination at Love Canal, she observed that officials "seemed to take refuge in being very firm . . . and in never admitting the uncertainty that was, of course, inherent in the situation for . . . from reassuring residents, such firm behavior made them think that the authorities were not to be trusted . . . 'I don't know' might have in fact been a perfectly correct response given the state of knowledge."

Discomfort with subjectivity in decision making may be because "many people have been led to believe that only 'objective' results are valuable, a belief that can be traced back to our pre-twentieth century scientific views, since it relies on the subject-object dichotomy" (Howard 1980). Decision analysis approaches can be expected to be most prone to raising subjectivity concerns, because they make the subjectivity in decision making explicit. Dyer (1978) notes that government agencies are uncomfortable with this because taking the position that information and values are subjective implies that reasonable bases may exist for disagreeing with agency decisions. Referring to OSHA, Dyer says, "They evidently feel compelled to develop a process of regulatory decision making that has the appearance of objectivity. In the highly political arena in which they must function, this may be a rational position for them to take, but it may not lead to the most socially desirable set of occupational health and safety regulations." Wynne (1984) warns that retaining the objective, tidy view of the scientific basis for regulatory decisions inevitably leads to the erosion of the credibility of science; "the formal public *expectations* of scientific knowledge and the informal, socially constructed reality, means that there are an infinite number of ways of embarrassing scientists in public policy areas by 'exposing' uncertainties in their knowledge."

Another problem with subjective probability judgments unsupported by frequency data is that they may be perceived as weak justifications for decisions and leave experts vulnerable to second-guessing. To illustrate, Slovic *et al.* (1975) suggest imagining the designers of the Edsel explaining in 1961 that their carefully constructed probability assessments indicated that it was likely to be a big seller.

Slovic (1978) notes that availability bias poses a serious barrier to open discussions of risk in the context of formal approaches. Naive observers of an analysis might be startled by the number of identified possible sources of disaster, some of which will be new and surprising to them. To illustrate, Slovic suggests, "Imagine an engineer explaining the basis for the estimated safety of disposing nuclear wastes in a salt mine by outlining the fault tree upon which the estimate was based .... Rather than reassuring the audience, such a presentation might have the opposite effect ('I didn't realize there were that many things that could go wrong')." Thus availability bias may hamper unbiased attempts to discuss low probability hazards without, at the same time, increasing the perceived probability of those hazards. From a statistical standpoint, convincing people that the catastrophe they fear is extremely unlikely is difficult under the best of conditions. Any mishap could be seen as proof of high risk, whereas demonstrating safety would require a massive amount of evidence (Slovic *et al.* 1979a).

A final compatibility issue concerns the technical complexity of formal approaches. Once analyses have been performed, they must be communicated to the various people who have a role in the risk-management process — industrialists, environmentalists, regulators, legislators, and voters. The technical complexity of formal approaches makes this a very difficult job. If people do not see, understand, or believe risk statistics, then distrust, conflict, and ineffective hazard management may well be the result.

Ethical Concerns

Two fundamental questions are central to the design of ethical social decision-making systems: (1) Which principles of ethics should be left to individual choice and which should be universally enforced? (2) Should ethical principles be stated in terms of ends or in terms of means? These questions appear in a variety of forms in the debate over social risk management. For example, "Under what conditions, if any, is someone in society entitled to impose a risk on someone else for the sake of supposed benefit to yet others?" (Highland 1978) and, "Should social risk decisions be the responsibility of a technical elite or approached through the democratic process?" (Fischhoff 1979). The principal source of ethical criticisms is the inability of formal approaches to provide solutions to these sorts of questions.

To be sure, the criterion by which an optimal decision is defined within a given approach may embody an ethical philosophy (e.g., cost-benefit theory and utilitarianism). The judgment, however, as to whether the particular ethical perspective is proper lies beyond the scope of formal approaches. To paraphrase Howard (1980), like mathematics, the logic of decision-making theory is amoral; questions of individual freedom versus government paternalism and end-state versus process ethics would exist even if formal decision-aiding approaches did not. The ethical danger consists of using formal approaches to obscure with technology what are in reality ethical problems.

Ethical problems that have been the source of criticisms of formal approaches include distributional equity, intergenerational effects, and anthropocentricity. These issues and the criticisms they lead to are discussed below.

Conflicts with Distributional Equity

All public policy decisions result in a distribution of benefits and burdens; some gain and others lose from a decision. The question is, how can we be sure that a decision derived from a formal approach will fairly distribute the costs and benefits of policies? This question cannot be answered objectively. To do so would require first an answer to the ethical question "What is a fair and equitable distribution?"

Cost-benefit analysis embodies the strongest ethical assumption and not surprisingly has been most subject to criticisms concerning its distributional impacts. As noted in Chapter 2, the efficiency criterion of CBA has been defended on the basis that it produces the largest "economic pie" which in principle could be distributed in such a way that everyone would be made better off. Redistribution of social profit resulting from cost-benefit decisions, however, rarely takes place in practice. (Even if it did, the redistribution process would not be costless.) Thus Howard (1980) asks, "Is not the adoption of an alternative that benefits some and hurts others not just a case of theft from those who have lost?"

A second argument for the efficiency criterion (also mentioned in Chapter 2) is that social costs are compensated by social benefits at a social level; that is, in the long run and on the average, the victims of a particular technological risk decision will most likely be the beneficiaries of other technological risk decisions of which they did not become victims. Flaws in accepting this as an ethical justification include: (1) risks in fact may not be reciprocal; (2) they may not balance out in the end; and (3) unless there is implicit consent to trade risks, we again have a situation which might be characterized as theft from those who have lost (Highland 1978).

Decision analysis approaches have tended to devote little attention to equity issues, although in theory an equity dimension could be incorporated into a DA value model.[81] Social welfare theory and related approaches adopting a social welfare criterion also provide flexibility for defining and accounting for social preferences over alternative distributions of risk and benefits. The problem is that mechanisms for doing so will still create conflict with the concept of individual rights. Pope (1979) traces this conflict to competing ethical systems:

> One goes back to John Locke. The purpose of government is to guarantee individual rights. Individuals own property and exchange it freely in markets. These transactions are voluntary; buyer and seller must agree on a price. That price is the value of the goods exchanged. One of the rights which is protected is the right not to sell, if agreement is not reached on a price. Society's concern is simply to make sure that transactions are in fact voluntary .... The social concept of value in the United States derives from Jeremy Bentham and utilitarianism .... Bentham was concerned not with maximizing individual rights, but with increasing the social good.

Regardless of the approach, once a value is assigned to an outcome, that outcome is placed on the market. Further, as Bogen (1980) observes, assigning a value of life not only places life on the market, it places all lives on the market:

> ... for some people, the [individual's right to avoid] increased health risks due to intentionally imposed, involuntarily assumed risks ... may have no selling price, or in effect, an infinite price ... such an infinite selling price is not intended to represent the value of a health loss per se, but rather is intended to represent the value of not compromising the Lockean principle that an individual has a right not to sell.

Even if one were willing to accept the Benthian view, the implications of the widely used willingness-to-pay measure of individual preferences may not lead to decisions that are to everyone's liking. An example is provided by the criticism that the use of market prices and willingness to pay sanctifies the existing distribution of wealth. Because people in their early sixties are generally wealthier than those in their thirties, willingness to pay may put relatively too much general tax money into programs that prolong the lives of people past middle age and thus deemphasize lifesaving programs for young adults with low incomes, large family responsibilities, and thus low average willingness to pay (Rhoads 1978). Illich (1977) has popularized the view that our willingness to pay may be too high, that we freely turn too many of our resources over to doctors and let them decide when and how we will die. Krieger (1970) has argued that basing social decisions on willingness to pay is likely to lead to less investment in protecting the environment because current federal environmental policy is such that the "rich get richer and the poor get poorer":

Rich people have the background and leisure to cultivate a taste in beautiful environments, and only they have the money to live in or near them. Rising property values in protected areas drive the poor out. If the pleasures of the poor were measured equally with those of the rich, then quicker than you can say "cost-benefit analysis," there would be parking lots, condominiums, and plastic trees.

Intergenerational Concerns

A special case of the distributional issue occurs in situations in which the possible consequences of a risk extend across generations. As noted in Chapter 2, the typical procedures for comparing future with current impacts is to discount future outcomes:

. . . concern for the future is usually assumed to diminish over time. We care more for our children than for their children, and so on. Consumers' sovereignty, then, dictates that the social welfare function should include a rate of time preference, a discount factor that counts consequence for less as they occur further into the future (MacLean 1979).

Discounting can produce uncomfortable implications, however, especially if the time period spans several generations. For example, if the discount rate were 5 percent, one case of poisoning today would be valued roughly equal to more than 10 billion cases 500 years from now. Thus, applying any significant discount rate produces strong implications for decisions yielding current economic costs and effects on second or third generations — for example, the depletion of supplies of hydrocarbons, the buildup of atmospheric carbon dioxide, and the disposal of nuclear waste.

Confronted with the implications of discounting, economists have presented a variety of supporting arguments, such as the opportunity costs of investment, improving living standards, technological progress, and the uncertainty of long-range planning. The fact remains, however, that many feel that these explanations fail to take future generations seriously enough. Howard (1980) describes people's reaction to his "galactic bargain":

Suppose that we were approached by galactic travelers with very advanced technology who offer us the following opportunity. They guarantee to double the standard of living of everyone on earth for one thousand years. At the end of that time they will arrange that humans will become painlessly sterile; when the last human is gone, the galactic travelers will occupy the planet. The question I now ask each person is whether they would accept this deal. I have not yet had a positive response.

Two reasons have been given to explain the breakdown in discounting: (1) equating one death today with a great many in the future may be inappropriate, because the opportunities may not exist to save a great many future lives at a reasonable cost; and (2) other social values come into play when we consider major cataclysms, such as our concern for the preservation of human civilization itself, apart from the individual lives saved (Raiffa *et al.* 1977). These considerations have led to proposals that for decisions with very long-term implications, a variable discount rate be used, perhaps declining with time to better reflect the relative values for future risks and benefits (English 1978).

Regardless of whether or how future risks and benefits are discounted, intergenerational impacts create an ethical issue in that future generations are not here to "vote" on the decisions that will affect their lives and life expectancies. Because of the potential irreversibility of consequences, it will not always be possible for each generation to revise ongoing programs to their own liking. The exclusion of preferences other than those of current participants in the decision-making process is a special case of the more general criticism of anthropocentricity.

Promotion of Anthropocentric Values

Several authors criticize formal decision-aiding approaches for promoting anthropocentric value systems based on maximizing self interest. Kelman (1980a), for example, argues against placing "instrumental" values on things that have an important "intrinsic" value. Friendship, for instance, has been given an intrinsic value by society. It would be considered gauche to offer a friend payment for a favor because it would imply that you had merely an instrumental value for the friendship. Because the values assigned by decision-aiding approaches are instrumental values, they ignore values that society deems important. For example, assigning a finite cost to pollution implies that polluting behavior can be a worthy action which should be encouraged, provided that it contributes enough to society. This conclusion, Kelman believes, ignores the motive of the behavior and loses some of the negative value that many would place on such activity.

Kelman also warns that repeated use of a decision calculus based on self-interest will have the effect of reinforcing self-interest as an index of value at the expense of altruism and spontaneity. Likewise, placing an explicit value on something that is rightfully "not for sale" tends to erode its special value. When something is priced, its perceived value is constantly being assessed, and a standing invitation exists to reconsider that original price. If people were faced with questions such as "how much would you sell your vote for if you could?" Kelman reasons the perceived value of the right to vote would be quickly eroded.

Similarly, while Tribe (1974) admits that analyses can accommodate whatever values individuals are capable of discerning, he questions whether "an institutional system or an analytic technique that treats all such values as manifestations of individual human preference will prove satisfactory." To illustrate, Tribe uses the example of plastic trees planted in a Los Angeles median strip:

> To insist on the superiority of natural trees in the teeth of a convincing demonstration that plastic ones would equally well serve human purposes may seem irrational. Yet the tendency to balk at the result of the analysis remains. There is a suspicion that some crucial perspective has been omitted from consideration, that the conclusion is as much a result of myopia as logic.

The omission, Tribe explains, is that "an inchoate sense of obligation toward natural objects is flattened into an aspect of self-interest." In other words, an ethical distinction exists between exploiting the environment as a resource in the human interest and managing it as a natural resource in its own right.

The ethical problem with formal approaches, in these critics' views, is that by articulating environmental goals wholly in economic terms, a system of discourse is created that structures human thought and feeling in ways that in the long run may erode the very sense of obligation to nature that provided the impetus for concern for the environment in the first place. Tribe would prefer that man increasingly come to regard nature as having a right to exist independent of man's frame of reference. "What is crucial to recognize is that the human capacity for empathy and identification is not static; the very process of recognizing rights in those higher vertebrates with whom we can already empathize could well pave the way for still further extensions as we move upward along the spiral of moral evolution" (Tribe 1974).

Stokey and Zeckhauser (1978) counter the criticism of anthropocentricity. While admitting that, considered abstractly, the rights of nonhuman entities may seem a valid criterion for policy choice, they point out that there is ample pragmatic support for an anthropocentric approach. All philosophical protestations to the contrary, unless human beings care about the rights of nature, nature will be injured: "[Nature's] rights are meaningless unless championed by people; neither the redwoods nor the bluebirds can speak for themselves. If this judgment strikes you as unduly hardnosed, look at the other side of the coin. How many voices are raised on behalf of that vanishing species, the small pox virus? And who speaks for the boll weevil?"

Formal approaches do not preclude the recognition of natural rights. The fault, Tribe agrees, derives not from a logical flaw in analysis but rather from "the ideological bias of the system in which such analysis is embedded, a system that has come to treat human will and its wants as the center around which reason as calculation must revolve." As a practical matter, as the environmental movement has progressed, proponents of environmental protection have been the subject of increasing pressures to make the worth of the values they expose explicit. Inevitably, this causes a move away from justifying environmentalism as an ethical tenet toward a utilitarian index for the worth of preserving the environment.

The primary ethical dilemma, as Bennett (1981) observes, is that "regardless of one's own philosophical standpoint, . . . for man to exist at all he must disrupt his environment to some degree. In practice bioethics is therefore an issue of to what extent we are morally justified in manipulating the environment for our own ends." Like many of the other criticisms that strike at the foundations of formal approaches, anthropocentricity has been associated with analysis not so much because it is a concern associated uniquely with formal approaches but because the clarity that formal logic casts on decision making tends to highlight the difficulty of our social decisions and our fundamental limitations as decision makers.

Chapter 4

Comparative Evaluation: Summary and Conclusions

As the reader has no doubt gathered from the discussion to this point, there exists no currently available formal decision-aiding approach for social risk decisions that is free from criticism. Furthermore, available approaches have different strengths and weaknesses. It is not possible, therefore, to identify any specific approach as in any absolute sense the "best."

While it is not the intent of this book to recommend a specific approach for a given type of social risk decision, an objective is to provide a rudimentary framework and supporting information for evaluating and selecting among approaches. The proposed framework consists of three components: (1) an identification and characterization of essential features of a risk problem that may be relevant to the decision to apply a formal approach, (2) a means of characterizing and distinguishing among available approaches, and (3) criteria for evaluating the use of an approach in a given situation. Chapters 1 and 2 of this book have discussed features and characteristics that distinguish risk problems and approaches. Although to an extent the various approaches were compared in terms of their susceptibility to criticisms in Chapter 3, we have not yet discussed explicit criteria for evaluating decision-aiding approaches. This last chapter discusses briefly such evaluation criteria and then describes the conceptual framework proposed for use in evaluating and selecting alternative approaches. Finally, some general observations and conclusions are summarized.

Criteria for Evaluating Decision-Aiding Approaches

Establishing evaluation criteria requires an understanding of the role or function of the thing being evaluated. For example, the criterion applied when selecting a pair of ski boots is very different from that used when purchasing hiking shoes, and someone who mistakenly thought ski boots were for hiking would have a very different opinion of their value than would a skier. Thus identifying criteria for evaluating decision-aiding approaches requires an understanding of the role of an approach within the decision process.

Role and Professed Value of Approaches

Many of the criticisms leveled at the use of formal decision-aiding approaches seem to assume that analysts view the approach as a rigid rule to supplant or replace the decision-making process. A few analysts do take the prescriptive nature of their approaches literally, believing, for example, that decision analysis or cost-benefit analysis can encompass all important value considerations while remaining neutral in deriving the implications of information. Generally, analysts make a much more modest claim: they do not recommend decisions; they simply present relevant information in useful ways that will aid decision makers. "Formal analysis is meant to serve as an aid to the decision maker and not as a substitute for him" (Keeney and Raiffa 1972). "We believe that the basic function of formal analysis should be to help policymakers gain insight into complex issues" (Raiffa *et al.* 1977). "The important element is not the bottom line, but the process of reaching it" (Slovic and Fischhoff 1982). "Even the most purportedly scientifically objective assessment is useful only as another input along with data from myriad other sources, interested and disinterested, informed and uninformed, rational and irrational" (Green 1980).[82]

The tendency for critics of decision-aiding approaches to confuse the approach with the actual process of arriving at a decision may be partly a result of terminology. As Majone (1980) observes, the role of analysis in assisting choice is generally discussed in a language that obliterates the difference between analysis and decision making. "The same categories — objectives, alternatives, prediction, control, optimization, and so on — are used in both cases, thus setting up a one-to-one correspondence between the stages of the decision process and the stages of the analytical process." In fact, analysis operates on a simplified model of the decision situation and not on the the real-world problem itself. Expressing the results in terms of conclusions concerning the real situation requires interpretation. Should the model be confused with reality, then the outputs of analysis will be assigned a significance beyond that which is justified.

Hammond *et al.* (1980) provide a summary of specific claims made for the value of decision-aiding approaches. Most of the claims fall into the category "aids clear thinking." For example, theoretical and applications literature claim that formal approaches promote a policy focus that helps "shift the decision maker's attention from specific actions to the values these actions serve," "helps separate judgments of probabilities from those of values," "forces full consideration of all consequences of a given action," "identifies what is important for making decisions and what is not important," and so forth. In addition, analysts argue that separating information from preference inputs "permits elected officials to function exclusively as policymakers and scientists to function exclusively as scientists," tends to produce insights that "aid in generating creative alternatives," and "improves communication" in such a way as to permit decision makers to "isolate and quantify differences of opinion among experts." It is also claimed that decision-aiding approaches can help individuals who have already made up their mind by supporting advocacy or defending a particular viewpoint. For example analysis "helps the decision maker defend and communicate his decision . . . makes the point that the decision was not frivolous and that all factors were considered."

Viewing a formal approach as an aid to decision making rather than as a way of making decisions emphasizes the importance of the analyst. With this view, the analyst largely plays the role of a facilitator, and the overall objective of a formal approach may be regarded as that of providing a framework for investigation that is likely to produce information of useful value in the decision process.

Two important implications may be drawn from this perspective. First, the skill and capabilities of the analyst are of prime importance in determining the success achieved from the use of a formal approach. Although decision theories provide a logic for decision making, they offer only limited guidance to the analyst who must choose among various procedures, methods of analysis, assumptions, and data. Relying on an understanding of the decision problem, theory, and basic concepts of analysis (e.g., quantification, iteration, and modeling — see chapter 2), the analyst must piece together an approach. Majone (1980) likens analysis to craft and notes that in both cases successful performance depends on an intimate knowledge of materials, tools, and processes and on a highly personal relationship between agent and task.

The second implication of viewing analysis as an aid rather than a replacement for the decision process is the shift from evaluation by results to process. Because the impact of a formal approach on the decision actually reached depends on a very complex, indirect, and poorly understood interaction, the process by which the approach is conducted is intrinsically important in determining success. Majone (1980) notes that this is typical of professional work. In the case of physicians, for example, "much, perhaps most, of what a physician does must be categorized as process, and process not even calculated to affect outcome."

The more general view of the role of formal approaches as an aid rather than a way of making decisions argues for a pragmatic perspective in identifying evaluation criteria. Such a perspective is taken in what follows.

Evaluation Criteria

Criteria for evaluating decision-aiding approaches may be categorized as internal or external (Majone and Quade 1980). Internal evaluation criteria involve considerations that are within the domain of analysis (i.e., mathematics) and relate to what Ford *et al.* (1979) have described as the "quality of the analysis" and the logical and technical "correctness" of the approach. External criteria are imposed by considerations normally thought to be outside the disciplines of analysis, particularly the desires and constraints imposed by decision makers and the public and the objective limitations of time and resources. Although both types of criteria are useful, one or the other often seems to be improperly elevated to primary importance. Analysts, for example, often emphasize high technical quality, assuming that a technically sound analysis is sufficient to guarantee its practical effectiveness. Examples of analyses of high technical quality that had no visible impact on policy demonstrate the error in this view. Sponsors of analyses, on the other hand, tend to judge the value of the approach solely by the favorable or unfavorable impressions of those whose opinions affect them. A comprehensive and balanced evaluation should include both types of considerations.

Logical soundness, completeness, and accuracy represent the major categories of internal criteria for evaluating decision-aiding approaches. Collectively, these account for the various theoretical and operational criticisms presented in Chapter 3 — criticisms claiming the impossibility of finding a socially optimal decision rule; destruction through decomposition; inherent incompleteness; inability to account for the costs of irreversibility; omissions and inaccuracies; difficulties of measuring benefits and costs; assessment, modeling, and analysis bias; propensity toward interference by the analyst; susceptibility to manipulation; and susceptibility to misuse and misinterpretation.

Whether or not an approach is logically sound is determined by the degree to which it can be justified in terms of theory and whether actual applications are likely to violate the fundamental assumptions of that theory. For an approach to be logically sound, conclusions must be derived via a defensible decision rule. Such a rule should be sensitive to the various aspects of the problem, reliable in the sense that independent applications to the same problem would produce the same results, justifiable in terms of theoretical arguments or empirical evidence, and unbiased in its recommendations by not giving undue weight to any specific interest or consideration (Fischhoff *et al.* 1981b). Considerations for assessing the logical soundness of approach would include whether the approach adheres to or violates basic principles of logic, whether differences in the quality or strength of evidence are properly reflected, the degree to which tests are undertaken to establish the credibility of the models used, and so forth.

Completeness addresses both whether in theory the method can account for all relevant problem aspects and whether, because of difficulties encountered in practice, the analyst who uses the approach is likely to omit certain information or considerations because they are hard to accommodate. Considerations addressed within this category include the degree to which multiple concerns are accounted for (environmental "intangibles," distributional effects, varying degrees of impacts), labile and conflicting social values are acknowledged, uncertainty in data and judgment are considered or assumed away, long-run impacts are represented, and the costs of irreversibility are reflected.

Accuracy of analysis includes both the precision or level of confidence attributable to results due to sensitivity to untested or untestable assumptions and the potential for biases or omissions resulting from errors of elicitation and estimation. Errors of elicitation can be the result of psychological factors, such as fatigue, boredom, or inattention; errors of estimation are associated with imprecision of procedures from a theoretical standpoint (Hammond *et al.* 1980).

External criteria are imposed by the realities of the decision-making process. They include the institutional and ethical criticisms of Chapter 3; namely, incompatibility with principles of democratic decision making, incompatibility with constitutional and legal principles, incompatibility with organizational structure and processes, incompatibility with social norms, conflicts with distributional equity, and promotion of anthropocentric values. The major categories of external criteria are practicality and acceptability.

Practicality means that the analysis can be conducted in the real-world, problem-solving environment using available resources and information. Basically, this requires that a pool of expertise be available to implement the approach and that the costs, time, and effort required are appropriate and compatible with the constraints of the situation. Practicality is obviously influenced by the availability of the required inputs and the extent to which the approach is flexible in its ability to use different types of input data. Availability of people must be considered when evaluating input requirements — applying an approach is a labor-intensive exercise involving both substantive experts who understand the risk and normative experts who specialize in the decision-aiding methodology. Also, depending on the approach, considerable time and information demands may be made on decision makers. The crisis atmosphere of most decision making and the reluctance of decision makers to spend funds for the uncertain benefits of good advice, which at best only increases one's chances of obtaining a favorable outcome, generally result in analysts being provided with less time and resources than they would like to conduct their analyses.

Acceptability largely relates to the desires of decision makers, public perception of the approach, and compatibility with existing institutions. The confidence, familiarity, and experience of decision makers with the approach will obviously be of major importance in determining acceptability. Characteristics influencing public perception would include understandability, the degree to which the concerns of the various participants in the process are responsibly and demonstrably addressed, and compatibility with soical norms, such as the degree to which the approach is viewed as fair, legal, and rational. The extent to which value judgments are clear and the evaluation process is transparent will affect public perception in these areas. Although clarity may be essential for acceptability, if an approach produces clear but unwanted conclusions, the offended parties may choose to discredit the approach rather than contest the assumptions leading to the undesired conclusion (Fischhoff *et al.* 1981b). Furthermore, some decision makers may prefer an approach that cloaks their decisions in ambiguity over one that is logically clear. They may feel uncomfortable with what others consider to be elements of a good analysis, such as extended time horizons, explicitly acknowledged uncertainty, and extensive reliance on outsiders.

Notice that the various evaluation criteria are not independent. Weaknesses in some areas (e.g., logical soundness) are likely to preclude strengths in others (e.g., acceptability). Furthermore, some of the goals may be in conflict. It may be easier to derive a logically sound decision rule by leaving out certain awkward issues and thereby sacrificing completeness. If no approach does, or even can, satisfy all of the criteria and if their respective strengths and weaknesses lie in different areas, then the choice of an approach will require tradeoffs.

Comparative Evaluation

The various types of decision-aiding approaches have already been compared in terms of their underlying rationales, standard procedures, and susceptibility to criticisms. A useful way of conceptualizing these comparisons is in the form of a matrix that evaluates the potential application of different types of approaches to different kinds of risk problems. Table 8 shows a matrix in which risk problems with various characteristics are represented by rows (using the risk problem taxonomy outlined in Table 4 of Chapter 1), and different categories of decisonaiding approaches are represented by columns. Table 6 (Chapter 2) summarizes the basic characteristics that might be associated with the labels used in Table 8. Because there is disagreement over the meaning of the various labels used to describe approaches (e.g., DA, social DA, supra DA, CBA, Paretian CBA), clear definitions must be established for the approach categories used in such an evaluation scheme. With such definitions, as shown in Figure 21, the approach categories might be regarded as providing a mutually exclusive, collectively exhaustive partition of the space of decision-making rationales. Comparing Figure 21 with Figure 20 shows that such attempts at defining and distinguishing approach cate-

Table 8. Conceptual framework for evaluating approaches

Risk problem characteristic*	CBA	Decision-maker CBA	Social DA	Supra DA	Clinical DA	Applied SWT
1.1.1.1						
1.1.1.2						
1.1.2.1						
1.1.2.2						
1.1.2.3						
1.1.3.1						
1.1.3.2						
1.1.4.1						
1.1.4.2						
1.1.5.1						
1.1.5.2						
1.1.5.3						
1.1.5.4						
1.1.5.5						
1.1.6.1						
1.1.6.2						
1.1.7.1						
1.1.7.2						
1.2.1.1						
1.2.1.2						
1.2.2.1						
1.2.2.2						
1.2.3.1						
1.2.3.2						
1.2.3.3						
1.2.4.1						
1.2.4.2						
1.2.5.1						
1.2.5.2						
1.2.6.1						
1.2.6.2						
1.3.1.1						
1.3.1.2						
1.3.2.1						
1.3.2.2						
1.3.3.1						
1.3.3.2						
1.3.3.3						
1.4.1.1						
1.4.1.2						
1.4.2.1						
1.4.2.2						
1.4.3.1						
1.4.3.2						
1.4.3.3						
1.4.4.1						
1.4.4.2						
1.4.4.3						
1.4.5.1						
1.4.5.2						

Evaluation Criteria
1. Logical soundness
2. Completeness
3. Accuracy
4. Practicality
5. Acceptability

* See Table 4

Risk problem characteristic*	CBA	Decision-maker CBA	Social DA	Supra DA	Clinical DA	Applied SWT
2.1.1.1						
2.1.1.2						
2.1.1.3						
2.1.2.1						
2.1.2.2						
2.1.3.1						
2.1.3.2						
2.1.4.1						
2.1.4.2						
2.2.1.1						
2.2.1.2						
2.2.2.1						
2.2.2.2						
2.2.3.1						
2.2.3.2						
2.2.3.3						
2.2.4.1						
2.2.4.2						
2.2.5.1						
2.2.5.2						
2.2.5.3						
2.2.6.1						
2.2.6.2						
2.2.6.3						
2.2.7.1						
2.2.7.2						
2.3.1.1						
2.3.1.2						
2.3.2.1						
2.3.2.2						
2.3.3.1						
2.3.3.2						
3.1.1.1						
3.1.1.2						
3.1.2.1						
3.1.3.1						
3.1.3.2						
3.2.1.1						
3.2.1.2						
3.2.1.3						
3.2.2.1						
3.2.2.2						
3.2.3.1						
3.2.3.2						
3.3.1.1						
3.3.1.2						
3.3.2.1						
3.3.2.2						
3.3.3.1						
3.3.3.2						

Evaluation Criteria
1. Logical soundness
2. Completeness
3. Accuracy
4. Practicality
5. Acceptability

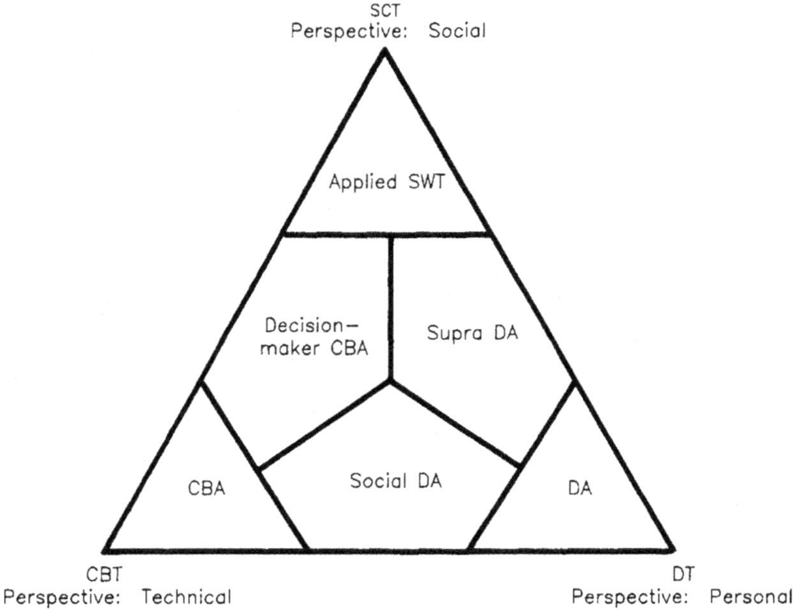

Fig. 21. Categories of decision-aiding approaches defined to span the space of underlying theories and perspectives.

gories are likely to produce some inconsistencies with the way in which the labels are currently being used.

The columns in Table 8 do not, of course, uniquely distinguish an approach. As illustrated in the analytic taxonomy of Table 7 in Chapter 2, there are a very large number of analytic choices that must be made in selecting procedures, each of which also differentiates an approach. Even if these more detailed choices are ignored and the comparative evaluation is conducted at the level of approach types, it would be difficult to present an exhaustive evaluation in the format of Table 8. The matrix contains approximately one hundred distinct decision situations, and there are six categories of approaches. That results in roughly 600 cells in the matrix, each of which represents the potential application of a specific type of approach to a risk problem possessing certain characteristics. Although evaluating each combination of approach and risk problem characteristic according to each of the five evaluation criteria listed above would require 3,000 separate evaluations, it is conceptually straightforward to reorganize the comparisons and criticisms of Chapters 2 and 3 to provide "first-cut" evaluations for any cell. To illustrate, the sections below abstract from previous sections summary evaluations for the various approach categories (assuming definitions consistent with Table 6 and Figure 21).[83]

Logical Soundness

Cost-benefit analysis, decision analysis, and applied social welfare theory represent the operationalization of cost-benefit theory, decision theory, and social choice theory; they appeal, therefore, to these theories for their logical justification. As described in Chapter 2, CBT, DT, and SCT are each coherent, consistent theories that describe how information and values should be integrated to identify alternatives that are in some sense "best." CBT is founded on the concept of economic efficiency and reflects a technical, objective worldview. DT is based on axioms of individual choice and reflects an individualistic, subjective worldview. SCT is based on the concept of a social welfare function and reflects a worldview centered on equity and process. Decision-maker CBA, social DA, and supra DA are hybrid approaches that derive their perspectives and procedures from more than one of the decision-making theories. Decision-maker CBA appeals to both CBT and SCT for its rationale and as a result mixes both equity and efficiency criteria. Social DA is based on both CBT and DT and contains some subjective versus objective inconsistencies in its perspective. Supra DA relies on both DT and SCT. Consequently it tends to mix individual with group decision-making perspectives.

A critical determinant of the logical soundness attributed to the various approaches is the persuasiveness of their underlying behavioral assumptions. To the extent that market prices do not reflect preferences, people reject the axioms of decision theory, or equity assumptions cannot be justified, the logical appeal of the approaches is diminished. Thus CBA will be less attractive for situations in which there are considerable differences in the quality of relevant information; DA is less useful if there is no decision maker willing to provide or delegate responsibility for providing subjective information and preference judgments; and SWT loses its appeal in situations where a comparison of the strengths of preferences of impacted parties is not regarded as the central issue for the decision. Because all approaches rely on decomposition, enumeration, and quantification, they are all subject to charges that information and values cannot be effectively separated, that all possible consequences cannot be anticipated, and that important factors cannot be quantified.

Completeness

Formal approaches embody few restrictive assumptions about how problems are to be defined. In each case, a decision model is constructed whose structure and content is largely determined by the substantive information provided to the analyst by the client, decision maker, or other "experts." The completeness of any analysis will thus be limited by the breadth and depth of vision possessed or achieved by those experts and the creativity and resourcefulness of the analyst. Furthermore, because the sources and perspectives identified for providing information differ among the approaches, the nature of the omissions that occur is likely to depend on the approach that is applied.

Obviously, approaches are less effective in situations were the factors they tend to omit are crucial. Thus, CBA will be less effective in situations where little or no data are available for quantifying important uncertainties or where important outcomes exist that do not have immediate, tangible economic implications. DA will be less effective if decision makers are likely to define the problem in habitual ways that overlook options or overemphasize certain factors. Social DA, because of its development of an extrapersonal decision model through the participation of a large number of substantive experts, can be expected to provide a more complete accounting of available information. Similarly, supra DA and SWT, with their assessment and integration of the preferences of stakeholders, are likely to produce a more complete representation of social preferences. For situations where distributional equity is an important issue, decision-maker CBA or another approach other than a standard CBA will permit explicit accounting of considerations beyond efficiency.

In general, DA approaches allow a more comprehensive accounting of factors than do CBA approaches. In principle, at least, any concept in the problem definition may be translated into operational terms under DA. Social DA approaches are especially flexible: they allow economic and statistical estimates to be used when they are available, while permitting subjective judgment as a source when they are not.

Accuracy

As discussed in Chapter 3, inaccuracies in approach applications may result from a variety of sources, including omissions, the difficulty of measuring costs and benefits, biases, and manipulation. CBA approaches will be most accurate in assessing well-defined projects, but they are less useful in situations where there are significant market distortions (due to taxes or externalities) or when information or preferences are changing rapidly. DA approaches, because they rely heavily on the assessment of subjective judgments, may be expected to yield less accurate results in situations where respondents have difficulty providing the required assessments (e.g., when experts do not have coherent, articulated views on a topic). SWT and supra DA approaches, which require measures of strengths of preferences of stakeholders, will be subject to inaccuracies in situations where strong motivational biases exist for stakeholders to adopt a strategy of misrepresenting their preferences.

Iteration, as described in Chapter 2, is an important component of analysis. Since errors are to be expected in complex problems, it is important that analyses be subjected to peer review. Weaknesses should be corrected and improvements incorporated into the analysis. Too often the relatively small additional resources required to correct the inaccuracies that a careful review would reveal are not made available. Furthermore, there is a natural tendency for analysts and clients stubbornly to defend their assumptions against all attacks rather than for them to view critiques as indications that their approach has successfully illuminated the problem and focused debate.

Practicality

Despite their logical appeal, neither CBA, DA, nor SWT was specifically designed for application to social decisions involving risk. Consequently, each shows some strain (in the form of conceptual difficulties or in resorting to ad hoc procedures) when forced to accommodate aspects of risk problems that are unlike those of the traditional problems for which these approaches originally evolved. For example, CBA is best suited for go/no-go decisions with immediate, predictable consequences and for which responsive markets exist and consumers are well informed. When CBA is used for other kinds of problems, the data requirements necessitated by the objective perspective of CBA create significant problems for estimating the impacts of proposed changes and placing dollar values on physical measures of health and other environmental improvements (because determining how much people are willing to pay for those improvements is difficult). The inputs for DA approaches, are, in principle, more easily obtained, but DA presumes the existence of a decision maker empowered and willing to speak on behalf of society. Assuming such individuals can be found, considerable time and information demands may be placed on them. Although DA approaches are much more flexible in their ability to handle difficult measurement problems (e.g., value of a life), workable and validated procedures are not available for addressing all problem aspects (e.g., joint time-risk preferences and assessing future values).

An important characteristic of formal approaches is their flexibility in level of detail and time and resource requirements. Uncompromising, thorough approaches can be expected to be the most complete and accurate. However, a full-blown approach can also be expensive and time consuming. Fast, limited analyses not only place greater reliance and demands on experts, they are also likely to require the most highly skilled analysts.

Acceptability

Critical acceptability issues include transfer of power and the explicitness of formal approaches. Critics worry that formal approaches will place too much societal decision-making power in the hands of a technical elite comprised of those who perform analyses and interact with the analyst. CBA provides the most constraining decision-making rule by requiring that decisions be based on a balancing of aggregate costs and benefits. Thus, CBA provides little opportunity for stakeholders to contribute to the analysis, except perhaps in defining the problem (e.g., identifying alternatives). At the same time, however, CBA avoids the necessity of a decision maker providing subjective value judgments. CBA is thus likely to appeal to some regulators because it appears to them as a value-free guide to decision making. In reality, of course, CBA embodies strong value judgments.

Decision analysis and other approaches that permit subjective inputs are more amenable to public participation because anyone's perspective can be represented. On the other hand, clinical DA, supra DA, SWT, and decision-maker CBA all require explicit, potentially controversial value comparisons to be established. Social DA demands that equivalent dollar values be assigned to human health and other nonmonetary outcomes and requires unambiguous statements from experts regarding scientific uncertainties. It also highlights the risks and uncertainties associated with alternatives and therefore may appear less attractive in situations in which decision makers are reluctant to admit the subjectivity or uncertainty inherent in their decisions. Furthermore, such approaches may be more prone to raise suspicions that analysts are playing sophisticated numbers game.

In general, DA approaches will be most acceptable in situations as an aid to individual decision makers concerned with achieving rationality and consistency in decisions for which they have complete authority. High-level agency decisions that are less subject to public scrutiny and that produce important consequences — for example, the setting of programmatic priorities — are especially suitable for DA. In these cases decision makers will appreciate the ability to incorporate sensitive information and personal value judgments, such as estimates of the future actions of opponents, and preferences for the distribution of benefits and costs to various identifiable groups. In situations in which the analysis is to be used to convince others, public officials may need to be intentionally vague on controversial issues in order to avoid criticism and retain sufficient room for maneuvering. Under these circumstances the less personal perspective adopted by social DA and CBA may be more acceptable.

Formal approaches have already established a niche in most institutions. Regulatory agencies, industry, labor unions, professional organizations, and consumer groups have all found it useful to commission at least an occasional analysis. CBA seems to have achieved a slightly broader acceptance to date, although this may simply reflect its seniority to DA and other approaches.

Selecting a Decision-Aiding Approach

Although an evaluation of the various approach categories in terms of the five criteria may provide some idea of how well each currently does in practice from various perspectives, it does not tell us which to choose. None of the options appears to surpass the others in all respects. Thus the criteria that are most important in a particular situation must be determined. The most appropriate approach for one problem might be rejected in another situation for which its particular strengths (e.g., completeness) are less essential and its weaknesses (e.g., political acceptability) more so.

With current practice, the analyst rarely treats the selection or the design of an approach for a specific risk problem as a strategic decision. Instead, analysts tend to select the approaches with which they are most familiar and then tactically modify the procedures and methods of analysis as required to meet the technical requirements of the problem. Clearly a more deliberate consideration of the options available is needed. A conceptual framework for approach evaluation and selection such as that represented in Table 8 may be of value in this respect.

Thus the first step to approach selection is problem characterization. The analyst must discern the technical characteristics of the problem that influence the difficulty of conducting the analysis using different approaches, understand the aspects of the decision process that influence the acceptability and applicability of approaches, and appreciate the perceptual characteristics that will influence people's reactions to the likely outputs of the analysis. In other words, the risk problem should be located in the relevant technical, organizational/process, and perceptual spaces. Obviously, relating the problem under consideration to other risk problems has benefits beyond approach selection. It should help in the broader problem of designing strategies for risk management, including, for example, the identification of effective means for educating the public and building trust and mutual respect among stakeholders.

The second step is to consider alternative approaches. The spectrum of available procedures and techniques is much wider than the options that are likely to occur at first glance. Analysts and sponsors of formal analysis must resist the temptation to choose the approaches with which they are most familiar. The selection and development (and ultimately the interpretation of the results) of an approach should be conducted with attention given to specific evaluation criteria, including logical soundness, completeness, accuracy, practicality, and acceptability. Stated differently, the analyst should consider explicitly the strengths and weaknesses of various approach types as they might be entered in the relevant rows of Table 8 and deliberately select analytic procedures through conscious choices such as those shown in Table 6. Such systematic consideration in the design of an approach is likely to produce a significant improvement in the quality of analysis for social risk decisions.

Some simple rules-of-thumb and practical advice for analyzing risk decisions can easily be deduced from viewing approach selection as a conscious choice to be made on the basis of problem characteristics. Although these may sound trivial, unsuccessful applications can often be traced to the analysts' failure to heed one or more of these simple rules. First, it is essential that analysts develop a thorough understanding of the risk problem before designing the analysis. Without such an understanding, much time can be wasted producing inaccurate abstractions of what the system is like. Second, analytic resources should be devoted to the aspects of the problem that make it difficult for decision makers. Whether the approach places more emphasis on procedures for representing values or knowledge should depend on whether the principal source of difficulty is estimating what the outcomes of the decision will be or analyzing preferences over those outcomes. Third, the approach should utilize the "best" available information for the decision. This might be directly relevant empirical data, indirectly relevant data, the cause-effect theories of experts, or direct subjective judgments. Fourth, the threats to good analysis raised by critics of formal approaches should be countered by good analytic procedures. Thus, for example, if the novelty of the risk causes concern that pathways to disaster will be overlooked, then extra attention should be devoted to augmenting memory and imagination through creating prompts such as lists of archetypal or historical accident scenarios which might serve to stimulate the identification of all of the components that need to be considered. Fifth, producing an approach that is most valuable may require trading off completeness for defensibility, accuracy for simplicity, or empirically-based objectivity for relevance. Finally, because of the inherent subjectivity and judgment required in all approaches, the assumptions made in the analysis should be clearly communicated to decision makers and other users of analysis. Analysts should routinely identify each area of inference where scientific uncertainty is confronted, and they should state the analytic choices made in each area. In this way analysis can routinely be subjected to both scientific and political scrutiny.

Concluding Observations and Comments

The metadecision problem of deciding among decision-making approaches is difficult because the options are not directly comparable: each approach embodies an alternative concept of rational decision making. If applied competently, each does best what it sets out to do. Rather than attempt to solve the metaphysical question "What is the best definition of rationality?" we have chosen to ask "Which approach best serves the needs of a specific problem?" To help in this respect, five evaluation criteria have been presented: logical soundness, completeness, accuracy, practicality, and acceptability. Determining the relative importance of these criteria in the context of a given social risk problem underlies the choice of a decision-aiding approach.

Many of the results of this research are directly relevant to an even broader and more hotly debated question, namely, the usefulness of decision-aiding approaches in sociotechnological risk problems. Sharply different views have been expressed on this issue: to some, analysis should be avoided; to others, it should be embraced regardless of circumstances. Clearly, the value of a decision-aiding approach depends critically on a wide range of factors, including the technical aspects of the risk problem, the decision-making institutions and people involved, and the characteristics of the approach applied.

Decision-aiding approaches may indeed represent the application of science to decisions; but the process of applying an approach is too ill-defined, too poorly understood, and depends too much on the analyst and other participants to qualify as a traditional science. Despite their logical appeal, application of formal approaches often have a more ad hoc flavor than the underlying theory would suggest. Critics question whether all relevant considerations can be accommodated and worry about power being concentrated in an intellectual elite. To turn society's difficult decisions involving risk over to analysts convinced of the superiority of formal analysis not only would run counter to the basic principles on which our society is based but also would result almost certainly in less equitable, efficient, and accceptable decisions.

An important factor in judging formal approaches, however, is to distinguish criticisms of the tool itself from those addressing its adequacy or effectiveness for a specific use. In most, perhaps all, situations a decision-aiding approach should be regarded as a process for deepening understanding and clarifying issues. It is not appropriate for use as a formal device for making decisions. This is not to say that someone looking for a hard and fast rule for making social decisions involving risk could do better than to use one of the decision-aiding approaches discussed in this book, but rather that no formal rule exists that does not have serious deficiencies. Most especially, to argue that social decisions should be made in such a way that people do not bear technological risks makes no sense whatsoever. Society never has been and never will be made risk-free. All strategies proposed to deal with technological risks, including the elimination of technology, have their own inherent risks.

Criticisms of formal decision-aiding approaches serve a value not only in identifying the dangers of using formal analysis as a decision-making mechanism, but also in identifying similar but perhaps less obvious problems with existing decision-making procedures that do not entail formal analysis. While it is true that formal approaches are subject to uncertainties, errors, and subtle manipulations, so is every other method for arriving at decisions. Leaving difficult decisions entirely to a technical elite, the intuitive judgments of regulators, the courts, or even the democratic process suffers from the same or even more serious problems. Formal approaches, in fact, may be expected to be somewhat less susceptible to these problems than the more informal approaches to decision making, for they are designed to highlight the ingredients that go into a choice. If presented in a professional manner, they lend themselves to the introduction of alternative sets of assumptions that enable the policy maker and his critics to see whether different conclusions would emerge. The complexity of social decisions involving risks demands that they be illuminated from all possible perspectives if those decisions are to be made in the best way possible. Formal approaches can provide an important, although sometimes uncomfortable, pragmatic perspective. For example, they force us to recognize that society's resources are limited, that above some level of expenditure additional programs designed to reduce risks to health and safety will be counterproductive because of economic and political effects, and that resources expended inefficiently do result, at least in principle, in less health and safety than might otherwise be obtained.

The most serious danger associated with the use of formal decision-aiding approaches is not, in my opinion, their technical, theoretical, or operational limitations, but rather that they may obscure responsibility through the appearance that method bears responsibility for results. Approach formulation is creative, iterative, and poorly understood. Yet it is critical to the conclusions that will result. Thus the analyst bears a great personal responsibility for the conclusions of the analysis. Analysts have an obligation to state clearly the assumptions and analytic choices made in performing the analysis. Furthermore, no approach can capture all of the considerations important to a decision. In particular, decisions about risk almost always involve conflicts of interests. Although analysis can help identify alternatives that are "best" according to some specified criterion, it cannot objectively resolve conflicts of interests. Therefore, regardless of whether a formal approach is or is not used, decision makers must bear responsibility for any decisions taken.

Formal analysis has the potential but cannot be guaranteed to improve decision making. Analysis can clarify a decision; it can also obscure it in technical and mathematical abstraction. Analysis can also sometimes exacerbate controversy by enabling protagonists to understand fully that their interests are opposed to those of others and by making explicit the critical assumptions on which any decision is based. Although explicitness may increase controversy, it might also provide greater focus and relevance and elevate the level of discussion by reducing the amount of uninformed and unconstructive rhetoric. The positive value derived from using formal analysis can be enhanced through the careful selection of the approach and the methods for presenting the results. Analysts need to become more aware of the options that are available and the importance of social and institutional realities when choosing from among these options.

The debate between opponents and proponents of formal decision-aiding approaches is useful to the extent that it leads to improved methods for helping us make difficult decisions on where and how to allocate the scarce resources available in our finite world. Unfortunately, this debate sometimes slips into name-calling, histrionics, posturing, and grandstanding — behavior that is not merely unprofessional but decidedly counterproductive. Our mutual goal should be to maximize agreement between those who apply formal approaches and those who are concerned that such applications improve and do not detract from our ability to achieve our ultimate social goals. This book has attempted to review available decision-aiding approaches, enlighten the reader on the controversies surrounding their use, and provide some guidance by which a more careful selection of an approach might increase the value derived from analysis.

Notes

1. In one application, e.g., Keeney (1980a) has defined the risk of an action as a probability distribution over the set of individual probabilities of dying (p_i, $i = 1, \ldots, N$), where p_i is the probability that the ith individual will die prior to the end of the next time period due to the action taken. In the widely publicized Reactor Safety Study (Rasmussen Report) sponsored by the Nuclear Regulatory Commission (1975), the risk of a nuclear reactor was described by a (complementary cumulative) probability distribution over numbers of fatalities per year reference reactor.

2. Additional factors influencing risk perception are listed later in this chapter and in Chapter 3.

3. Examining historical data, Ferreira and Slesin (1976) found that for accidents involving between 5 and 1,000 deaths the frequency of accidents falls nearly as fast as the cube of the number of fatalities.

4. This observation was also made by the Committee on Risk and Decision Making of the National Research Council (1982).

5. Citing the results of research indicating that large amounts of saccharin caused bladder cancer in rats, the FDA ordered a ban on the artificial sweetener beginning in mid-1977. Congress, however, responded to widespread public objection by twice postponing its implementation. In the meantime epidemiological studies failed to confirm an effect on humans, and the FDA rescinded its ban.

6. These agencies deal most directly with risks to health, safety, and the environment. Other regulatory bodies, such as the Securities and Exchange Commission and the Federal Trade Commission are also concerned with risks, but with risks of a different sort.

7. The Administrative Procedures Act of 1946 sets forth required procedures for agencies to follow when they engage in rule making and is generally applicable to all regulatory agencies. Because the Act antedates most of the regulatory responsibilities created by the major health, safety, and environmental laws, it has little effect on the methodological and administrative specifics regarding the use of formal analysis in establishment of regulations (Bowonder 1980).

8. In some cases it is a bit more complicated than this, but the effect is the same. E.g., the head of the FDA is appointed by and serves at the pleasure of the Secretary of the Department of Health and Human Services. The head of OSHA is an Assistant Secretary of Labor, who requires Senate confirmation. CPSC is governed by five commissioners, who make major regulatory decisions by majority vote. The commissioners are appointed for fixed terms by the President with Senate confirmation.

9. These difficulties are reflected in the novel legal theories occasionally developed by plaintiffs in cases of technological risk. Jasanoff and Nelkin (1981), e.g., cite several cases, including *Honicker v. Hendrie*, in which the plaintiff's lawyer argued that the harmful effects of radiation justified the immediate closure of all nuclear power plants. Legal support for this position was derived from an array of national and international sources of law, principles adopted during the Nuremberg trials, covenants of the United Nations, and provisions of the U.S. Constitution. Not surprisingly, the Supreme Court viewed the case as overstepping the dividing line between adjudication and policy making. In other cases, however, demonstrators at nuclear power plants have on occasion successfully defended themselves by arguing that the crime of trespass is a "lesser of two evils": that trespassing is a legitimate means of dramatizing the greater evil associated with the dangers of nuclear power.

10. See, e.g., *Essex Chemical Corporation v. Ruckelshaus* 486 Federal Reporter, 2nd series, 427, D.C. Cir., 1971.

11. *Reserve Mining Company v. EPA*, 514 Federal Reporter, 2d series, 492, 7 ERC 1618, 8th Cir., 1975.

12. *U.S. v. Carroll Towing Co.*, 159 Federal Reporter, 2d series, 169, 173, 2d Cir., 1947.

13. *Environmental Defense Fund v. EPA*, 636 Federal Reporter, 2d series, ERC 1267, 1283, D.C. Cir., 1980.

14. *Ethyl Corporation v. EPA*, 541 Federal Reporter, 2d series, 1, ERC 1785, D.C. Cir., 1976.

15. *Kerr-McGee Nuclear Corporation v. NRC*, 673 Federal Reporter, 2d series, ERC 1124, 10th Cir., 1982. For other examples, see "In the Matter of Consolidated Edison Company of New York, Power Authority of the State of New York," Docket Nos. 50-247-SP, 50-286-SP (ASLBP-83-39), 18 NRC 811, 1983, and "In the Matter of Long Island Lighting Company, Shoreham Nuclear Power Station, Unit 1," Docket No. 50-322-OL (ALAB-788), 20 NRC 1102, 1984.

16. *Scott v. Food and Drug Administration*, 728 Federal Reporter, 2d series, 322, 6th Cir., 1984.

17. *Gulf South Insulation v. United States Consumer Product Safety Commission*, 701 Federal Reporter, 2d series, 1137, 5th Cir., 1983.

18. *Asbestos Information Association v. OSHA*, 727 Federal Reporter, 2d series, 415, 5th Cir., 1984.

19. Out of the eight laws administered by the EPA in Table 3, two (the Toxic Substances Control Act and FIFRA) empower the agency to require manufacturers to provide risk assessments of their products before they can be marketed and to regulate those products found to be hazardous. Similarly, the Federal Food, Drug, and Cosmetic Act allows the FDA to require manufacturers to perform risk assessments demonstrating the safety and effectiveness of all new drugs and the safety of unexempted food additives and substances such as drug and pesticide residues. Even when the agency has the authority to require such assessments, however, statutory time deadlines and limited staff may reduce the effectiveness of this ability. E.g., in the EPA's premarketing notification procedures for

pesticides, the agency is required to make its decisions within 90 days of receipt of notification (although this statutory deadline is not always met). A typical notification involves the submission of very limited toxicological data, and EPA rarely demands further information. The agency often must infer risk levels from the chemical's physical structure and chemical properties.

20. National Environmental Policy Act, 42 U.S. Code, Section 4321, 1969. NEPA is best known for requiring the preparation of environmental impact statments for every "major federal action significantly affecting the quality of the human environment." This requirement has led to the preparation of more than a thousand impact statements per year since 1970 (Andrews 1982).

21. The courts have, however, interpreted NEPA as requiring a quantitative comparison of costs and benefits in cases where other methods lack adequate detail (Anderson 1973).

22. Outer Continental Shelf Lands Act, 43 U.S. Code, Section 2605(a), 1978.

23. Toxic Substances Control Act, 15 U.S. Code, Section 2605(a), 1976.

24. Federal Environmental Pesticide Control Act, 7 U.S. Code, Sections 136(bb), 13a(c)(5)(A—D), 1976.

25. In only a few laws has Congress explicitly required formal quantitative analyses to be conducted in performing the balancing of risks and benefits. E.g., the Pipeline Safety Act of 1979, P.L. 96—129, 1979, which instructs the Department of Transportation (DOT) to balance the "reasonableness of any proposed standards" against the "extent to which the standards will contribute to public safety," required two risk assessments to be performed. One section required DOT to perform a study of the risks associated with the production, transportation, and storage of liquefied natural gas and to include recommendations to Congress for appropriate action. Another section required DOT to perform a cost-benefit analysis of the adequacy and effectiveness of existing pipeline safety regulations and several additional specific safety-related proposals.

26. Agency discretion in choosing a particular balancing method is constrained to some extent, of course, by the legal challenges that may be made by the affected parties. Baram and Field (1983) note that legal challenges to an agency's use of a particular approach to support balancing decisions fall into one of three areas: the agency's choice does not conform to its statutory mandate as intended by Congress; the use of the approach was flawed in some significant way; or agency choice and use, though adequate, produced results which do not adequately support its final decision.

27. Technology designed to reduce one set of risks often produces others. For example, U.S. automobile manufacturers used catalytic converters in the early 1970s to meet emission standards for regulated pollutants. Although the catalytic converter decreases regulated gaseous emissions, it tends to result in increased particulate emissions, including sulfuric acid (Kawecki 1978). Furthermore, the production and installation of materials used in safety devices involve additional risk. It has been estimated, for example, that the production of certain kinds of safety equipment can be expected to result in one death per $30 million of equipment (the average for all manufacturing) (Bates *et al.* 1982).

28. *Industrial Union Department, AFL-CIO v. Hodgson*, 499 Federal Reporter, 2d series, 467 D.C. Cir., 1974.

29. *American Iron and Steel Institute v. OSHA*, 577 Federal Reporter, 2d series, 465, 3rd Cir., 1978.

30. *American Petroleum Institute v. OSHA*, 581 Federal Reporter, 2d series, 493, 5th Cir., 1978.

31. *American Textile Manufacturers Association v. Donovan*, 452 U.S. 490, 1981.

32. *Kerr-McGee Nuclear Corp. v. NRC*, 673 Federal Reporter, 2d series, 1124, 10th Cir., 1982.

33. The EPA, e.g., has been doing this since 1981 as part of their Integrated Environmental Analysis Program. Regulatory options for technology-only statutes, as well as balancing statutes, have been compared using cost-effectiveness analysis, a variation of cost-benefit analysis which omits the final step of assigning monetary values of outcomes.

34. In some cases "to the extent feasible" language is used in conjunction with a requirement that costs be taken into account, for example, as to the Safe Drinking Water Act. In such cases a balancing requirement has been construed.

35. *International Harvester v. OSHRC*, 628 Federal Reporter, 2d series, 982, 988, 7th Cir., 1980. *RMI Co. v. Secretary of Labor*, 561 Federal Reporter, 2d series, 82, 6th Cir., 1979.

36. Food, Drug, and Cosmetics Act, 21 U.S. Code, Section 348d(c)(3)(A), 1976.

37. See 48 Federal Register 15706 and 48 Federal Register 33112.

38. EPA's reliance on the results of this application, which was severely criticized by industry reviews as being poorly executed, has been at issue in court challenges to the revised standard. EPA's original promulgation notice in the Federal Register suggested that the results were a factor in its decision, but the agency subsequently maintained that the analysis did not play a role in the decision process. Although the agency has been sued on the basis that its decision was arbitrary and without adequate scientific basis, the revised standard has survived court challenges.

39. *American Textile Manufacturers Institute v. Marshall*, U.S. Supreme Court Docket No. 79—1429 and *National Cotton Council of America v. Marshall*, U.S. Supreme Court Docket No. 79—1583.

40. Executive Order 11821, 39 Fed. Reg. 41501, November 29, 1974, amended by Executive Order 11949, 42 Fed. Reg. 1017, December 31, 1976.

41. Executive Order 12044, 43 Fed. Reg. 12661, March 23, 1978.

42. Despite the requirements of Carter's order, regulatory agencies were not entirely negative in their judgment of it. The administrator of EPA defended it as an "encouraging beginning" to resolving regulatory conflicts and shaping decisions in a manner that reflects overall policy objectives (Costle 1980).

43. Executive Order 12291, 46 Fed. Reg. 13193, February 1981.

44. Public Citizen Health Research Group et al. v. Tyson, D.C. Cir., Docket No. 84—1252, July 1986.

45. House of Representatives, "Risk Analysis Research and Demonstration Act of 1982," HR 6159, 97th Congress, June 24, 1982, and "Risk Analysis Research and Demonstration Act of 1983," HR 4192. 98th Congress, October 21, 1983.

46. Recognition of the necessarily creative elements of decision-aiding approaches explains what might otherwise appear as a curious omission in the literature. Authors often provide good accounts of the historical foundations and theoretical justifications for an approach but present little or no discussion of the specific procedures to be used in its practical application. Conversely, the applications literature generally describes the technical procedures used without pointing out the logical justification for those procedures in terms of underlying methodological principles.

47. Most normative approaches also contain descriptive elements in the form of empirically observed or expressed values and beliefs. Behavior prescriptions are based on consistency requirements.

48. For an interesting example of the debate between objectivist and subjectivist schools, see the series of letters to the editor by Abramson (1981) and Kaplan and Garrick (1981) published in the journal *Risk Analysis.*

49. Other normative decision theories also exist. See, e.g., Arrow (1951). This study restricts attention to the most widely accepted and familiar normative theories of decision making.

50. This is also called the Kaldor-Hicks criterion.

51. The historical reference is reproduced in D. Bernoulli, "Exposition of a New Theory on the Measurement of Risk" (L. Sommer, trans.), *Econometrica* 22 (1954): 23–26.

52. Other axioms (see Howard 1976) are

Continuity. If A is preferred to B and B is preferred to C, then there must be some lottery between A and C (i.e., an uncertain situation in which either A or C will be received) for which the decision maker is indifferent between that lottery and receiving B for sure. (B is then referred to as the "certain equivalent" of the lottery.)

Substitutability. A lottery and its certain equivalent are in all situations interchangeable without affecting preferences.

Monotinicity. If A is preferred to B, then in all evaluations of lotteries involving only A and B, the decision maker must prefer those with the highest probability of yielding A.

Decomposability. The value assigned by a decision maker to a lottery should not depend on whether the uncertainty is resolved in stages or all at once, but only on the probabilities and possible outcomes.

53. Both subjective probability theory and utility theory contain a number of directly relevant principles. E.g., Bayes' theorem specifies an optimal rule for revising subjective probabilities in the light of new evidence. Multiattribute utility provides a theory for making multiple value dimensions commensurable in terms of a common scale of value (see, e.g., Anderson *et al.* 1981; Keeney and Raiffa 1976; Gardiner and Edwards 1975; Edwards and Newman 1982).

54. Simple majority rule, e.g., does not satisfy this requirement. Majorities can prefer candidate A to candidate B and candidate B to candidate C, but at the same time a majority can prefer candidate C to candidate A. Runoff elections among leading candidates prevent intransitivities from being observed in the election process.

55. Also referred to as benefit-cost analysis, risk-benefit, analysis, and other similar permutations. For introductory discussions, see Prest and Turvey (1965), Peters (1968), Layard (1972, 1974), Bohm (1974), Mishan (1976), Stokey and Zeckhauser (1978), Abelson (1979), Drummond (1980), Thompson (1980a), and Gramlich (1981).

56. For introductions, see Howard (1968a, 1968b), Raiffa (1968), Schlaifer (1969), Brown *et al.* (1974), and Holloway (1979).

57. The procedures applied within an approach might be further characterized according to the tasks involved in developing each of the six elements. Each element requires the representation of data derived from some source in terms of numeric scales and mathematical relationships, followed by analysis. In general, then, the process of developing these elements may involve steps directed at conditioning the sources of data, assessing the data, structuring those data, and analysis. Other partitions have also been proposed (e.g., Hammond *et al.* 1980: Ch. 14).

58. A social DA application conducted as part of a methodology development effort for the EPA provides an illustration of these six elements (Smith *et al.* 1982). Actions consisted of seven alternative national ambient carbon monoxide (CO) air quality standards. The method of characterizing the alternatives was in terms of assumed hourly average CO concentrations by exposure district resulting under each standard in each of four cities (Chicago, Los Angeles, Philadelphia, and St. Louis). The selection criterion established by the analysts was defined to be the least stringent standard producing an expected health impact below some maximum "acceptable" level. The measure for assessing health impact level was the total number of "equivalent days of restricted activity (EDRAs)," defined as a weighted sum of specific health effects experienced by four sensitive population groups. Associating health impacts with alternatives involved using a model designed to account for the varying exposures of individuals (by accounting for the hourly variability of CO concentrations by geographic location and the movement of population groups among these locations) and the increased incidence of various health effects that could be attributed to these exposures. Analysis of the model consisted of encoding judgmental probabilities and using decision trees to derive probability distributions over the total number of EDRAs occurring under each standard in each city.

59. Alternatively it has been argued that no explicit consideration of risk attitudes is needed with CBA because they are automatically incorporated into the market data: if people are risk averse, they will pay more for safer goods, making those prices rise (Fischhoff *et al.* 1981a). The validity of this argument depends, of course, on the extent to which a free market exists and takes risk issues into accounts.

60. Citing these statistics the EPA has inferred that a "statistical life" should be valued at between $400,000 and $7,000,000 and has recommended that this range be used as a guideline for preparing regulatory impact analyses under E. O. 12291 (EPA 1983a).

61. Howard's model (1979) illustrates that individuals may rationally place infinite values on their lives when the chances of dying are large (e.g., refusing a gamble involving a 0.5 chance of death for any amount of money) and yet willingly accept a finite amount of money in return for a small increase in the probability of death.

62. Both Howard's and Keeney's procedures are somewhat difficult to apply in practice; as a result, DA and SWT approaches have tended to use the more standard procedures associated with CBA. When used as part of a decision-aiding approach such structured processes are models in the sense that they are approximations to the decision-making process. Those with authority for the decision can choose to accept or reject their results.

63. As more and more applications are conducted, existing models appropriate for various elements of the risk chain will become more readily available. It may be expected, therefore, that it will be increasingly more efficient to adapt an existing model than to construct a new model. Currently, however, various problems of model integration, such as lack of required input data and the difficulties often associated with obtaining access to the models developed by other researchers, usually limit the use of existing models in decision-aiding approaches.

64. More precisely, the arrows in an influence diagram indicate to the analyst the terms that must be present in the expansion of joint probability distributions over the variables expressed as functions of conditional probabilities.

65. In the extreme case where prior information is noninformative or where new data add only insignificantly to existing knowledge, Bayes theorem yields results identical to or very close to those produced under a classical analysis. Thus the practical value of a Bayesian analysis derives in the in-between cases in which both existing (or generic) and new information must be taken into account.

66. Indeed, probabilities are regarded by decision analysts as appropriately obtained only through elicitation: extrapolating from frequency counts to prediction requires judgment, because to rely solely on past data implies the belief that the conditions surrounding the event are identical to those existing in the past.

67. When using direct methods to assess rare events (probabilities less than about 0.05), Lichtenstein (see Morgan *et al.* 1979) reports that subjects do better using odds rather than probabilities. Savage (1971) argues that indirect methods are generally superior to direct methods because they avoid explicit mention of probabilities.

68. The expansion rule states that a joint probability can be written as the product of a marginal and conditional probability. E.g., if $P(B)$ is the probability of event B occurring and $P(A|B)$ is the conditional probability of event A occurring given that event B occurs, the probability of both events A and B occurring, $P(A, B)$ is

$$P(A, B) = P(A|B)P(B)$$

The expansion rule follows directly from the definition of conditional probability.

69. Rowe (1977a) provides some useful benchmarks along this scale. An objective measure of probability that an event will occur may be determined by direct measurement of repeated trials of the event. When the number of trials is large, the estimate of probability is properly considered objective because it represents an empirical estimation. On the other hand, if probability estimates are made on the basis of one or a few trials or by conjecture, then the probability estimate is subjective. This view is consistent with classical definitions of objective and subjective probability. Between these extremes lies another estimate that Rowe refers to as "synthesized probability." In this case the probability of an event is not measured directly, but rather is extrapolated through simulation using models of causal systems that are expected to behave similarly.

A similar range between objective and subjective sources exists in estimating consequent value. An objective estimate of consequent value consists of a consequence that is directly measured along a value scale; e.g., the medical costs borne by a particular population. Subjective consequence value, on the other hand, occurs when the consequence value to a particular risk agent depends completely on the risk agent's personal value system and situation; e.g., the amount of money necessary to compensate (make indifferent) the victim of an accident. Between these two extremes is a value estimate, which Rowe terms "observable consequence value," in which the behavioral response of groups in society to risk consequences is ascertained and measured by studying actual behavior; e.g., the amount of money people spend on safety.

70. Rowe *et al.* (1980), however, were unable to find evidence of such behavior in fairly extensive empirical studies dealing with preferences for levels of visibility.

71. Incompleteness of analysis is not solely caused by the finite nature of analysis and knowledge; it is compounded by human fallibilities and procedural limitations.

72. As a consequence of inadequate training and control room design, operators repeatedly misdiagnosed the condition of the reactor and took inappropriate corrective actions (Doniger 1980). Although the specific sequence of events occurring at Three Mile Island was not included among the scenarios evaluated in the "Reactor Safety Study," the general category of accidents involving operator error was considered.

73. Evidence suggests that regulation stimulates commercially superior technology in established, mature industries (Utterback 1978).

74. See the following section for a discussion of the difficulty of measuring costs and benefits.

75. According to cognitive psychologists, a heuristic is a rule of thumb for simplifying a complex problem. If an individual attempts to solve a problem through heuristics, he simplifies the search for a solution by making educated guesses as to where in the search space the solution is likely to be found. Unlike algorithmic search, which is based on calculation or exhaustive consideration of all possibilities and is guaranteed to find a solution if there is one, heuristic search merely increases the chances of finding a solution within a reasonable time frame. Simon (1957) views heuristics as necessitated by the limited information-processing capacity of humans.

76. Delay cannot, of course, be counted on as the inevitable consequence of careful analysis. Often analyses will point to preferred alternatives, will clarify the costs of delay, or may establish that improving the technical information base will have little or no effect on the decision in question (Redmond 1978).

77. It has also been pointed out that a false sense of imprecision without numbers may give the impression that less is known than is really known (National Research Council 1982).

78. E.g., health-only statutes. See Chapter 1.

79. See Chapter 1.

80. "Doubts Linger on Cyclamate Risks," *Eugene Register-Guard,* January 14, 1976.

81. Keeney (1981), however, has found a tension between the decision analysts' concept of risk aversion and equity. Starting from the axioms of decision theory, he shows that it is inconsistent for an individual to prefer both more equitable distributions of risks over society's members and to be risk averse regarding numbers of fatalities.

82. MacLean (1982b), however, points out that there is a fine line between prescribing a decision and presenting relevant information in the form judged by analysts to be most

useful to decision makers. "The more information relevant to a decision is brought into a risk assessment, the more one leaves only formal considerations, like consistency and transitivity of preferences, standing between estimation and a decision . . . if someone says he is only showing . . . what it should be rational to do, that he is not prescribing a choice, then he is disingenuous in claiming that his analysis is a modest one."

83. The organization of the following sections is similar to and borrows some of its insights from a more general comparative evaluation by Fischhoff *et al.* (1981b).

References

Abelson, P. 1979. *Cost-Benefit Analysis and Environmental Problems.* Westmead, England: Saxon House.
Abrams, N. E. and R. S. Berry. 1977. 'Mediation: A Better Alternative to the Science Court.' *Bulletin of the Atomic Scientists* April: 50.
Abramson, L. R. 1981. 'Some Misconceptions about the Foundations of Risk Analysis' and 'A Rejoinder to Kaplan and Garrick.' *Risk Analysis* 1(4): 229—230, 235.
Ahern, W. 1980. 'California Meets the LNG Terminal.' *Coastal Zone Management Journal* 7: 185—221.
Allais, M. 1953. 'La Psychologie de l'Homme Rationale devant le Risque: Le Théories et l'Expérience.' *Journal of the Statistics Society* (Paris) 94: 47—73.
Allison, G. T. 1971. *Essence of Decision.* Boston: Little, Brown.
American Nuclear Society and Institute of Electrical and Electronics Engineers. 1983. 'PRA Procedures Guide: A Guide to the Performance of Probabilistic Risk Assessments for Nuclear Power Plants.' NRC Report NUREG/CR—2300, NTIS. January.
Anderson, B. F., D. H. Deane, K. R. Hammond, G. H. McClelland, and J. C. Shanteau. 1981. *Concepts in Judgment and Decision Research.* New York: Praeger.
Anderson, F. R. 1973. *NEPA in the Courts.* Baltimore: Johns Hopkins University Press.
Anderson, L. G. and R. F. Settle. 1977. *Benefit-Cost Analysis: A Practical Guide.* Lexington, Mass.: Lexington Books.
Andrews, R. N. L. 1982. 'Politics: Cost-Benefit Analysis as Regulatory Reform.' In D. Swartzman, R. A. Liroff, and D. G. Croke, eds. *Cost Benefit Analysis and Environmental Regulations: Politics, Ethics, and Methods.* Washington, D.C.: Conservation Foundation.
Apostolakis, G. 1981. 'Bayesian Methods in Risk Assessment.' In J. Lewins and M. Becker, eds. *Advances in Nuclear Science and Technology.* Vol. 13. New York: Plenum.
Arrow, K. J. 1951. 'Alternative Approaches to the Theory of Choice in Risk-Taking Situations.' *Econometrics* 19(4): 404—437.
——. 1974. *Essays in the Theory of Risk Bearing.* New York: Elsevier.
——. 1977. 'Current Developments in the Theory of Social Choice.' *Social Research* 44(4): 607—622.
Aschenbrenner, K. M. 1978. 'Singlepeaked Risk Preferences and Their Dependability on the Gambles Presentation Mode.' *Journal of Experimental Psychology: Human Perception and Performance* 4: 513—520.
Ashford, N. A. 1978. 'Federal Control of Toxic Substances in the Workplace and General Environment: Legal, Regulatory, and Scientific Complexities.' Center for Policy Alternatives, Massachusetts Institute of Technology.
——, G. R. Heaton, Jr., and W. C. Priest. 1979. 'Environmental Safety Regulation and Technological Innovation.' In C. T. Hill, ed. *Federal Regulation and Chemical Innovation.* Washinton, D.C.: American Chemical Society.
Australian Department of Environment. 1980. *Report 9 — Economic Evaluation of Eppalock Soil Conservation Study.* Australian Government Publishing Service.

Ayres, R. 1979. 'Comments.' In U.S. Congress, Senate Committee on Environmental and Public Works, Subcommittee on Environmental Pollution, 96th Congress, 1st Session, Serial No, 96—H4, p. 27.

Bailey, M. J. 1980. *Reducing Risks to Life*, p. 2. Washington, D.C.: American Enterprise Institute for Public Policy Research.

Baram, M. S. 1980. 'Cost-Benefit Analysis: An Inadequate Basis for Health, Safety, and Environmental Regulatory Decisionmaking.' *Ecology Law Quarterly* 8: 473—531.

—— and R. I. Field. 1983. 'Law, Risk Management, and the Influence of the Psychological and Decision Sciences.' Paper presented at the NATO—NSF Advanced Study Institute on Technology Assessment, Environmental Impact Assessment and Risk Analysis: Contributions from the Psychological and Decision Sciences, Les Arcs, France. August.

—— *et al.* 1977. 'Federal Energy Regulation and the Use of Cost-Benefit Analysis: Legal and Public Policy Considerations.' *Part I: Adoption of Cost-Benefit Analysis by Government Agencies.* Report to Environmental Policy Office, Brookhaven National Laboratory.

Bar-Hillel, M. 1980. 'The Base Rate Fallacy in Probability Judgments.' *Acta Psychologica* 44: 211—233.

Barrager, S. M., B. R. Judd, and D. W. North. 1976. 'The Economic and Social Costs of Coal and Nuclear Electric Generation.' Decision Analysis Department, SRI International, Menlo Park, Calif. March.

Bates, D., I. Burton, and G. Butler. 1982. 'On the Application of Risk — An Overview Paper.' In J. T. Rogers and D. V. Bates, eds. *Risk: A Symposium on the Assessment and Perception of Risk to Human Health in Canada, October 18—19, 1982.* Toronto: Royal Society of Canada.

Behn, R. D. and J. W. Vaupel. 1983. *Quick Analysis for Busy Decision Makers.* New York: Basic Books.

Bennett, G. 1981. 'Environmental Decision Faking.' *New Scientist* August 20: 468.

Bentkover, J. D., V. T. Covello, and J. Mumpower, eds. 1986. *Benefits Assessment: The State of the Art.* Dordrecht: D. Reidel.

Bernick, K. 1982. 'The Inflation Impact Statement Program and Executive Branch Coordination.' Cited by R. A. Liroff in D. Swartzman, R. A. Liroff, and K. G. Croke eds. *Cost-Benefit Analysis and Environmental Regulations: Politics, Ethics, and Methods.* Washington, D.C.: Conservation Foundation.

Bodily, S. E. 1979. 'A Delegation Process for Combining Individual Utility Functions.' *Management Science* 25(10): 1035—1041.

Bogen, K. T. 1980. 'Public Policy and Technology Risk.' *Journal of Law and Technology* 24(1): 37—74.

——. 1981. 'Coordination of Regulatory Risk Analysis: Current Framework and Legislative Proposals.' Report No. 81—209 SPR. Congressional Research Service, Library of Congress. June 19.

Bohm, P. 1974. *Social Efficiency: A Concise Introduction to Welfare Economics.* Stockholm: Macmillan Press.

Borch, K. 1968. *The Economics of Uncertainty.* Princeton: Princeton University Press.

Boulding, K. E. 1966. 'The Ethics of Rational Decision.' *Management Science* 12: 161.

Bowonder, B. 1980. 'Issues in Environmental Risk Assessment.' *Journal of Environmental Systems* 10(4): 305—331.

Brickman, R., S. Jasanoff, and T. Ilgen. 1985. *Controlling Chemicals: The Politics of Regulation in Europe and the United States.* Ithaca: Cornell University Press.

Brock, H. W. 1979. 'A Game Theoretic Account of Social Justice.' *Theory and Decision* 10: 239—265.

Brown, R. V. In press. 'Decision Analysis Tools.' In K. B. Levitan, ed. *Government Infostructures.* Westport, Conn.: Greenwood.

——, A. S. Kahn, and C. R. Peterson. 1974. *Decision Analysis for the Manager.* New York: Holt, Rinehart, and Winston.

Buede, D. M. 1979. 'Decision Analysis: Engineering Science or Clinical Art?' Technical Report TR 79—2—97, Decisions and Designs, Inc., McLean, Va. November.

Bunn, D. W. 1984. *Applied Decision Analysis.* New York: McGraw-Hill.

Callahan, D. 1980. 'Statement of Daniel Callahan, Director. The Hastings Center, Hastings-on-Hudson, NY.' In *Risk/Benefit Analysis in the Legislative Process.* U.S. Senate Committee on Science and Technology, No. 71. Washington, D.C.: GPO.

Carter, E. E. 1972. 'What are the Risks in Risk Analysis.' *Harvard Business Review* July—August: 72—82.

Chacko, G. K. 1976. *Applied Operations Research/Systems Analysis in Hierarchical Decision-Making II.* Amsterdam: North-Holland.

Chankong, V. and Y. Y. Haimes, 1983. *Multiobjective Decision Making: Theory and Methodology.* New York: North-Holland.

Chapman, L. J. and J. P. Chapman, 1969. 'Illusory Correlation as an Obstacle to the Use of Valid Psychodiagnostic Signs.' *Journal of Abnormal Psychology* 74: 271—280.

Churchman, C. 1961. *Prediction and Optimal Decision: Philosophical Issues of a Science of Values.* Englewood Cliffs, N.J.: Prentice-Hall.

Cochran, T. B. 1976. 'Comments by Thomas B. Cochran, National Resources Defense Council, Inc..' Section M in S. M. Barrager, B. R. Judd, and D. W. North, eds. 'The Economic and Social Costs of Coal and Nuclear Electric Generation.' Decision Analysis Department, SRI International, Menlo Park, Calif. March.

Conrad, J. 1981. 'Society and Problem-Oriented Research: On the Socio-Political Functions of Risk Assessment.' In H. Kunreuther, ed. *Risk: A Seminar Series.* Laxenburg, Austria: International Institute for Applied Systems Analysis.

Conservation Foundation. 1984. 'Risk Assessment and Risk Control.' In *State of the Environment: An Assessment at Mid-Decade*, pp. 261—318. Washington, D.C.: Conservation Foundation.

Costle, D. 1980. 'Panel III: Managing the Regulatory Process.' *Administrative Law Review* 32: 256—260.

Covello, V. T. 1985. 'Social and Behavioral Research on Risk: Uses in Risk Management Decisionmaking.' In V. T. Covello, J. L. Mumpower, P. J. M. Stallen, and V. R. R. Uppuluri, eds. *Environmental Impact Assessment, Technology Assessment, and Risk Analysis: Contributions from the Psychological and Decision Sciences.* Berlin: Springer Verlag.

Cox, L. A., Jr. 1986. 'Theory of Regulatory Benefits Assessment: Econometric and Expressed Preference Approaches.' In J. D. Bentkover, V. T. Covello, and J. Mumpower, eds. *Benefits Assessment: The State of the Art* Dordrecht: D. Reidel.

Cumming, R. B. 1981. 'The Urge to Shout.' *Risk Analysis* 1(3): 165.

Cummings, R. G., L. A. Cox, Jr., and A. M. Freeman III. 1986. 'General Methods for Benefits Assessment.' In J. D. Bentkover, V. T. Covello, and J. Mumpower, eds. *Benefits Assessment: The State of the Art.* Dordrecht: D. Reidel.

David, E. E. 1975. 'One-Armed Scientists?' *Science* 189: 891.

De Finetti, B. 1968. 'Probability Interpretations.' *International Encyclopedia of the Social Sciences.* Vol. XII, pp. 496—505, New York: Macmillan.

DeMuth, C. 1980. 'The White House Review Programs.' *Regulation* January/February: 27.

Dodge, C. H. and R. L. Civiak. 1981. 'Risk Assessment and Regulatory Policy,' p. 1. Library of Congress, Issue Brief No. IB81019. September 24.

Doniger, D. 1980. 'Statement of David D. Doniger, Natural Resources Defense Council.' In *Comparative Risk Assessment, Hearings before the Subcommittee on Science and Technology*, No. 129, May 14—15. Washington, D.C.: GPO.

Douglas, M. and A. Wildavsky. 1982. *Risk and Culture.* Berkeley: University of California Press.

Drummond, M. F. 1980. *Principles of Economic Appraisal in Health Care.* Oxford: Oxford University Press.

Dyer, J. 1978. *Applications of Decision Analysis to Occupational Health and Safety*

Problems. Los Angeles: UCLA.
—— and R. Miles. 1976. 'An Actual Application of Collective Choice Theory to the Selection of Trajectories for the Mariner Jupiter/Saturn Project.' *Operations Research* 24: 220—224.
—— and R. Sarin, 1978a. 'Cardinal Preference Aggregation Rules for the Case of Certainty.' In S. Zionts, ed. *Proceedings of the Conference on Multiple Criteria Problem Solving*. New York: Springer-Verlag.
—— and ——. 1978b. 'Measurable Preference Aggregation Rules.' Discussion paper No. 71, Management Science Study Center. Los Angeles: UCLA.
Eads, G. C. and M. Fix. 1984. *Relief or Reform? Reagan's Regulatory Dilemma*, p. 123. Washington, D.C.: Urban Institute Press.
Edwards, W. 1954. 'The Theory of Decision Making.' *Psychological Bulletin* 51(4): 380—417.
——. 1977. 'How to Use Multi-Attribute Utility Measurement for Social Decision Making.' *IEEE Transactions on Systems, Man, and Cybernetics* SMC—7: 326—339.
—— and J. R. Newman. 1982. *Multiattribute Evaluation*. Beverly Hills, Calif.: Sage.
English, J. M. 1978. 'A Perceptual Time Scale for Determination of a Discount Function.' In C. Van Dam, ed. *Trends in Financial Decision-Making*. Boston: Martinus Nijhoff.
Environmental Protection Agency. 1980. 'Economics at EPA.' Science Advisory Board, Subcommittee on Economic Analysis, July 22.
——. 1983a. *Guidelines for Performing Regulatory Impact Analyses*, Office of Policy Analysis, EPA—230/01—84—003, Washington, D.C. December.
——. 1983b, *Proposed Process for Evaluation and Control of Toxic Pollutants*, External Staff Draft. March 23.
Fares, R. J., R. E. Burt, D. L. Britt, J. D. Kinsman, and S. A. Schoepke. 1983. 'Preliminary Guidebook for Conducting a Regulatory Impact Analysis for Health, Safety, and Environmental Regulations.' General Research Corp. McLean, Va. August.
Ferreira, J. and L. Slesin. 1976. 'Observations on the Social Impact of Large Accidents.' Operations Research Center, Technical Report No. 122, Massachusetts Institute of Technology, Cambridge, Mass.
Fischhoff, B. 1977. 'Cost-Benefit Analysis and the Art of Motorcycle Maintenance.' *Policy Sciences* 8: 177—202.
——. 1979. 'Informed Consent in Societal Risk-Benefit Decision.' *Technological Forecasting and Social Change* 13: 347—357.
—— and L. A. Cox, Jr. 1986. 'Conceptual Framework for Regulatory Benefits Assessment.' In J. D. Bentkover, V. T. Covello, and J. Mumpower, eds. *Benefits Assessment: The State of the Art*. Dordrecht: D. Reidel.
——, P. Slovic, and S. Lichtenstein. 1977. 'Knowing with Certainty: The Appropriateness of Extreme Confidence.' *Journal of Experimental Psychology: Human Perception and Performance* 3: 552—564.
——, G. Hohenemser, R. E. Kasperson, and R. W. Kates. 1978a. 'Handling Hazards: Can Hazard Management Be Improved?' *Environment* 20(7).
——, P. Slovic, and S. Lichtenstein. 1978b. 'Knowing What You Know: Measuring Labile Values.' Decision Research, Eugene, Or.
——, ——, and ——. 1980a. 'Knowing What You Want: Measuring Labile Values.' In T. S. Wallsten, ed. *Cognitive Processes in Choice and Decision Behavior*, pp. 117—142. Hillsdale, N.J.: Erlbaum.
——, ——, and ——. 1980b. 'Lay Foibles and Expert Fables in Judgments about Risk.' Decision Research, Eugene, Ore. June.
——, ——, and ——. 1981a. '"The Public" vs. "The Experts": Perceived vs. Actual Disagreement about Risks.' Presented at the First Annual Meeting of the Society for Risk Analysis, Washington, D.C. June 1—3.
——, S. Lichtenstein, P. Slovic, S. L. Derby, and R. L. Keeney. 1981b. *Acceptable Risk*. New York: Cambridge University Press.

Fisher, D. W. and D. von Winterfeldt. 1978. 'Setting Standards for Chronic Oil Discharges in the North Sea.' *Journal of Environmental Management* 7: 177—199.

Flowerdes, A. D. J. 1972. 'Choosing a Site for the Third London Airport: The Roskill Commission's Approach.' In R. Layard, ed. *Cost-Benefit Analysis.* Middlesex, England: Penguin Books.

Food Safety Council. 1980. 'Proposed System for Food Safety Assessment.' Final Report of the Committee of the Food Safety Council, Washington, D.C. June.

Ford, C., R. L. Keeney, and C. W. Kirkwood. 1979. 'Evaluating Methodologies: A Procedure and Application to Nuclear Power Plant Siting Methodologies.' *Management Science* 24(1): 1—10.

Foster, C. D. and M. E. Beesely. 1963. 'Estimating the Social Benefit of Constructing an Underground Railway in London.' *Journal of the Royal Statistical Society Series A* 126 (Part I).

Freeman, A. M. III. 1979a. 'The Benefits of Air and Water Pollution Control: A Review and Synthesis of Recent Estimates.' Report prepared for the U.S. Council on Environmental Quality. December.

——. 1979b. *The Benefits of Environmental Improvement: Theory and Practice.* Baltimore: Johns Hopkins University Press.

——. 1982. *Air and Water Pollution Control: A Benefit Cost Assessment.* New York: Wiley.

——. 1986. 'Estimating the Benefits of Environmental Regulations.' In J. D. Bentkover, V. T. Covello, and J. Mumpower, eds. *Benefits Assessment: The State of the Art.* Dordrecht: D. Reidel.

Fussell, J. B. and G. R. Burdick. 1977. *Nuclear Systems Reliability Engineering and Risk Assessment.* Philadelphia: Society for Industrial and Applied Mathematics.

Gardiner, P. C. and W. Edwards. 1975. 'Public Values: Multiattribute Utility Measurement for Social Decision Making.' In M. F. Kaplan and S. Schwartz, eds. *Human Judgment and Decision Processes.* New York: Academic Press.

Gass, S. I. and R. L. Sisson. 1975. *A Guide to Models in Governmental Planning Operations.* Potomac, Md.: Sauger Books.

General Accounting Office. 1982. 'Improved Quality, Adequate Resources, and Consistent Oversight Needed if Regulatory Analysis Is to Help Control Costs of Regulation.' GAO/PAD—83—6. November 2.

——. 1984. 'Cost-Benefit Analysis Can Be Useful in Assessing Environmental Regulations, Despite Limitations.' GAO/RCED—84—62. April 6.

Gibbard, G. S. 1973. *Analysis of Groups: Contributions to Theory, Research, and Practice.* San Francisco: Jossey—Bass.

Goicoechea, A., D. R. Hansen, and L. Duckstein. 1982. *Multiobjective Decision Analysis with Engineering and Business Applications.* New York: Wiley.

Goodin, R. E. 1980. 'The Ethics of Social Risks.' *Science and Public Policy* October: 318—327.

Goodin, W. R. 1981. 'Advanced Air Quality Modeling.' *Engineering Bulletin* 58: 21—36.

Gori, G. B. 1980. 'The Regulation of Carcinogenic Hazards.' *Science* 208 (April 18): 257.

Graham, J. D. and J. W. Vaupel. 1981. 'Value of a Life: What Difference Does It Make?' *Risk Analysis* 1(1): 89—95.

Gramlich, E. M. 1981. *Benefit-Cost Analysis of Government Programs.* Englewood Cliffs, N.J.: Prentice-Hall.

Green, H. P. 1975. 'Legal and Political Dimensions of Risk-Benefit Methodology.' In D. Okrent, ed. *Risk-Benefit Methodology and Application: Some Papers Presented at the Engineering Foundation Workshop, September 22—26, 1975, Asilomar, California.* Los Angeles: UCLA.

——. 1979. 'The Risk-Benefit Determination.' *American Industrial Hygiene Association Journal* 40 (December): 1200—1206.

——. 1980. 'Statement of Harold P. Green, Attorney, Professor of Law, George

Washington University, Washington, D.C.' In *Risk/Benefit Analysis in the Legislative Process*. Washington, D.C.: U.S. Senate Printing Office.

Green, J. R. and J.-J. Laffont. 1979. *Incentives in Public Decision Making*. New York: North-Holland.

Hammond, K. R. and L. Adelman. 1976. 'Science, Values and Human Judgment.' *Science* 194 (October 22): 389—396.

——, G. H. McClelland, and J. Mumpower. 1980. *Human Judgment and Decision Making*. New York: Praeger.

Hampton, J. M., P. G. Moore, and H. Thomas. 1973. 'Subjective Probability and Its Measurement.' *Journal of the Royal Statistical Society Series A* 136: 21—42.

Hapgood, F. 1978. 'Risk-Benefit Analysis: Putting a Price on Life.' *Atlantic*: 33—38.

Harris (Louis) and Associates. 1980. 'Risk and a Complex Society.' Public opinion survey conducted for Marsh and McLennan, New York.

Harsanyi, J. C. 1955. 'Cardinal Welfare, Individualistic Ethics, and Interpersonal Comparisons of Utility.' *Journal of Political Economy* 63 (August): 309—321.

——. 1975. 'Nonlinear Social Welfare Functions.' *Theory and Decision* 6: 311—332.

——. 1977a. 'Mortality and the Theory of Rational Behavior.' *Social Research* 44(4): 623—667.

——. 1977b. *Rational Behavior and Bargaining Equilibrium in Games and Social Situations*. Cambridge: Cambridge University Press.

Highland, J. H. 1978. 'Risk-Benefit Analysis in Regulating Decision Making.' Presented at the Internation Conference on Health Hazards of Asbestos Exposure, New York. Washington, D.C.: Environmental Defense Fund.

Hirshleifer, J. 1975. 'The Economic Approach to Risk-Benefit Analysis.' In D. Okrent, ed. *Risk-Benefit Methodology and Application: Some Papers Presented at the Engineering Foundation Workshop, September 22—26, 1975, Asilomar, California*. Los Angeles: UCLA. December.

——, T. Bergstrom, and E. Rappaport. 1974. 'Applying Cost-Benefit Concepts to Projects Which Alter Human Mortality,' p. 5. UCLA—ENG—7478, School of Engineering and Applied Science, UCLA. November.

Hobbs, B. F. 1979. 'Analytical Multiobjective Decision Methods for Power Plant Siting: A Review of the Theory and Applications.' Division of Regional Studies, Brookhaven National Laboratory, Upton, New York.

Holdren, J. P. 1976. 'The Nuclear Controversy and the Limitations of Decision Making by Experts.' *Bulletin of the Atomic Scientists* 32(3): 20—22.

Holling, C. S. 1977. 'The Curious Behavior of Complex Systems: Lessons from Ecology.' In H. A. Linstone and W. H. C. Simmonds, eds. *Futures Research: New Directions*, pp. 107—108. Reading, Mass.: Addison-Wesley.

——. 1981. 'Resilience in the Unforgiving Society.' Working Paper R—24, Vancouver Institute of Resource Ecology. March.

Holloway, C. 1979. *Decision Making Under Uncertainty: Models and Choices*. New York: Prentice-Hall.

Hooker, C. A., J. J. Leach, and E. F. McClennen, eds. 1978. *Foundations and Applications of Decision Theory*. Vols. I and II. Boston: D. Reidel.

Hoos, I. R. 1972. *Systems Analysis in Public Policy: A Critique*. Berkeley: University of California Press.

Howard, R. A. 1968a. 'The Foundations of Decision Analysis.' *IEEE Transactions on Systems Science and Cybernetics* SSC—4(3): 211—219.

——. 1968b. 'Decision Analysis: Applied Decision Theory.' In *Proceedings of the Fourth Conference on Operational Research*. New York: Wiley.

——. 1971. *Dynamic Probabilistic Systems — Volume I: Markov Models*. New York: Wiley.

——. 1975. 'Social Decision Analysis.' *Proceedings of the IEEE* 63: 359—371.

——. 1976. 'Risk Preference.' In R. A. Howard, J. E. Matheson, and K. L. Miller, eds. *Readings in Decision Analysis*. Menlo Park, Calif.: SRI International.

——. 1979. 'Life and Death Decision Analysis.' In *Lawrence Symposium on Systems and Decision Sciences*. North Hollywood: Western Periodicals.

——. 1980. 'An Assessment of Decision Analysis.' *Operations Research* 28 (January): 4—27.

—— and J. E. Matheson. 1984. 'Influence Diagrams.' Private report (January 1981). In R. A. Howard and J. E. Matheson, eds. *Readings on the Principles and Applications of Decision Analysis*. Vol. II. Menlo Park, Calif.: Strategic Decisions Group.

——, ——, and D. W. North. 1972. 'The Decision to Seed Hurricanes.' *Science* 176 (June): 1191—1202.

——, M. W. Merkhofer, A. C. Miller, and S. N. Tani. 1975. 'A Preliminary Characterization of a Decision Structuring Process for the Task Force Commander and His Staff.' Technical Report MSC—4030, SRI International, Menlo Park, Calif. December.

Hull, J. C. 1977. 'Dealing with Dependence in Risk Simulations.' *Operational Research Quarterly* 28(1): 201—213.

Hurter, A. P., Jr., G. S. Tolley, and R. G. Fabian. 1982. 'Methods — Benefit-Cost Analysis and the Common Sense of Environmental Policy.' In D. Swartzman, R. A. Liroff, and K. G. Croke, eds. *Cost-Benefit Analysis and Environmental Regulations: Politics, Ethics, and Methods*. Washington D.C.: Conservation Foundation.

Hutt, P. B. 1984. 'Legal Considerations in Risk Assessment under Federal Regulatory Statutes.' In J. Rodricks and R. Tardiff, eds. *Assessment and Management of Chemical Risks*, pp. 83—94. ACS Symposium Series, American Chemical Society.

Hwang, C. L. and A. S. M. Masud. 1979. *Multiple Objective Decision Making Methods and Application: A State-of-the-Art Survey*. New York: Springer-Verlag.

Illich, I. 1977. *Medical Nemesis: The Expropriation of Health*. New York: Bantam.

Jackson, C., *et al.* 1976. 'Benefit-Cost Analysis of Automotive Emission Reductions.' General Motors Research Publication 2265. October.

Jasanoff, S. and D. Nelkin. 1981. 'Science, Technology, and the Limits of Judicial Competence.' *Science* 214 (December 11): 1211—1215.

Just, R. E., D. L. Hueth, and A. Schmitz. 1982. *Applied Welfare Economics and Public Policy*. Englewood Cliffs, N.J.: Prentice-Hall.

Kahneman, D. and A. Tversky. 1973. 'On the Psychology of Prediction.' *Psychology Review* 80: 237—351.

—— and ——. 1977. 'Intuitive Prediction: Biases and Corrective Procedures.' Decision Research, Technical Report PTR—1042—77—6, Eugene, Ore.

—— and ——. 1979. 'Prospect Theory.' *Econometrica* 47(2): 263—292.

Kaplan, S. and B. H. Garrick. 1981. 'Some Misconceptions about Misconceptions: A Response to Abramson.' *Risk Analysis* 1(4): 231—233.

Kawecki, J. M. 1978. 'Emission of Sulfur-Bearing Compounds from Motor Vehicle and Aircraft Engines: A Report to Congress.' EPA—600/9—78—028, Biospherics Inc., Rockville, Md. August.

Keeney, R. L. 1976. 'A Group Preference Axiomatization with Cardinal Utility.' *Management Science* 23(2): 140—145.

——. 1977. 'The Art of Assessing Multiattribute Utility Functions.' *Organizational Behavior and Human Performance* 19(2): 267—310.

——. 1980a. 'Equity and Public Risk.' *Operations Research* 28(3): 527—534.

——. 1980b. *Siting Energy Facilities*. New York: Academic Press.

——. 1981. 'Evaluation of Mortality Risks from an Organizational Perspective.' Presented at the Conference on the Value of Life, Geneva. March 30—April 1.

—— and C. Kirkwood. 1975. 'Group Decision Making Using Cardinal Social Welfare Functions.' *Management Science* 22(4): 430—437.

—— and H. Raiffa. 1972. 'A Critique of Formal Analysis in Public Decision Making.' In

A. W. Drake, R. F. Keeney, and P. M. Morse, eds. *Analysis of Public Systems*. Cambridge, Mass.: MIT Press.

—— and ——. 1976. *Decisions with Multiple Objectives, Preferences and Value Tradeoffs*. New York: Wiley.

——, R. Kulkarni, and K. Nair. 1979. 'A Risk Analysis of an LNG Terminal.' *Omega* 7: 191—205.

Kelman, S. 1980a. 'Economic Incentives and Environmental Policy: Politics Ideology, and Philosophy.' Kennedy School of Government, Harvard University,Cambridge, Mass.

——. 1980b. 'Occupational Safety and Health Administration.' In J. Q. Wilson, ed. *The Politics of Regulation*, pp. 165—168. New York: Longman.

——. 1982. 'Ethics — Cost-Benefit Analysis and Environmental, Safety, and Health Regulation: Ethical and Philosophical Considerations.' In D. Swartzman, R. A. Liroff, and K. G. Croke, eds. *Cost-Benefit Analysis and Environmental Regulations: Politics, Ethics, and Methods*. Washington, D.C.: Conservation Foundation.

Kirkwood, C. W. 1972. 'Decision Analysis Incorporating Preferences of Groups.' Technical Report No. 47, Operations Research Center, M.I.T., Cambridge, Mass.

——. 1974. 'Group Decision Making under Forced Uncertainty.' Technical Report 74—5, Department of Industrial and Operations Engineering, University of Michigan, Ann Arbor.

——. 1979. 'Pareto Optimality and Equity in Social Decision Analysis.' *IEEE Transactions on Systems, Man and Cybernetics* SMC—9 (2): 89—91.

—— and R. K. Sarin. 1980. 'Preference Conditions for Multiattribute Value Functions.' *Operations Research* 28: 225—231.

Kraft, M. E. 1982a. 'Risk Analysis in the Legislative Process: Congress and Risk Managment Decisionmaking.' Technical Report No. 82—1398—2, J. H. Wiggins Company, Redondo Beach, Calif. April.

——. 1982b. 'The Use of Risk Analysis in Federal Regulatory Agencies: Problems and Prospects.' Technical Report No. 82—1398—3, J. H. Wiggins Company, Redondo Beach, Calif. June.

Krieger, M. H. 1970. 'Six Propositions on the Poor and Pollution.' *Policy Sciences* 1: 311—318.

Krutilla, J. and O. Eckstein. 1958. *Multiple Purpose River Development*. Baltimore: Johns Hopkins University Press.

Lathrop, J. W. 1980. 'The Role of Risk Assessment and Facility Siting: An Example from California.' Working Paper WP. 80—150. Laxenburg, Austria: International Institute for Applied Systems Analysis. October.

—— and J. Linnerooth. 1983. 'The Role of Risk Assessment in a Political Decision Process.' In P. Humphries, O. Svenson, and A. Vari, eds. *Analysing and Aiding Decision Processes*. Amsterdam: North-Holland.

Lave, L. B. 1972. 'Risk, Safety, and the Role of Government.' In *Perspectives on Benefit-Risk Decision Making*, pp. 96—108. Report of a colloquium conducted by the Committee on Public Engineering Policy, National Academy of Engineering, Washington, D.C.

——. 1980. 'Economic Tools for Risk Reduction.' In R. C. Schwing and W. A. Albers, Jr., eds. *Societal Risk Assessment: How Safe is Safe Enough*. New York: Plenum.

Lawless, E. W., M. V. Jones, and R. M. Jones. 1984. 'Comparative Risk Assessment: Toward an Analytical Framework.' Midwest Research Institute, Kansas City, Mont. January.

Layard, R., ed. 1972. *Cost-Benefit Analysis: Selected Readings*. Middlesex, England: Penguin.

——. 1974. *Cost-Benefit Analysis*. New York: Penguin.

Leape, J. P. 1980. 'Quantitative Risk Assessment in Regulation of Environmental Carcinogens.' *Harvard Environmental Law Review* 4: 96—97.

Levine, A. 1982. *Love Canal: Science, Politics and People.* Boston: Heath.

Lichtenstein, S. and B. Fischhoff. 1977. 'Do Those Who Know More Also Know More About How Much They Know?' *Organizational Behavior and Human Performance* 20: 159—183.

——, P. Slovic, and L. D. Phillips. 1977. 'Calibration of Probabilities: The State of the Art.' In H. Jungermann and G. deZeeuw, eds. *Decision Making and Change in Human Affairs.* Amsterdam: D. Reidel.

——, ——, B. Fischhoff, M. Layman, and B. Combs. 1978. "Judged Frequency of Lethal Events.' *Journal of Experimental Psychology: Human Learning and Memory* 4: 551—578.

Lind, R. C. 1972. 'The Analysis of Benefit-Risk Relationships: Unresolved Issues and Areas for Further Research.' In *Perspectives on Benefit-Risk Decision Making*, report of a colloquium conducted by the Committee on Public Engineering Policy, National Academy of Engineering, Washington, D.C.

Linnerooth, J. 1979. 'The Value of Human Life: A Review of the Models.' *Economic Inquiry* 17: 52—74.

Linstone, H. A. 1981. 'The Multiple Perspective Concept, with Applications to Technology Assessment and Other Decision Areas.' *Technological Forecasting and Social Change* 20: 275—325.

Liroff, R. A. 1982. 'Federal Experience — Cost-Benefit Analysis in Federal Environmental Programs.' In D. Swartzman, R. A. Liroff, and K. G. Croke, eds. *Cost-Benefit Analysis and Environmental Regulations: Politics Ethics, and Methods.* Washington, D.C.: Conservation Foundation.

Liska, A. E., ed. 1975. *The Consistency Controversy.* New York: Wiley.

Lovins, A. B. 1977. 'Cost-Risk-Benefit Assessments in Energy Policy.' *George Washington Law Review* 45(5): 911—943.

Lowrance, W. W. 1976. *Of Acceptable Risk*, p. 129. Los Altos, Calif.: Kaufmann.

——. 1980. 'The Nature of Risk.' In R. C. Schwing and W. A. Albers, Jr., eds. *Societal Risk Assessment: How Safe is Safe Enough*? New York: Plenum.

MacGregor, O. and P. Slovic. 1984. 'Perceived Acceptability of Risk Analysis as a Decision-Making Approach.' Decision Research, Report SY—1, Eugene, Ore.

MacLean, D. 1979. 'A Moral Requirement for Energy Policies.' In *Energy and the Future.* Totawa, N. J.: Rowman and Littlefield.

——. 1982a. 'Risk and Consent: Philosophical Issues for Centralized Decisions.' *Risk Analysis* 2(2): 59—67.

——. 1982b. 'Values at Risk.' Prepared for the Working Group on Risk and Consent, draft paper, Center for Philosophy and Public Policy, University of Maryland, College Park, Md.

Majone, G. 1980. 'An Anatomy of Pitfalls.' In G. Majone and E. S. Quade, eds. *Pitfalls of Analysis.* International Institute for Applied Systems Analysis. Chichester, England: Wiley.

—— and E. S. Quade. 1980. 'Introduction.' In G. Majone and E. S. Quade, eds. *Pitfalls of Analysis.* International Institute for Applied Systems Analysis, Chichester, England: Wiley.

Makridakis, S. and S. C. Wheelwright, eds. 1979. *Forecasting.* New York: North-Holland.

Manne, A. S. and R. G. Richels. 1978. 'A Decision Analysis of the U.S. Breeder Reactor Program.' *Energy* 3: 747—767.

March, J. 1978. 'Bounded Rationality, Ambiguity and the Engineering of Choice.' *Bell Journal of Economics* 9 (Spring): 587—608.

Markey, H. T. 1980. 'Statement of Hon. H. T. Markey, Chief Judge of the U.S. Court of Customs and Patent Appeals, Washington, D.C.' In *Risk/Benefit Analysis in the Legislative Process*. U.S. Senate Committee on Science and Technology, No. 71. Washington, D.C.: GPO.

Marraro, C. H. 1982. 'Revising the Ozone Standard.' In L. Lave, ed. *Quantitative Risk Assessment in Regulation*. Washington, D.C.: Brookings Institution.

McKean, R. N. 1958. *Efficiency of Government through Systems Analysis*. New York: Wiley.

Merkhofer, M. W. 1977. 'Methodology for Back-Contamination Risk Assessment for a Mars Sample Return Mission.' SRI Final Report, SRI International, Menlo Park, Calif. April.

—— and W. M. Saade. 1978. 'Decision Flexibility in a Learning Environment.' SRI International, Menlo Park, Calif.

—— *et al.* 1984. 'Risk Assessment and Risk Assessment Methods: The State-of-the Art.' A report prepared for the National Science Foundation. Boston, Mass. December.

Merrill, R. A. 1985. 'Legal Impediments to the Use of Risk Assessment by Regulatory Agencies.' in D. G. Hoel, R. A. Merrill, and F. P. Perera, eds. *Risk Quantitation and Regulatory Policy*. Banbury Report 19. New York: Cold Spring Harbor Laboratory.

Milburn, T. W. and R. S. Billings. 1976. 'Decision-Making Perspectives from Psychology.' *American Behavioral Scientist* 20 (September — October): 111—126.

Miller, A. C. and T. R. Rice. 1983. 'Discrete Approximations of Probability Distributions.' Management Science 29(3): 352-362.

Mishan, E. J. 1967. *The Costs of Economic Growth*. London: Staples.

——. 1976. *Cost-Benefit Analysis*. New York: Praeger.

—— and T. Page. 1979. 'The Methodology of Cost-Benefit Analysis: With Particular Reference to the Ozone Problem.' Social Studies Working Paper 249, California Institute of Technology, Pasadena.

Moll, K. S., E. Baum, F. Capener, G. Dresch, and R. Wright. 1975. 'Hazardous Wastes: A Risk-Benefit Framework Applied to Cadmium and Asbestos.' SRI International, Menlo Park, Calif.

Morgan, M. G. 1981. 'Choosing and Managing Technology-Induced Risk.' *IEEE Spectrum* December: 53—59.

——. 1982. 'Uncertainty and Quantitative Assessment.' In J. V. Rodricks and R. G. Tardiff, eds. *Assessment and Management of Chemical Risks*. Washington, D.C.: American Chemical Society.

——, M. Henrion, and S. C. Morris. 1979. 'Expert Judgments for Policy Analysis: Report of an Invitational Workshop Held at Brookhaven National Laboratory, 1979 July 8—11.' Report BNL 51358, Brookhaven National Laboratory, Upton, New York.

——, ——, J. Nair, T. Mulin, and C. Wiecha. 1984. 'A Generic "PREHEED" on Characterizing and Dealing with Uncertainty in Health and Environmental Risk Assessment.' Carnegie-Mellon University, Pittsburgh, Penn.

Morris, F. A. and E. Duvernoy. 1984. 'The Statutory Basis of Risk Assessment.' In R. A. Waller and V. T. Covello, eds. *Low-Probability High-Consequence Risk Analysis*. New York: Plenum.

Moss, T. and B. Lubin. 1981. 'Risk Analysis: A Legislative Perspective.' In C. R. Richmond, P. J. Walsh, and E. D. Copenhavem, eds. *Health Risk Analysis — Proceedings of the Third Life Sciences Symposium*. Philadelphia: Franklin Institute Press.

National Academy of Sciences. 1977. 'Considerations of Health Benefit-Cost Analysis for Activities Involving Ionizing Radiation Exposure and Alternatives.' EPA 520/4—88—003, U.S. Environmental Protection Agency.

National Planning Association. 1978. 'Social Decision Making for High Consequence, Low Probability Occurrences.' Washington, D.C. October.

National Research Council. 1977. 'Decision Making in the Environmental Protection

Agency.' Committee on Environmental Decision Making, National Academy of Sciences, Washington, D.C.

——. 1982. *Risk and Decision Making: Perspectives and Research.* Committee on Risk and Decision Making. Washington, D.C.: National Academy press.

——. 1983. *Risk Assessment in the Federal Government: Managing the Process.* Committee on the Institutional Means for Assessment of Risks to Public Health. Washington, D.C.: National Academy Press.

National Safety Council. 1979. 'Accident Facts.'

Newell, A. and H. A. Simon. 1972. *Human Problem Solving.* Englewood Cliffs, N.J.: Prentice-Hall.

Nuclear Regulatory Commission. 1975. 'Reactor Safety Study: An Assessment of Accident Risks in U.S. Commercial Nuclear Power Plants.' WASH 1400 (NUREG–75/014). Washington, D.C. October.

Office of Management and Budget. 1980. 'Improving Government Regulations.' Executive Office of the President. November.

Office of Technology Assessment. 1981. *Assessment of Technologies for Determining Cancer Risks from the Environment.* Washington, D.C.: GPO.

Otway, H. J. and D. von Winterfeldt. 1982. 'Beyond Acceptable Risk: On the Social Acceptability of Technologies.' *Policy Sciences* 14: 241–246.

Otway, M. 1980. 'A Perspective on Risk Perception: Confession of a Disillusioned Analyst.' Unpublished manuscript prepared for Risk Perception Workshop held at Decision Research, Eugene, Ore. December 11–13.

Owen, D. L. 1979. 'The Concept of Influence and Its Use in Structuring Complex Decision Problems.' Ph.D. dissertation. Department of Engineering-Economic Systems. Stanford University, Stanford, Calif. June.

Page, T. 1978. 'A Generic View of Toxic Chemicals and Similar Risks. *'Ecology Law Quarterly* 7(2): 207–243.

Parducci, A. 1974. 'Contextual Effects: a Range-Frequency Analysis.' In E. C. Carterette and M. P. Friedman, eds. *Handbook of Perception.* Vol. II. New York: Academic Press.

Parish, R. M. 1976. 'The Scope of Benefit-Cost Analysis.' *Journal of the Economic Society of Australia and New Zealand* 52: 302–314.

Park, C. N. and R. D. Snee. 1983. 'Quantitative Risk Assessment: State-of-the Art for Carcinogenesis.' *American Statistician* 34(4): 427–441.

Parkinson, C. N. 1979. *Parkinson's Law.* New York: Ballantine.

Parry, T. W. and P. W. Winter. 1981. 'Characterization and Evaluation of Uncertainty in Probabilistic Risk Analysis.' *Nuclear Safety* 22: 379–391.

Patwardhan, A., R. Kulkarni, and D. Tocher. 1980. 'A Semi-Markov Model for Characterizing Recurrence of Great Earthquakes.' *Bulletin of the Sociological Society of America* February: 323–347.

Perrone, N. 1976. 'A Mathematical Model to Predict Skull Fracture under Impact Loads.' No. 760768 Society of Automotive Engineers, Warrendale, Penn.

Peters, G. H. 1968. 'Cost Benefit Analysis and Public Expenditure.' Eaton Paper 8, 2nd ed. Institute of Economic Affairs, London.

Philipson, L. L. and J. D. Gasca. 1982. 'Risk Assessment Methodologies and Their Uncertainties — Volume I: A Review of Risk Estimation Approaches.' Technical Report No. 80–1398. J. H. Wiggins Company, Redondo Beach, Calif.

——, H. E. Lambert, and H. E. Roland. 1981. 'A First Look at the Applicability of the Digraph-Fault Tree Process to Transportation Accident Occurrence Modeling.' Proceedings of the Fifth International System Safety Conference, Denver, Colo. July 27–31.

Phillips, L. D. and H. Thomas. 1973. 'Assessing Probability and Utility.' Royal Economic Society Conference on Decision Analysis, Lancaster.

Polanyi, M. 1962. *Personal Knowledge.* London: Routledge & Kegan Paul.

Pope, C. D. 1979. 'The Problem of Symmetry in Assessing Toxic Substance Risks, Bentham vs. Locke.' Paper presented at the Technical Information Project, Inc., Conference on "Toxic Substances: Decisions and Values." Washington, D.C. February 19—20.

Prest, A. R. and R. Turvey. 1965. 'Cost-Benefit Analysis: A Survey.' *Economic Journal* 75: 683—735.

Quade, E. S. 1980. 'Pitfalls in Formulation and Modeling.' In G. Majone and E. S. Quade, eds. *Pitfalls of Analysis*. International Institute for Applied Systems Analysis. Chichester, England: Wiley.

Raiffa, H. 1968. *Decision Analysis: Introductory Lectures on Choices under Uncertainty*. Reading, Mass.: Addison—Wesley.

—— and R. Zeckhauser. 1981. 'Reporting of Uncertainties in Risk Analysis.' In T. McCurdy and H. Richmond. *Conceptual Approaches to Health Risk Assessment*. Washington. D.C.: U.S. Environmental Protection Agency. May.

——, W. B. Schwartz, and M. C. Weinstein. 1977. 'Evaluating Health Effects of Societal Decisions and Programs.' In *Decision Making in the Environmental Protection Agency, Selected Working Papers*. Vol. IIb. National Research Council, National Academy of Sciences, Washington, D.C.

Rappaport, E. 1974. 'Economic Analysis of Life-and-Death Decision Making.' Appendix 2 in Report No. ENG 7478, School of Engineering and Applied Science, UCLA. November.

Rasmussen, N. C. 1975. 'Rasmussen on Reactor Safety.' *IEEE Spectrum* 12 (August): 46—55.

Ravetz, J. R. 1980. 'Public Perception of Acceptable Risks.' In M. Dierkes *et al.* eds. *Technological Risk: Its Perception and Handling in the European Community*. Cambridge, Mass.

Redmond, W. H. 1978. 'Values in Forecasting and Planning.' *Journal of Long Range Planning* II (June): 22—25.

Rhoads, S. E. 1978. 'How Much Should We Spend to Save a Life?' *Public Interest* 51 (Spring): 74—92.

Ricci, P. F. and L. S. Molton. 1981. 'Risk and Benefit in Environmental Law.' *Science* 214 (December 4): 1096—1110.

Richmond, H. M. 1981. 'A Framework for Assessing Health Risks Associated with National Ambient Air Quality Standards.' *Environmental Professional* 3(3—4): 225.

Rivlin, A. M. 1971. *Systematic Thinking for Social Action*. Washington, D.C.: Brookings Institution.

Roback, H. 1972. 'Politics and Expertise in Policy Making.' In *Perspectives on Benefit-Risk Decision Making*. Report of a colloquium conducted by the Committee on Public Engineering Policy. Washington, D.C.: National Academy of Engineering.

Robertson, L. S. and P. L. Zaldor, 1977. *Driver Education and Fatal Crash Involvement of Teenage Drivers*. Washington, D.C.: Insurance Institute for Highway Safety.

Rosen, S. 1974. 'Hedonic Prices and Implicit Market: Product Differentiation in Pure Competition.' *Journal of Political Economy 82*: 34—55.

Rosenberg, M. 1981. 'Presidential Control of Agency Rulemaking: An Analysis of Constitutional Issues Which May Be Raised by Executive Order 12291,' pp. 38 and 42. Committee Print 97—0, Committee on Energy and Commerce, U.S. House of Representatives.

Rothenberg, J. 1975. 'Cost-Benefit Analysis: A Methodological Exposition.' In M. Guttentag and E. Streuning, eds. *Handbook of Evaluation Research*. Vol. II. Beverly Hills, Calif.: Sage.

Rowe, W. D. 1977a. 'Governmental Regulation of Societal Risks.' *George Washington Law Review* 45(5): 944—968.

——. 1977b. *An Anatomy of Risk*. New York: Wiley.

——. 1979. 'Risk/Benefit Determination.' *American Industrial Hygiene Association Journal* 40 (December): 1200—1206.

—— and I. E. Broder. 1979. 'Risk Assessment in Catastrophe Management Policies.' Final Report, NSF P. O. No. 79—5P—1028. Americn University Institute for Risk Analysis, Kogod College of Business Administration. Washington, D.C. July.

——, R. C. d'Arge, and D. S. Brookshire. 1980. 'An Experiment on the Economic Value of Visibility.' *Journal of Environmental Economics and Management* 7: 1—19.

Runchal, A. K., M. W. Merkhofer, E. Olmsted, and J. Davis. 1984. 'Probability Encoding of Hydrologic Parameters.' Analytic and Computational Research, West Los Angeles, Calif.

Sabatier, P. A. 1977. 'Regulatory Policy-Making: Toward a Framework of Analysis.' *Natural Resources Journal* 17 (July): 415—460.

Sagan, L. A. 1984. 'Problems in Health Measurements for the Risk Assessor.' In P. Ricci *et al.*, eds. *Technological Risk Assessment.* The Hague: Martinus Nijhoff.

Sage, A. P. and E. B. White. 1980. 'Methodologies for Risk and Hazard Assessment: A Survey and Status Report.' *IEEE Transactions on Systems, Man, and Cybernetics* SMC—10(8): 425—446.

Savage, L. J. 1971. 'The Elicitation of Personal Probabilities and Expectations.' *Journal of the American Statistical Association* 66: 783—801.

——. 1972. *The Foundations of Statistics.* Rev. ed. New York: Dover.

Schelling, T. C. 1966. 'The Life You Save May Be Your Own.' In S. B. Chase, Jr., ed. *Problems in Public Expenditure Analysis.* Washington, D.C.: Brookings Institution.

Schlaifer, R. O. 1969. *Analysis of Decisions under Uncertainty.* New York: McGraw-Hill.

Schneider, H. and G. Beier. 1974. 'Experiment and Accident: Comparison of Dummy Tests.' SAE Technical Paper Series No. 741177. Society of Automotive Engineers. Warrendale, Penn.

Schoemaker, P. J. 1980. *Experiments on Decisions under Risk, the Expected Utility Hypotheses.* Boston: Martinus Nijhoff.

Schulze, W. D., D. S. Brookshire, E. G. Walther, K. Kelly, M. A. Thayer, R. L. Witworth, S. Ben-David, W. Malm, and J. Molenar. 1981. 'The Benefits of Preserving Visibility in the National Parklands of the Southwest.' Vol. VIII. *Methods Development for Environmental Control Benefits Assessment.* Final report to the U.S. EPA, Grant No. R805059010. Office of Exploratory Research, Office of Research and Development.

Schweitzer, G. E. 1982. 'Monitoring to Support Risk Assessments at Hazardous Waste Sites.' In F. Long and G. Schweitzer, eds. *Risk Assessment at Hazardous Waste Sites.* New York: American Chemical Society.

Sen, A. K. 1970. *Collective Choice and Social Welfare.* San Francisco: Holden-Day.

Simon, H. A. 1957. *Models of Man: Social and Rational.* New York: Wiley.

Sinden, J. A. and A. C. Worrel. 1979. 'Unpriced Values: Decisions without Market Prices.' Yale University, New Haven, Conn.

Slovic, P. 1978. 'Judgment, Choice and Societal Risk Taking.' In K. R. Hammond, ed. *Judgment and Decision in Public Policy Formation*, AAAS Selected Symposium Series. Boulder, Colo.: Westview Press.

—— and B. Fischhoff. 1982. 'How Safe Is Safe Enough? Determinants of Perceived and Acceptable Risk.' in Gould and Walker, eds. *Too Hot to Handle.* New Haven: Yale University Press.

—— and S. Lichtenstein. 1971. 'Comparison of Bayesian and Regression Approaches to the Study of Information Processing Judgment.' *Organizational Behavior and Human Performance* 6: 649—744.

——, B. Fischhoff, and S. Lichtenstein. 1975. 'Cognitive Processes and Societal Risk Taking.' In D. Okrent, ed. *Risk-Benefit Methodology and Application: Some Papers Presented at the Engineering Foundation Workshop, September 22—26, 1975, Asilomar, California.* Los Angeles: UCLA. December.

——, ——, and ——. 1979a. 'Rating the Risks.' *Environment* 21(3).

——, ——, and ——, 1979b. 'Risk Assessment: Technical and Behavioral Issues.' Prepared for the Congressional Forum on Risk-Benefit Analysis in the Legislative Process, Washington D.C. July 25. Decision Research, Eugene, Ore.

——, ——, and ——. 1981. 'Perceived Risk: Psychological Factors and Social Implications.' In F. Warner and D. H. Slater, eds. *The Assessment and Perception of Risk*. London: Royal Society.

——, S. Lichtenstein, and B. Fischhoff. 1982. 'Images of Disaster: Perception of Acceptance of Risks from Nuclear Power.' In G. Goodman, ed. *Impacts of Risks of Energy Strategies*. New York: Academic Press.

——, B. Fischhoff, and S. Lichtenstein. 1983. 'Characterizing Perceived Risk.' In R. W. Kates and C. Hohenemser, eds. *Technological Hazard Management*. Cambridge, Mass.: Oelgeschlager, Gunn and Hain.

Smith, A. E., P. C. McNamee, and M. W. Merkhofer. 1982. 'Development of Decision Analysis Methodology for Health Risk Assessment: An Illustrative Application to Alternative Air Quality Standards for Carbon Monoxide.' Final Report 3432, SRI International, Menlo Park, Calif.

Smith, H. 1958. *The Religions of Man*, p. 86. New York: Harper & Row.

Smith, V. K. 1986. 'A Conceptual Overview of the Foundations of Benefit-Cost Analysis.' In J. D. Bentkover, V. T. Covello, and J. Mumpower, eds. *Benefits Assessment: The State of the Art*. Dordrecht: D. Reidel.

—— and W. J. Vaughan. 1980. 'The Implications of Model Complexity for Environmental Management. *Journal of Environmental Economics and Management* 7: 184–208.

Sohn, M. and R. Litan. 1981. 'Regulatory Oversight Wins in Court.' *Regulation* July–August: 17–24.

Spetzler, C. S. and C.-A. S. Staël von Holstein. 1975. 'Probability Encoding in Decision Analysis.' *Management Science* 22: 340–358.

Staël von Holstein, C.-A. S. and J. E. Matheson. 1979. 'A Manual for Encoding Probability Distributions.' SRI Project 7078, SRI International, Menlo Park, Calif.

Stafford, R. T. 1981. 'Remarks of U. S. Senator Robert T. Stafford.' Conference on Compensation for Toxic Substances Pollution, The Environmental Law Institute, Virginia Environmental Endowment, University of Virginia Law School, Charlottesville, Va. November 23.

Stainaker, R. L. 1983. 'Crash Victim Simulation — A First Step in Child Auto Safety.' No. 831661, Society of Automotive Engineers, Warrendale, Penn.

Starr, C. 1978. 'Risk-Benefit Analysis and Full Disclosure.' Presented at the American Association for the Advancement of Science. Washington, D.C. February 13.

Stokey, E. and R. Zeckhauser. 1978. *A Primer for Policy Analysis*. New York: Northon.

Strauch, R. 1980. 'Risk Assessment as a Subjective Process.' P–6460. Rand Corporation, Santa Monica, Calif.

Sugden, R. and A. Williams. 1978. 'Practical Cost-Benefit Analysis.' Oxford, England.

Tani, S. N. 1978. 'Decision Analysis of the Synthetic Fuels Commercialization Program.' National Computer Conference Proceedings, SRI International Preprint, SRI International, Menlo Park, Calif.

Thaler, R. and S. Rosen. 1976. 'The Value of Saving of Life: Evidence from the Labor Market.' In N. Terleckys, ed. *Household Production and Consumption*. New York: Columbia University Press.

Thomas, K. 1981. 'Comparative Risk Perception: How the Public Perceives the Risks and Benefits of Energy Systems.' In F. Warner and D. H. Slater, eds. *The Assessment and Perception of Risk*. London: Royal Society.

Thompson, M. S. 1980a, *Benefit-Cost Analysis for Program Evaluation*. Beverely Hills, Calif.: Sage.

——. 1980b. 'An Outline of the Cultural Theory of Risk.' Working Paper No. 80—177. International Institute for Applied Systems Analysis, Laxenburg, Austria.

——. 1982. *Decision Analysis for Program Evaluation*. Cambridge, Mass: Ballinger.

Throsby, D. C. and P. W. Abelson. 1976. 'Economic Evaluation of Net National Benefits from Sandmining on Moreton Island,' pp. 2—148 to 2—177. In *Moreton Island Environmental Impact Study and Strategic Plan*. A. A. Health and Partners.

Tomatis, L., C. Agthe, H. Bartsch, J. Huff, R. Montesano, R. Saracci, E. Walker, and J. Wilbourn. 1978. 'Evaluation of the Carcinogenicity of Chemicals: A Review of the Monograph Program of I. A. R. C..' *Cancer Research* 38: 877.

Torrance, G. W., M. H. Boyle, and S. P. Horwood. 1982, 'Application of Multiattribute Utility Theory to Measure Social Preferences for Health States.' *Operations Research* 30(6): 1043—1069.

Travis, C. C. *et al.* 1983. 'Exposure Assessment Methodology and Reference Environments for Synfuel Risk Analysis.' ESD Publication No. 2232, Oak Ridge National Laboratory, Oak Ridge, Tenn.

Tribe, L. H. 1972. 'Policy Science: Analysis or Ideology?' *Philosophy and Public Affairs* 2: 66—110.

——. 1974. 'Ways Not to Think about Plastic Trees: New Foundations for Environmental Law.' *Yale Law Journal* 83(7): 1315—1348.

Turner, C. F. and E. Krauss. 1978. 'Fallible Indicators of the Subjective State of the Nation.' *American Psychologist* 33: 456—470.

Tversky, A. 1973. 'Availability: A Heuristic for Judging Frequency and Probability.' *Cognitive Psychology* 5: 207—232.

—— and D. Kahneman. 1974. 'Judgment under Uncertainty: Heuristics and Biases.' *Science* September 26: 1124—1131.

—— and ——. 1981. 'The Framing of Decisions and the Psychology of Choice.' *Science* 211 (January 30): 453—458.

—— and S. Sattagh. 1981. *Evidential Impact of Base Rates*. Technical Report 4, Stanford University, Department of Psychology, Stanford, Calif. May 15.

U.S. Congress. 1979. *Cost-Benefit Analysis: Wonder Tool or Mirage*. Committee on Interstate and Foreign Commerce, Subcommittee on Oversight and Investigations, 96th Congress, 2nd Session, Committee Print 96—IFC62. Washington, D.C.: GPO.

U.S. House of Representatives. 1982. 'Risk Analysis Research and Demonstration Act of 1982.' Committee on Science and Technology, Report No. 97—625. June 24.

Utterback, J. M. 1978. 'Creating a Climate for More Innovative Responses to Regulation.' Paper presented at the 32nd National Conference on the Advancement of Research, Denver Research Institute, University of Denver, Colo.

Van Horn, A. J. and R. Wilson. 1976. 'The Status of Risk-Benefit Analysis.' Discussion paper, Energy and Environmental Policy Center, Harvard University, Cambridge, Mass. December.

Vaughan, W. J. and C. Russell. 1982. *The National Benefits of Water Pollution Control: Fresh Water Recreational Fishing*. Baltimore: Johns Hopkins University Press.

Vesely, W. E. 1984. 'Engineering Risk Analysis.' In P. F. Ricci *et al.* eds. *Technological Risk Assessment*. The Hague: Martinus Nijhoff.

—— *et al.* 1981. 'Fault Tree Handbook.' NUREG/CR—0492, U.S. Nuclear Regulatory Commission. January.

von Foerster, H. 1972. 'Responsibilities of Competence.' *Journal of Cybernetics* 2(2): 1.

von Neumann, J. and O. Morgenstern. 1947. *Theories of Games and Economic Behavior*. Princeton: Princeton University Press.

von Winterfeldt, D. and W. Edwards. 1982. 'Costs and Payoffs in Perceptual Research.' *Psychological Bulletin* 91 (3): 609—622.

—— and ——. 1986. *Decision Analysis and Behavioral Research*. New York: Cambridge University Press.

Wasson, P. C. and P. N. Johnson-Laird. 1972. *Psychology of Reasoning, Structure and Content*. Cambridge, Mass.: Harvard University Press.

Watson, S. R. 1979. 'Decision Analysis as a Replacement for Cost/Benefit Analysis.' CUED/F—CAMS/TR197. Cambridge University, Cambridge, England.

Weathermax, R. K. 1975. 'Virtues and Limitations of Risk Analysis.' *Bulletin of the Atomic Scientists* 31(10): 767—769.

Weidenbaum, M. 1978. 'On Estimating Regulatory Costs.' *Regulation* May—June: 17.

White, D. J. 1975. *Decision Methodology*. New York: Wiley.

Wildavsky, A. 1980. 'Statement of Professor Aaron Wildavsky, Department of Political Science, University of California at Berkeley, Berkeley, California.' In *Risk/Benefit Analysis in the Legislative Process*. U.S. Senate Committee on Science and Technology, No. 71. Washington, D.C.: GPO.

Williams, C. and G. Crawford. 1980. *Analysis of Subjective Judgment Matrices*. Santa Monica, Calif.: Rand Corporation.

Wilson, J. Q. 1973. 'On Pettigrew and Armor: An Afterword.' *Public Interest* 30: 132—135.

Wilson, R. 1975. 'Examples in Risk-Benefit Analysis.' *Chemtech* No. 6 (October): 604—607.

Winkler, R. L. 1970. 'Consensus of Subjective Probability Estimates.' *Management Science B* 15(2): 61—76.

Wismans, J., J. Maltha, and J. Van Wiljk. 1982. 'MADYMO — A Crash Victim Simulation Computer Program for Biomechanical Research and Optimization of Designs for Impact Injury Prevention.' AGARD, Cologne. April.

Wynne, B. 1984. 'Commentary on "Uncertainty, Ignorance, and Policy," by Jerome R. Ravetz.' International Institute for Applied Systems Analysis, Laxenburg, Austria.

Zeckhauser, R. 1975. 'Procedures for Evaluating Lives.' *Public Policy* 23: 419—464.

—— and D. Shepard. 1976. 'Where Now for Saving Lives?' *Law and Contemporary Problems* 40(4): 5—45.

Applications of Decision-Aiding Approaches

To better illustrate the commonalities and differences among quantitative decision-aiding approaches, this appendix contains three case studies. Each illustrates an application of one or more specific approaches to a particular problem involving technological risk. The first is essentially a cost-benefit analysis, the second a decision analysis, and the third demonstrates the application of several versions of applied social welfare theory.

The studies are actual or slightly modified versions of professional applications of decision-aiding approaches. Although representative, one should not infer conclusions concerning the relative usefulness of the broad categories (CBA, DA, or applied SWT) that they represent. None of the case studies utilizes anywhere near the full range of sophisticated data collection, modeling, and analysis procedures that proponents of each type of approach often advocate. Furthermore, since the descriptions are abbreviated summaries, they do not provide the sort of detailed discussions of analytic assumptions and approximations normally associated with the documentation of a professional study. Without such information (and a clear description of the context of the analysis) it is impossible to reach a reasoned judgment about the value of an approach. The descriptions are meant solely to provide the reader with several concrete examples of the different perspectives and procedures that might be adopted by analysts in the application of decision-aiding approaches.

Appendix A

A Cost-Benefit Analysis of Mobile-Source Sulfur Oxide Emission Control

This case study describes an analysis of a situation for which previous research had provided a useful data base and for which a balancing of costs and benefits was perceived to be a legitimate basis for decision making. Although hypothetical, this case study closely follows an application originally described by Merkhofer *et al.* (1978) and Kawecki (1978).

Decision Situation

In Los Angeles the city council must decide whether to pass legislation that would place stringent limits on the sulfur content of gasoline sold in the area. If passed, the legislation is expected to improve the city's air quality. However, the legislation would require additional refining by gasoline producers and result in higher fuel prices for automobile drivers. Should the legislation be passed?

Background

Sulfur is one of the most abundant and complex of air pollutants. In the United States alone, nearly 30 million tons of sulfur oxides are emitted into the atmosphere annually. Most of these emissions are in the form of gaseous sulfur dioxide (SO_2) resulting from the burning of fossil fuels. However, both during emission and while in the atmosphere, SO_2 can undergo chemical reactions to form sulfate particulate matter generally referred to as sulfates (SO_4). Epidemiological and toxicological studies indicate that at least some of these oxidation products are more toxic than SO_2 and are the likely cause of the adverse health and welfare effects generally associated with sulfur emissions.

Although motor vehicles are responsible only for about 2 percent of sulfur emissions nationwide, these emissions have recently become of special concern. Beginning with the 1975 model year, automobile manufacturers began installing oxidation catalysts on cars as a means of reducing emissions of carbon monoxide (CO) and hydrocarbons (HC). Without a catalytic converter, most sulfur contained in gasoline passes out of the vehicle's tailpipe as SO_2. A converter, however, tends to convert a portion of the SO_2 directly into sulfur trioxide, which rapidly combines with water vapor in the exhaust to form sulfuric acid aerosol. Sulfuric acid is known to be one of the more harmful forms of sulfate.

Sulfur emissions are of particular concern in the Los Angeles basin, an area of high automotive traffic density and meteorological and geographic characteristics which are extremely favorable to the buildup of atmospheric pollutants. The abundant sunshine and photochemical smog tend to promote an unusually rapid and complete conversion of SO_2 to SO_4. Frequent temperature inversions, low average wind speeds, and the surrounding mountains restrict air movement, while limited rainfall precludes frequent clearing of sulfates from the atmosphere. Furthermore, forecast shortages of relatively clean natural gas and increased burning of sulfur containing fuel make it likely that sulfur emissions from stationary sources will increase over time.

To reduce the risks associated with anticipated increasing levels of sulfates in the Los Angeles basin, it has been proposed that regulations be established that would mandate that sulfur be removed from gasoline. Although this would require the oil industry to invest in new refinery equipment, reducing the level of sulfur in fuel would reduce proportionately both SO_2 and sulfate emissions.

Analysis

To support decision making on this issue, an analysis of the desulfurization option was commissioned. The project team established for the effort consisted both of cost-benefit analysts and individuals with expertise in relevant technical and health issues. The year in which the analysis was conducted was 1977. A description of the approach and the results follows.

Specification and Characterization of Alternatives

The basic option for the analysis, desulfurization of gasoline, was firmly specified by the clients. Since total elimination of sulfur was judged to be economically infeasible, desulfurization down to some maximum allowable level was considered. In Southern California the sulfur content of fuel was estimated to average between 310 parts per million (ppm) for unleaded and 600 ppm for regular leaded gasoline. The desulfurization option was taken to be a reduction in gasoline sulfur content to 100 ppm.

Decision Criterion

The criterion established for evaluating the desulfurization option was a comparison of its estimated costs and benefits. To simplify the evaluation, costs and benefits were computed and compared for a typical year, 1980.

Outcome Measures

Through a series of discussions both within the project team and with the clients of the study, specific outcome variables for measuring costs and benefits were selected. Costs were taken to be the incremental annualized capital investment charges and operating cost associated with desulfurization. Benefits consisted of improved health and visibility. Additional benefits were identified, such as reduced sulfur-caused damage to materials (especially galvanized steel and paint coatings on steel) and plant and animal losses (due, e.g., to the contribution of SO_2 to acid precipitation and the acidification of water bodies). However, to permit a defensible analysis within the limited time and resources available and because these other benefits were judged to be of less significance for Los Angeles, only health and visibility impacts were included.

Visibility was measured in miles. To identify variables for measuring health impact, epidemiological studies were reviewed. Such studies suggest that short-term exposures (24 hours or longer) to elevated levels of sulfates tend to aggravate preexisting heart and lung disorders in elderly patients, aggravate asthma, and perceptibly increase daily mortality. Repeated short-term exposures or elevations of annual average exposures seem to increase the incidence of lower respiratory disease in children and increase the risk of chronic respiratory disease in adults. Thus, health outcome measures selected for the analysis were person-days aggravation of heart and lung disease symptons, asthma attacks, deaths, cases of children's lower respiratory disease, and chronic respiratory disease.

Estimation of Costs

Estimation of the costs of the desulfurization option was relatively straightforward. A review of gasoline sources for the area indicated that most gasoline processed by refining crude oil would require desulfurization to attain a 100 ppm sulfur level.[1] Therefore incremental annualized costs were estimated assuming the installation, where necessary, of additional refining equipment to permit fluid catalytic cracking and coking units to attain a 100 ppm sulfur level. Because such costs would vary widely among refineries, capital investment and operating costs were estimated individually for several representative refinery situations. Capital investment included direct process investment for desulfurization, hydrogen manufacture (used in the desulfurization process), sulfur recovery facilities, associated off-site costs, interest during construction, and working capital. Manufacturing costs included return on direct capital investment at a 12 to 15 percent after-tax basis, fuel (energy) costs, and labor costs, and included hydrotreating the feedstock of refinery catalytic cracking units.

Total computed costs, expressed in 1977 dollars, ranged from nearly 0 to almost 4 cents per gallon, depending on processing configuration, the sulfur content of the crude oil processed by the refinery, its capacity, and the availability of hydrogen. Extrapolating according to the volumes of gasoline produced by various refineries facing various costs, an overall average was computed of approximately 2 cents per gallon of total gasoline produced.

Estimation of Benefits

Estimating the health and visibility benefits associated with the desulfurization option required a much more involved process. Figure A1

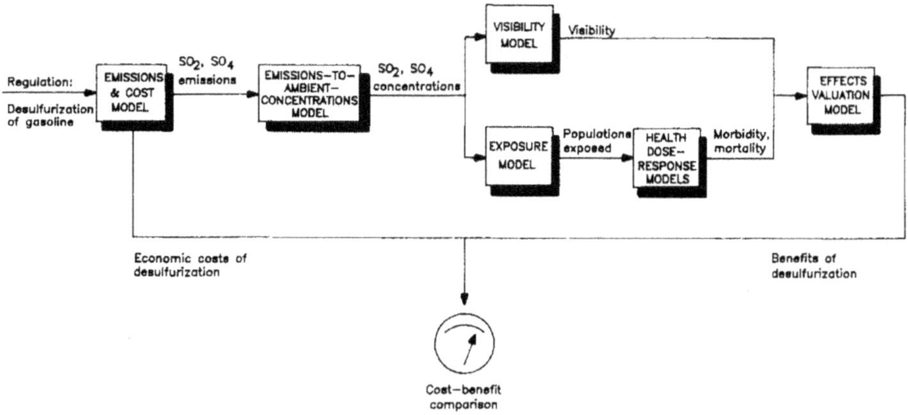

Fig. A1. Models constructed to estimate costs and benefits of desulfurization.

illustrates the models that were constructed for the analysis, including those developed to estimate benefits. First, an emissions model was developed to forecast the impact of gasoline desulfurization on the geographic pattern and composition of sulfur emissions. Then, an emissions-to-atmospheric concentrations model was developed to estimate the impact of changing emissions patterns on ambient concentrations of SO_4. Ambient concentrations were then input directly to a visibility model designed to determine the contribution of SO_4 to visibility degradation. To estimate health effects, an exposure model was used to convert ambient concentrations into exposure profiles for the sensitive population groups who are at risk. Then, a health effects dose-response model was used to convert exposure conditions into estimated numbers of health effects. Finally, a valuation model assigned dollar values to changes in visibility and health effects so that an equivalent economic value of benefits might be estimated.

Ambient concentrations of SO_4

For the purpose of estimating the impact of desulfurization on ambient concentrations of SO_4, various sources of sulfur emissions were identified and located geographically. Relatively simple plume models for the dispersion and conversion of SO_2 and SO_4 were then used to determine the contribution of each source to ambient concentrations.

The modeling region chosen was a square approximately 53 miles on each side overlying the central portion of the Los Angeles basin, as shown in Figure A2. The region was subdivided by a grid into 169 cells each roughly 4 miles on a side.[2]

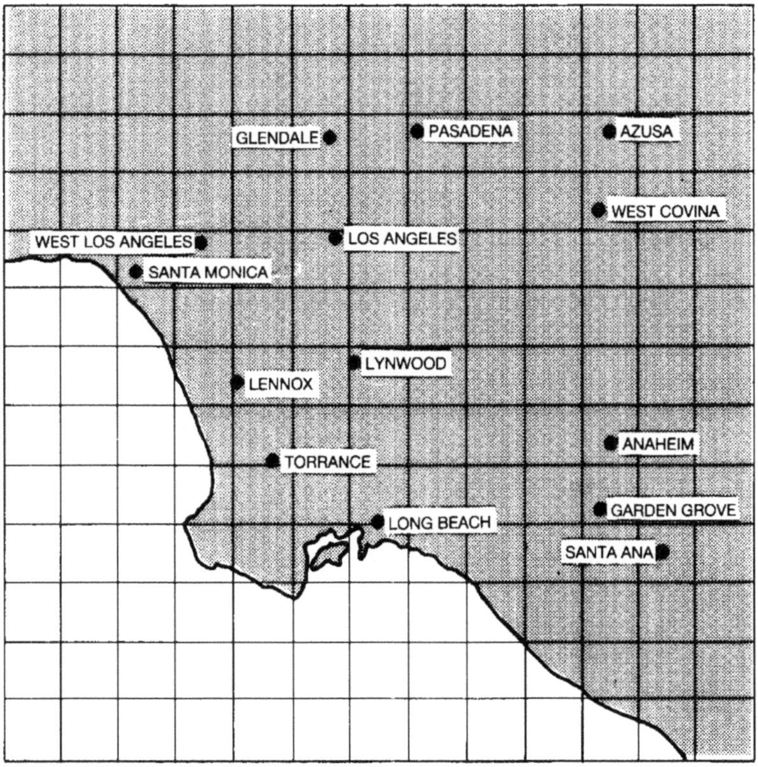

Fig. A2. Modeling region and cells defined for estimating SO_4 levels.

Automotive sources

Automobile emissions were assumed to be proportional to traffic density. Estimates of the spatial distribution of freeway and surface street traffic for 1969 provided by Roth *et al.* (1974) were used to estimate the daily vehicle miles traveled within each cell, as shown in Figures A3 and A4. These data were then scaled to 1980 using vehicle traffic growth rates by cell observed between 1969 and 1974 (Cass 1978).

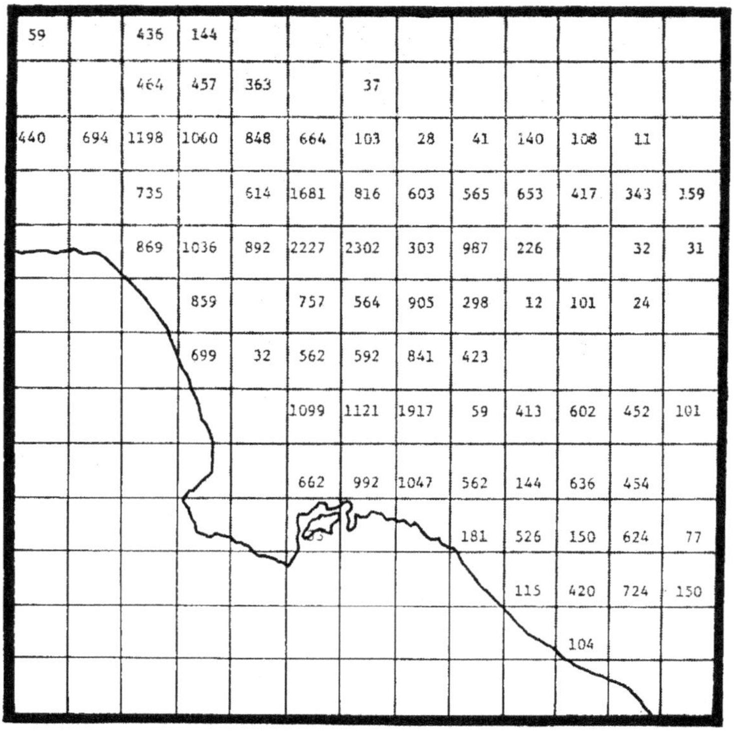

Fig. A3. Spatial distribution of freeway traffic, 1969.

To estimate the SO_2 and SO_4 automobile emissions within each cell, average rates for the 1980 on-the-road vehicle fleet were estimated. Estimates were performed separately for surface and freeway driving (because emission rates depend on automobile speed) and assumed a vehicle age and use distribution given in Nordsieck (1975). All sulfur not converted to SO_4 was assumed to be emitted as SO_2. The resulting average emission rates, which account for estimated improvements in fuel economy, catalytic converter technology necessary to meet more stringent CO and HC standards for 1979 and 1980, and gradual deterioration of catalysts with age, are summarized in Table A1. Total automobile emissions were obtained by multiplying the emission rates by estimated traffic by cell.

To obtain the impact of automobile emissions on ambient SO_4 concentrations, emissions within cells were approximated as point sources located at the centers of the cells. For any given direction of wind, each

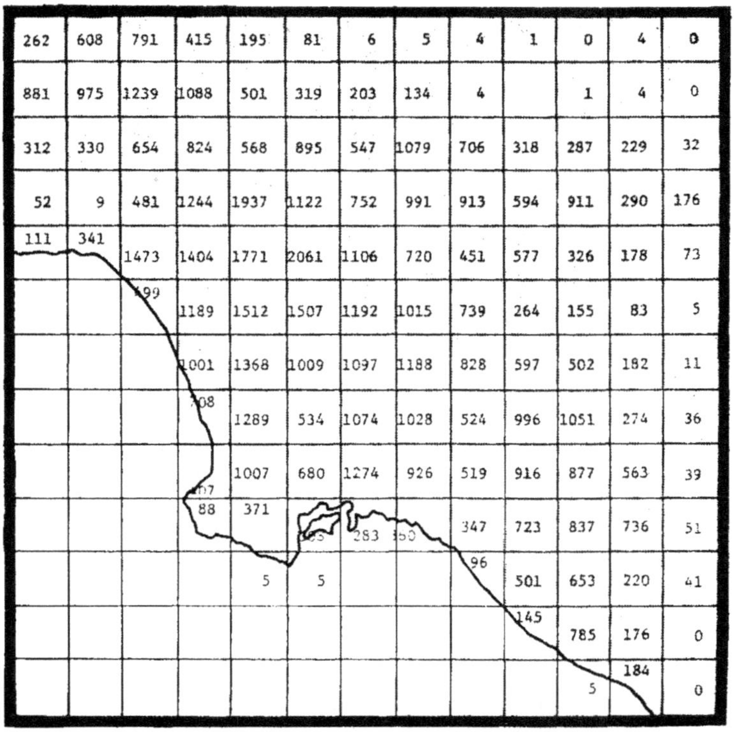

Fig. A4. Spatial distribution of surface street traffic, 1969.

Table A1. Average fleet SO_4/SO_2 emission rates for 1980 estimated under base case and desulfurization of gasoline assumptions

	Base case		Desulfurization of gasoline to 100 ppm	
	Freeway	Surface	Freeway	Surface
SO_4 emissions	11.7	5.3	2.34	1.06
SO_2 emissions	149	195	30	39

point source was assumed to emit a plume the shape of a half elliptical cone with emissions uniformly distributed over each cross section downwind from the source. The dimensions of the plumes were chosen so as to approximate downwind concentrations estimated using the binormal continuous plume dispersion model of Pasquill and Gifford (reported by Turner 1970) for a source at ground level under atmospheric stability class C. Conversion of SO_2 to SO_4 was assumed to take place within the plume according to a first-order rate reaction. Both SO_2 and SO_4 were assumed to be removed to the ground at constant rates, beginning at the time of emission.

Summer and winter concentrations were estimated separately, to account for increased sunlight-induced photochemical conversion of SO_2 in the summer months and the strongly seasonal wind pattern that tends to produce winds from the southwest during the summer and variable winds during the winter. Concentrations were first estimated for average wind speeds in each of the 16 azimuthal directions. Wind speed and the percentage of time the wind blows from each direction were obtained for each cell through a linear interpolation of monthly wind data obtained for the Los Angeles International Airport, the Burbank Airport, and Long Beach. Resulting summer and winter concentrations were then integrated over wind direction and averaged to obtain estimated total 1980 sulfate concentrations attributable to automobiles under both the base case and desulfurization scenarios. The results, shown in Figure A5 for the base case assumptions, range between 0 and 3 micrograms per cubic meter ($\mu g/m^3$).

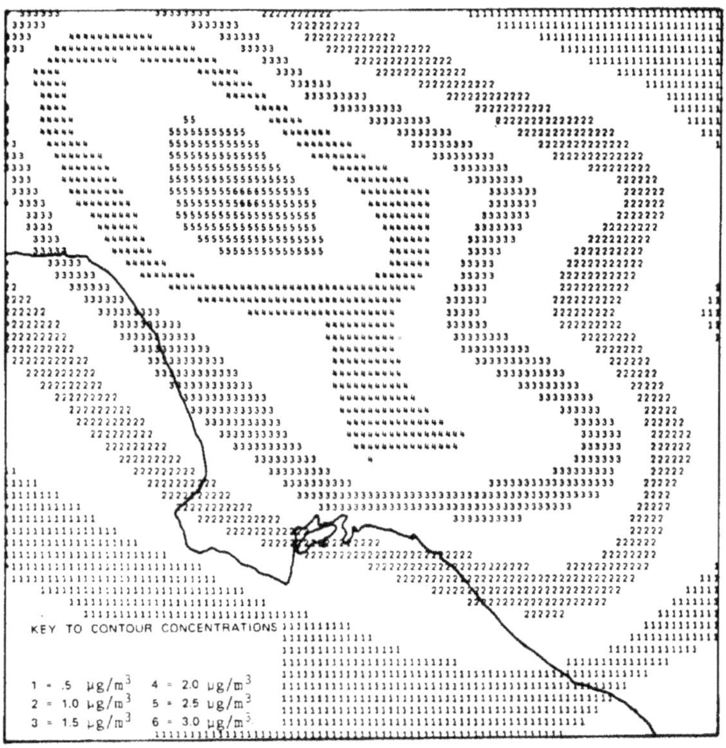

Fig. A5. Incremental automotive annual average sulfate concentrations predicted by the emissions-to-ambient-concentrations model. (Contours indicate areas of approximately constant concentrations.)

To help verify the predictive ability of the automotive emissions-to-atmospheric concentrations model, the model was exercised using lead as a tracer. Auto traffic density data were first reduced to account for vehicles not powered by gasoline. Suspendable lead emission factors for 1969 were derived from an analysis by Cass (1975) which was based on the lead content of gasoline, vehicle fuel economy, and the fate of lead in gasoline. Running the model (without atmospheric conversion) produced the estimated ground-level suspended lead concentrations shown in Figure A6. Actual annual average suspended lead concentrations measured at nine sites in 1969 are also shown. Comparison of estimated and measured values is good, considering the many approximations employed.

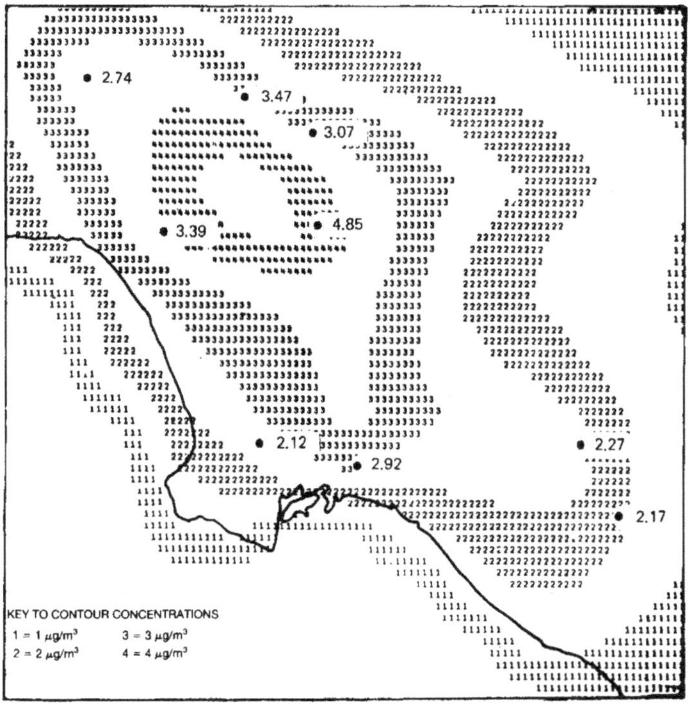

Fig. A6. Predicted and measured annual average suspended lead concentrations.

Stationary sources

Because of nonlinearities in dose-response relationships, the reduction in health effects per unit change in sulfate concentration depends on total sulfate levels. Similarly, evidence suggests that the visibility improvement obtained per unit decrease in sulfate concentration also depends on total sulfate levels. Because of these nonlinearities, it was necessary to predict total sulfate concentrations in 1980 in order to estimate the benefits in that year of reducing automobile sulfur emissions.

Summer and winter emission rates for stationary sulfur oxide sources were obtained by aggregating a month-by-month inventory of stationary sources compiled by Cass (1978). Figures A7 and A8 show the geographical locations of the major stationary sources in the basin (power plants, sulfur recovery operations, sulfuric acid plants, and petroleum coke kilns). To reduce computational complexity, only the largest stationary sources (those accounting for 90 percent of emissions) were included.

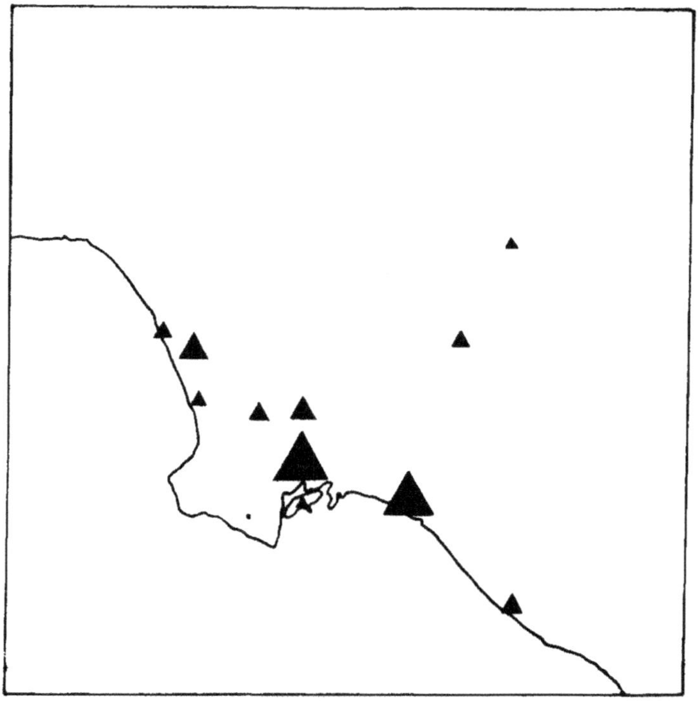

Fig. A7. Major stationary SO_x sources, summer 1973. (Triangle size proportional to emission rate.)

Each of the major stationary sources was modeled as a point source that emits a plume containing SO_2 and SO_4 such that, in a period shortly following emission, concentrations become uniformly distributed from the ground to a constant mixing layer height. The mixing layer height assumed was determined by the effective stack height of the source: chemical plants, petroleum refining and production operations, and residential fuel combustion were categorized as having low effective stack height; electric utilities and petroleum coke kilns were categorized as having high effective

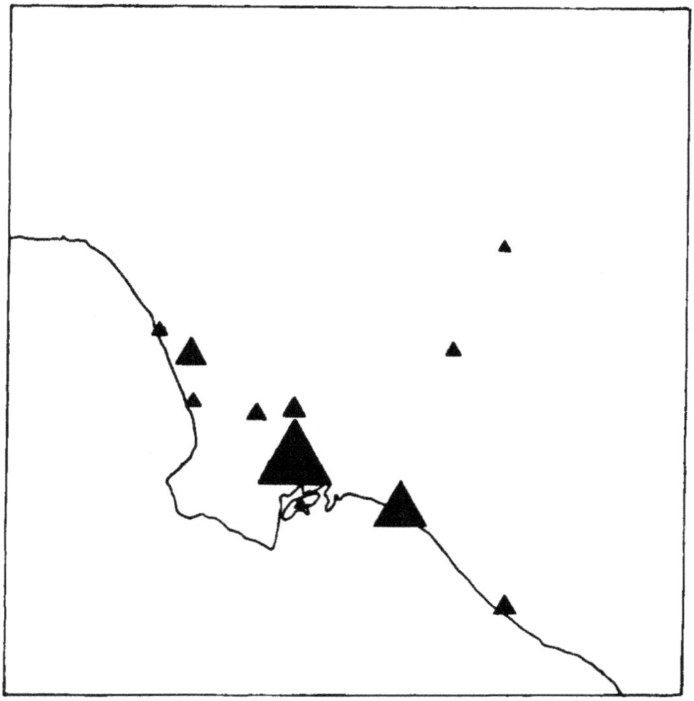

Fig. A8. Major stationary SO_x sources, winter 1973. (Triangle size proportional to emission rate.)

stack height. Plumes were assumed to be wedge-shaped with SO_2 and SO_4 uniformly distributed across each cross section. As in the case of automobile emissions, conversion of SO_2 to SO_4 and deposition of SO_2 and SO_4 to the ground were asumed to occur at constant rates per unit of time.

Average summer and winter concentrations were computed using the same procedure used for automotive emissions: concentrations were estimated for each average wind speed in each of the 16 azimuthal wind directions, assuming plume angles of 22 1/2 degress. Computed concentrations were then added in proportion to the percentage of time the wind blows in each direction.

Total SO_4 concentrations from all sources

To obtain estimates of total SO_4 concentrations, mobile sources other than automobiles and background sources must be added to the automotive and stationary source components computed above. Furthermore, it is important to calibrate the overall emissions-to-ambient concentrations model against empirical measurements. This was accomplished using ground-level SO_4 concentrations measured in the Los Angeles basin in 1973.

The model was first used to predict 1973 mobile source contributions to SO_4. Automotive emissions for 1973 were taken from Cass (1978), who used Roth's (1974) 1969 vehicle traffic estimates and accounted for vehicle traffic growth to 1973, vehicle fuel economy, and summer/winter sulfur content of gasoline. Emissions components for nonautomotive mobile sources (heavy trucks, aircraft, and ships) were then added (also taken from Cass). Running the model with these emissions produced contour plots for incremental mobile source SO_4 concentrations. These mobile concentrations were then added to similarly estimated stationary source contributions to obtain estimates of total 1973 summer and winter ambient SO_4 contributed by mobile and major stationary sources.

The SO_4 estimates yielded by the combined model neglect sources outside of the grid, minor manmade sources on the grid (that together account for 10 percent of stationary emissions), and natural sources. The principal natural source of atmospheric sulfur oxides is the oxidation of hydrogen sulfide or dimethyl sulfide gas that results from decaying vegetation (Kellogg *et al.* 1972). Near the ocean, significant sulfate is also emitted as part of sea spray. Natural background levels of atmospheric sulfate in the Los Angeles basin have been estimated at about 4 $\mu g/m^3$ (Trijonis *et al.* 1975).

Concentrations of SO_4 resulting from sources other than stationary and mobile on-grid sources were accounted for in the analysis through a least-squares fit of predicted concentrations to measured concentrations. With nominal estimates of 2.88 and 1.44 percent/hour for the dry deposition of SO_2 and SO_4, respectively, average summer and winter mixing layer heights of 325 meters, and summer and winter conversion rates of 6 and 2 percent/hour, a best fit was obtained with respective summer and winter background concentrations of 3.3 and 2.0 $\mu g/m^3$.[3] Figures A9 and A10 show 1973 SO_4 levels predicted by the model when background concentrations are taken into account. Also shown are the summer and winter average values actually measured at various sites.[4]

To obtain predictions of total SO_4 concentrations in 1980, stationary and background contributions were estimated and then added to the automotive component computed above. To obtain 1980 stationary and background components, projected 1980 emissions by source category for stationary sources were obtained from Hunter and Helgeson (1976). In the absence of better information, emissions rates were assumed to change uniformly, irrespective of spatial distribution within each source category. Nonautomotive mobile emissions were projected to increase uniformly with a growth rate of 2 percent/year. Figure All shows the total predicted base-case 1980 annual average sulfate levels contributed by all sources.

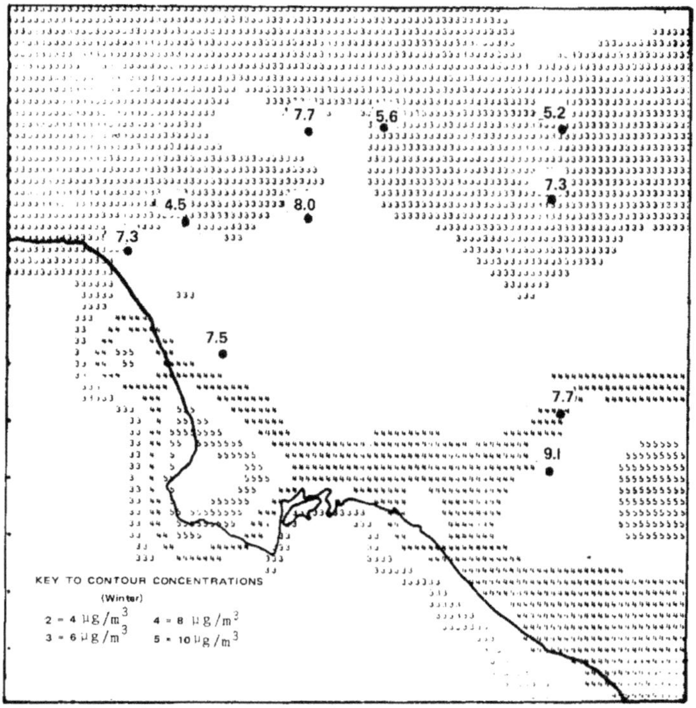

Fig. A9. Predicted and measured average SO_4 concentrations, for summer 1973. (Contours indicate areas of approximately constant concentrations. Actual measured values are shown to one decimal place adjacent to monitoring sites.)

Frequency distributions of 24-hour SO_4 concentrations

Daily SO_4 levels tend to follow log-normal frequency distributions. A least-squares fit (Cass 1978) to available data on 24-hour Los Angeles SO_4 concentrations by location was used to establish a relationship between annual average levels and the parameters of the log-normal distribution that achieved the best fit. This relationship was then used to estimate the 1980 frequency distribution for 24-hour sulfate levels within each geographic cell. Figure A12 show sample results for one location.

Impact on health

To determine the health impacts of the estimated SO_4 concentrations, the geographic distributions of various sensitive populations were determined and then used with dose-response relationships to predict reductions in the incidence of the various health effects.

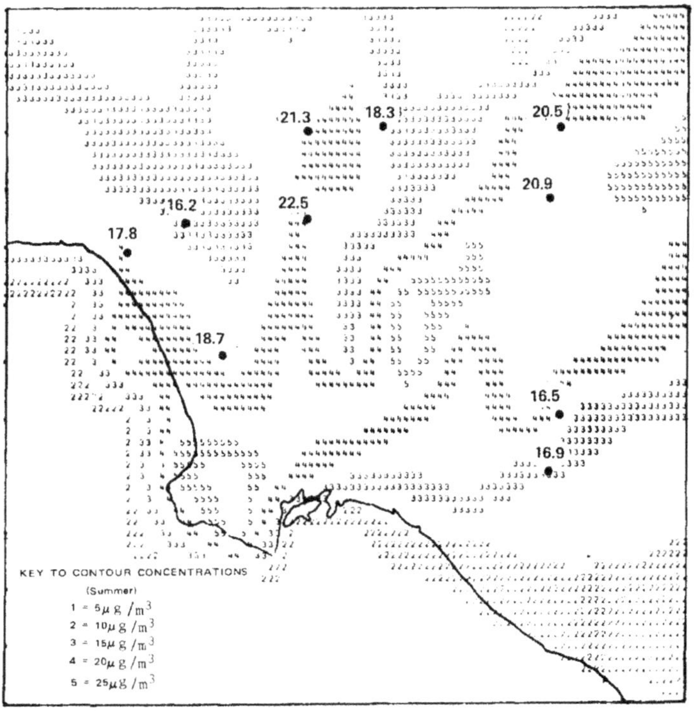

Fig. A10. Predicted and measured average SO_4 concentrations, winter 1973. (Contours indicate areas of approximately constant concentrations. Actual measured values are shown to one decimal place adjacent to monitoring sites.)

Populations at risk

Children, the elderly, asthmatics, and persons with chronic respiratory and heart disease tend to be more sensitive to air pollution. To quantify the health impact of changes in SO_4 concentrations, estimates were therefore made of the populations, age distributions, and health characteristics of individuals within each of the cells defined over the Los Angeles basin.

Population forecasts for 1980 were obtained by estimating 1974 populations by cell and then extrapolating to 1980 using annual growth rate data (Southern California Association of Governments 1976). The growth rate estimates used vary from cell to cell and reflected recent rapid growth in the southeast region and slow growth in the inner city. Elderly populations by cell were estimated using statistics in Horie and Chaplin (1977). Base rates for the various health effects were estimated using national or regional prevalence rates. Table A2 presents a summary of the estimated total numbers of health effects and provides additional references.

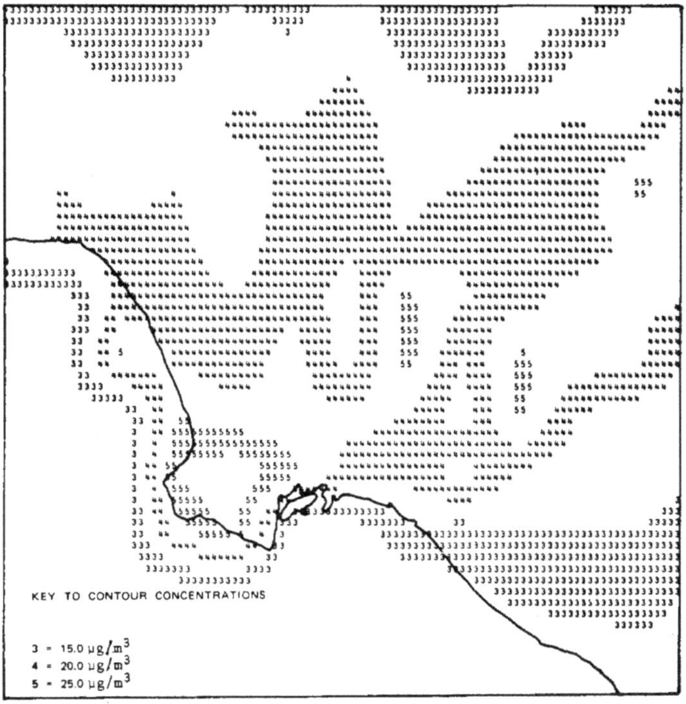

Fig. A11. Predicted total annual average sulfate concentrations, base case, 1980.

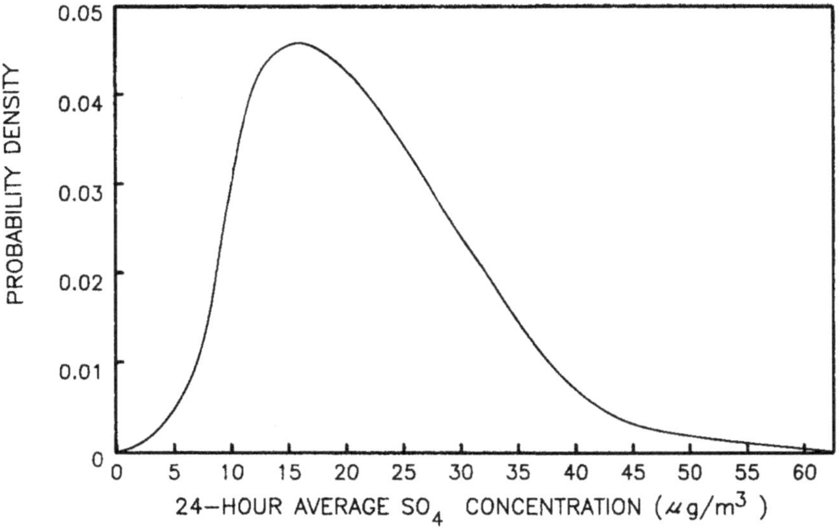

Fig. A12. Frequency distribution estimated for 1980 24-hour average SO_4 concentrations, downtown Los Angeles.

Table A2. Estimated total numbers of health effect cases, Los Angeles basin, 1980

Year	Estimated population	Mortality (thousands of deaths)	Heart and lung (thousands of person-days of aggravation)	Asthma (thousands of attacks)	Lower respiratory (thousands of cases)	Chronic respiratory (thousands of cases)
1980	8,660,000	65.8	15,879	2,150	176	308

The figures cited were derived individually by cell and then summed. Data are based on population estimates from Cass (1978) and Southern California Association of Governments (1976) and health effects prevalence rates. Local mortality prevalence rates are from the Orange County Services Agency (1976) and the Los Angeles County Department of Health Services (1976). Other prevalence rates used are as follows: (1) prevalence of heart and lung disease in elderly persons, 0.27; average number of aggravated days per day-person, 0.20; (2) asthma prevalence rate, 0.036 (National Center for Health Statistics 1973); attacks per day per asthmatic, 0.02; (3) annual incidence rate for lower respiratory disease in children, 0.07 (National Center for Health Statistics 1977); (4) chronic respiratory disease prevalence rate, 0.02 for nonsmokers (62 percent) and 0.10 for smokers (38 percent).

Health dose-response functions

It is clear that a simple relationship between exposure to sulfate and health does not exist. Toxicological findings suggest that human response to the respiration of sulfate particles varies markedly among individuals and depends on the size of the particles, their chemical properties (such as *p*H), and conditions of exposure (such as atmospheric humidity and temperature). Furthermore, although some sulfate compounds clearly appear to produce adverse health effects, others seem benign.

Because lack of information prevented specification of a relationship sufficiently sophisticated to account for all of these complexities, simple linear threshold relationships were assumed. The approximation made by such relationships, which were suggested by epidemiological studies conducted in the 1970s, is that no adverse health effects occur below some threshold level of exposure, and that above this threshold the percentage of cases of the health effect which might be attributed to sulfate (the percent excess) increases linearly with concentration. By plotting the results of various epidemiological studies on a graph relating percent excess health effects to levels of suspended sulfate, best fits for linear threshold functions may be obtained, as illustrated in Figure A13. The results of such curve fitting exercises and the studies on which they are based are summarized in Table A3.[5]

Following Morgan *et al.* (1978) a statistical analysis was conducted to obtain a measure of the uncertainty inherent in the derived dose-response functions. By examining the standard errors in the various studies, it was

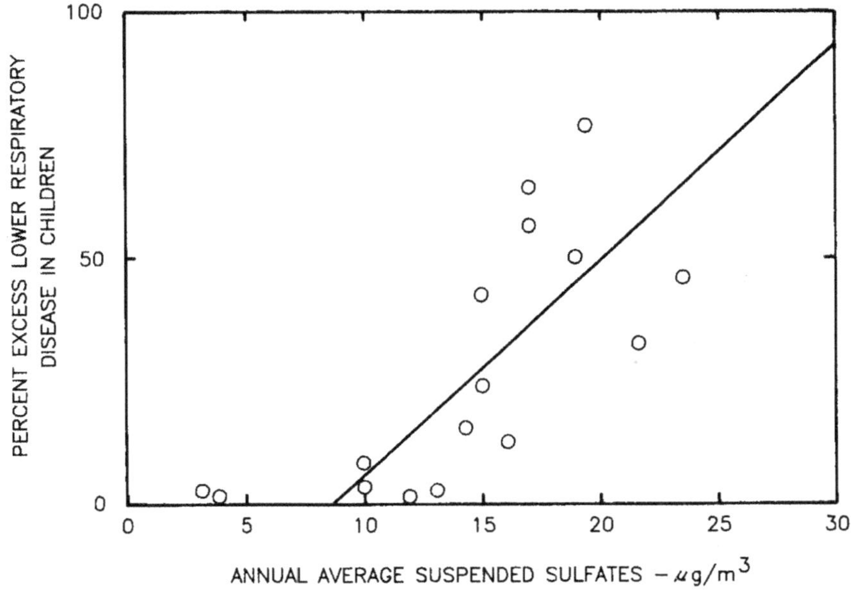

Fig. A13. Data and least-squares dose-response function for excess lower respiratory disease in children, from Federal Power Commission (1975).

found that they all had similar variability. Therefore, the results of the studies were interpreted as independent observations of the same linear threshold relationships, thereby permitting probability density functions (student's *t* distributions) to be associated with the thresholds and slopes, as illustrated in Figure A14.

Health effects estimates

As indicated in Table A3, daily mortality, aggravation of heart and lung disease, and aggravation of asthma were assumed to be related to short-term exposures, while excess lower respiratory disease and excess risk of chronic respiratory disease were related to long-term exposures. Thus, for each cell the percentage of cases of daily mortality, aggravation of heart and lung disease and of asthma attributable to sulfate were calculated by integrating the respective dose-response functions over the frequency distributions for 24-hour sulfate concentrations computed for that cell.[6] The percentage of cases of child's lower respiratory disease and chronic respiratory disease in adults was obtained directly from the dose-response curves and the predicted annual average concentrations.

Multiplying the computed percent excess cases by the numbers of observed cases of each health effect in each cell yielded the total number of cases of health effects attributable to SO_4 by cell. Totals were then obtained by summing health effects across cells. Subtracting the numbers

Table A3. Health effects dose-response functions*

	Least-squares threshold function		
Adverse health effect	Exposure duration	SO_4 threshold ($\mu g/m^3$)	Slope+
Increased daily mortality per 10^5 persons	24 hr or longer	12	0.319
Aggravation of heart and lung disease	24 hr or longer	4	1.31
Aggravation of asthma	24 hr or longer	3	2.05
Excess lower respiratory disease in children	Up to 10 years	9	4.57
Excess risk for chronic respiratory disease in adults++	Up to 10 years	12	12.1

* These dose-response relationships were developed in a study for the Environmental Protection Agency and appear in Finklea *et al.* (1975).

\+ Change in percent excess over base rate for population, per $\mu g/m^3$ change in suspended sulfate level.

++ For chronic respiratory disease, difficulties with available data required the unit of measurement to be excess risk rather than direct incidence of illness. Actually in its originally calculated form, separate dose-response functions were assessed for cigarette smokers and nonsmokers. The function described in the table is a weighted linear average based on the average prevalence of cigarette smoking in the adult population at risk.

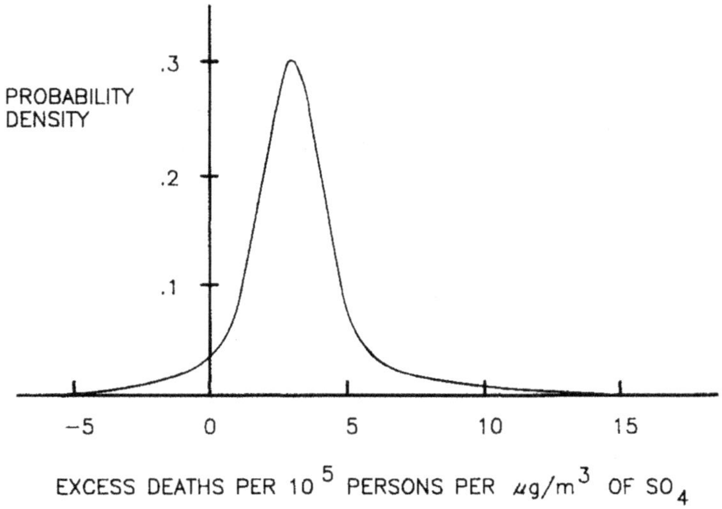

Fig. A14. Probability density function illustrating uncertainty on the slope of the assumed linear threshold dose-response function for excess mortality.

of computed health effects assuming no automotive sulfur from the numbers computed in the base case and desulfurization scenarios yielded the health impact of automotive sulfur emissions. The results are shown in Table A4.

Table A4. Estimated total number of health effect cases attributable to automotive sulfur emissions, 1980

Emission control assumption	Mortality (early deaths)	Heart and lung (thousands of person-days of aggravation)	Asthma (thousands of attacks)	Lower respiratory (thousands of cases)	Chronic respiratory (thousands of cases)
Base case	121	331	113	22	55
Desulfurization	23	66	23	4	11
Reduction in cases attributable to desulfurization	98	265	90	18	44

Impact on visibility

Much of the impairment of visibility in urban areas of the United States is caused by the scattering of light by airborne particulates. Evidence suggests that sulfates may be important contributors to visibility reduction by this means.[7] For the purposes of the analysis, visibility was defined as the meteorological range, the distance at which an ideal black object can just be seen against the horizon sky. The basic relationship between the visibility V and the composition of the atmosphere is then given by the Koschmeider equation (Middleton 1951):

$$V = \frac{0.0024}{b}$$

where b is the sum of the extinction coefficients of the various light-scattering and light-absorbing elements in the atmosphere. The contribution of sulfates to b is determined not only by the sulfate mass concentration but also by relative humidity; high humidity enhances the light scattering properties of sulfates.

A relationship between b, relative humidity, and the levels of various atmospheric pollutants was derived for downtown Los Angeles by Cass (1978), using data accumulated by the Los Angeles Air Pollution Control District between 1965 and 1974. Cass used this result to derive a relationship between a frequency distribution for the percent of time that visibility is less than any given value and assumed reductions in the level of sulfate concentration. For a given location, frequency distributions describing average daily visibility tend to have the shape of a log-normal distribution. Thus, the results of Cass' analysis and the assumption of a log-normal frequency distribution for daily average visibility make it possible to express the parameters of a log-normal distribution for daily visibility in downtown Los Angeles as a function of the fraction of atmospheric SO_4 removed.

For the analysis it was necessary to estimate what in 1977 were future, 1980, visibilities and to make these estimates for more areas in the basin than downtown Los Angeles. Unfortunately, no information was available for how the relationship between visibility and the concentrations of various light-scattering and -absorbing particulates might differ from location to location and year to year. Therefore, the simplifying assumption was made that the relative contributions of the various components to the extinction coefficient would be approximately the same, regardless of location or year. Thus, estimates of the frequency distributions describing 1980 visibilities could be obtained at any location given only total 1980 particulate level at that location relative to the level in downtown Los Angeles from 1965 to 1974, and given the fraction of the base-case SO_4 that is removed from the atmosphere.[8]

Figure A15 shows an example of the results obtained from this analysis. The figure shows the frequency distribution derived for 1980 visibilities in downtown Los Angeles under the base-case assumption and under the desulfurization option. As illustrated, desulfurization was estimated to

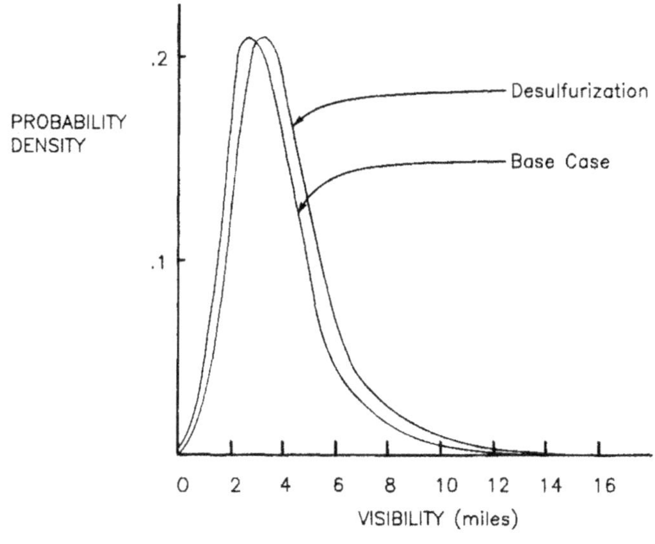

Fig. A15. Frequency distribution for visibility estimated under base case and desulfurization, downtown Los Angeles, 1980.

result in roughly a 10 percent improvement in average visibility at this location.

Valuation of Benefits

To permit a direct comparison with economic costs, the benefits of desulfurization must be expressed in monetary terms. This required that the adverse health effects and reduction in visibility associated with elevated levels of SO_4 be assigned dollar values.

Valuing health effects

The two categories of health effects, morbidity and mortality, were treated separately. To determine appropriate economic equivalents for morbidity, a willingness-to-pay survey was conducted. The survey consisted of mailing questionnaires to approximately 2,000 residents randomly selected from the greater Los Angeles area. The questionnaire began by requesting biographical and demographical data about the recipient (age, occupation, income, health status, length of residence, number in family). Willingness to pay was requested for symptoms (shortness of breath, chest pains, coughing, etc.) of specified severity (mild or severe) and duration (1, 7, and 90 days). Thus, willingness-to-pay information was obtained through requests of the form: "To avoid one day per year severe shortness of breath and chest pains, the most I would pay is $0, $0.50, $1, $2, $10, $20, $50, $120, $250, $1,000 per year (circle one)." Recipients were asked to react to symptoms rather than to diseases, because it was felt that individuals might not be familiar with specific diseases. A relatively large range of response values was allowed to provide sufficient spread for those with both high and low incomes and to account for the diversity of seriousness of health symptoms.

Each recipient of the questionnaire was supplied with a cover letter explaining the purpose of the survey (to determine the values people place on respiratory health) and a stamped return envelope. No explicit reference to air pollution was made in any of the materials. Two weeks after the inital mailing, postal cards were mailed to all persons in the sample reminding them to complete the questionnaire and thanking them for their help if they had already done so.

Out of the 2,000 questionnaires mailed, approximately 400 completed questionnaires were returned. Because of concern over the small size of the collected sample, the biographical and demographical data provided on the returned questionnaires was compared with census data to check for representativeness. Overall, the data suggested that the sample was reasonably representative of the toal population.

Figure A16 provides an example of the survey results. The figure shows the percentage of the sample reporting various maximum willingness-to-

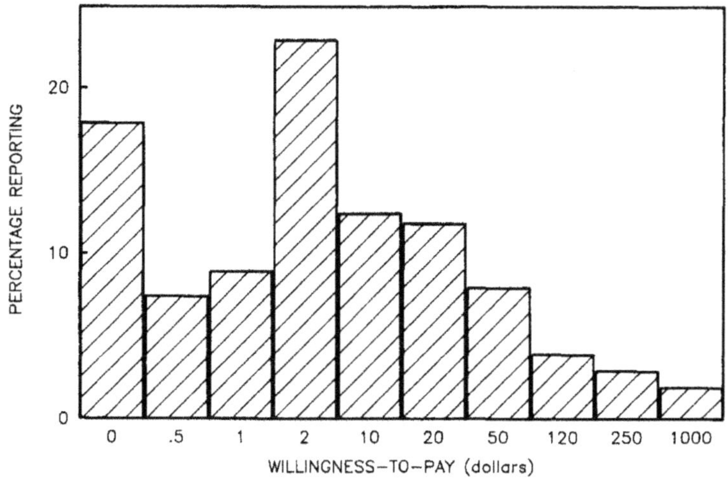

Fig. A16. Percent of sample reporting various willingness-to-pay values for reduction of one-day minor shortness of breath and chest pains.

pay values for reductions of minor shortness of breath and chest pains. Table A5 shows the average and median willingness-to-pay values computed for each health symptom. (Median values are the values for which half of the population would pay less and half would pay more). As expected, the percentages reporting a willingness to pay higher dollar amounts increased with severity and duration of the symptom.[9] The value reported per day, however, decreased with duration; for example, a week of suffering was not regarded as seven times as bad as one day. Severe illness was valued as roughly twice as bad as minor illness.

Table A5. Average and median willingness to pay for health symptoms of various severity and duration

		Average willingness to pay			Median willingness to pay		
		1 day	7 days	90 days	1 day	7 days	90 days
Shortness of breath	Minor	48.61	73.87	145.93	4.90	13.64	35.96
	Severe	79.15	136.12	251.84	10.92	35.93	97.80
Coughing/Sneezing	Minor	26.40	44.67	86.03	2.31	7.84	22.85
	Severe	45.77	72.29	147.48	6.95	19.90	50.56
Head congestion/ Eye, ear, throat irritation	Minor	32.50	41.51	90.37	3.80	9.58	25.14
	Severe	53.42	80.32	179.94	8.17	20.34	61.68

These data and the survey on which they is based are described in a report by the Interdisciplinary Center for Aeronomy and Other Atmospheric Sciences (1978).

Application of the results of the willingness-to-pay survey required that morbidity effects associated with elevated SO_4 exposures, such as asthma attacks and cases of child's lower respiratory disease, be translated into symptoms addressed by the survey. Therefore, approximate equivalencies between health effects and symptoms were established, based on opinions of health experts on the project team. For example, one asthma attack was assumed to produce roughly one day of severe shortness of breath; one case of child's lower respiratory disease was assumed to produce approximately seven days of severe coughing, minor head congestion and eye, ear, and throat irritation, and so forth. As a result of such assumptions, nominal dollar values were derived for each morbidity effect. Because average willingness-to-pay values were highly sensitive to a few unusually high reported values, less sensitive median values were used as the basis for computing the nominal estimates summarized in Table A6. Because of

Table A6. Nominal dollar equivalencies

One day aggravation of heart and lung disease symptoms	\$ 19
One asthma attack	12
One case child's lower respiratory disease	60
One case chronic respiratory disease	320

the lack of agreement over methods for valuing human life and the difficulties of obtaining meaningful willingness-to-pay responses in this area, mortality effects were not translated into monetary terms. However, to help clarify the significance of mortality benefits, effects on mortality were expressed in terms of life expectancy, using a simple linear relationship suggested by Schwing (1976). Schwing observed that plots of unadjusted mortality rates versus longevity across populations comes reasonably close to the relationship

$$\text{years} = 0.02 \times \text{crude death rate (as deaths/100,000)}$$

Because most deaths associated with high concentrations of SO_4 are thought to occur among chronically ill, elderly people (whose lives may be shortened only by a matter of days or weeks), expressing mortality as a decrease in average life expectancy may be more useful. Table A7 summarizes health benefits expressed in dollars and life expectancies.

Valuing visibility

Like health, visibility obviously has some value, and it seems likely that at least some people would be willing to decrease their material standard of living in order to improve their local visibility. Unfortunately, estimating a willingness to pay for local visibility improvements was regarded as too

Table A7. Estimated health benefits of desulfurization

Average increased life expectancy (weeks)	Decreased incidence of (millions of dollars)			
	Heart and lung disease symptoms	Asthma attacks	Child's lower respiratory disease	Chronic respiratory disease
1.2	5.0	1.1	1.1	14.1

complex a task to be attempted through a mail survey, and limited project resources prevented conducting the more sophisticated personal interview survey regarded as necessary to produce credible willingness-to-pay results. Very few attempts at estimating a dollar value for visibility had been made at the time at which the analysis was conducted, and there were apparently no studies of the economic value of visibility to the residents of Los Angeles.

To obtain a very rough estimate of the dollar value of visibility, it was necessary to rely on the results of a study conducted in Farmington, New Mexico (Blank *et al.* 1977). This study employed a bidding-game approach to determine how much households or visitors to the area would be willing to pay to avoid reductions in visibility, using photographs to illustrate different local conditions.

Application of the Farmington results to Los Angeles is problematic. Farmington lies in the picturesque Four Corners region of New Mexico. Its beauty derives largely from distant natural rock formations and mountains. Although the hills surrounding parts of the Los Angeles basin might also be regarded as scenic, visibilities there averaged about 5 miles, compared with 25 to 75 miles in Farmington. Despite these difficulties, the Farmington results were extrapolated for application to Los Angeles. Two models were considered.

Linear model

The simplest interpretation of the Farmington data is that the value of reducing visibility from a value V_1 to a value V_2 is proportional to the magnitude of the improvement:

$$k_1(V_1 - V_2$$

With this interpretation, the Farmington data yield an average value for k_1 of 70 cents per mile per person per year.[10]

Logarithmic model

The linear model predicts a social cost of 1-mile reduction in visibility that is the same whether the reduction occurs where prevalent visibility is 75 miles or 5 miles. An alternative assumption is that the value of visibility improvement is related to the percentage increase. Therefore, a logarithmic model was considered for which the value of improving visibility from V_1 to V_2 is given by

$$k_2 \log_{\text{base } 2}(V_1/V_2)$$

The factor k_2 may be interpreted as the social value per person of a 50 percent improvement in visibility. With the Farmington data, a value of \$24.50 per person per year was obtained for k_2.

Estimated social costs of visibility reduction

Total visibility costs were computed for each of the geographic cells defined over the Los Angeles basin assuming both the linear and logarithmic cost model. The total visibility costs obtained by summing across cells are shown in Table A8. For each scenario, the visibility costs were

Table A8. Visibility costs attributable to automotive sulfur emissions (millions of dollars)

Assumption	*Linear model*	*Logarithmic model*
Base case	\$0.8	\$12.4
Desulfurization	0.1	2.5
Visibility benefits	\$0.7	\$ 9.9

measured relative to the situation in which all SO_4 attributable to automobiles was removed from the atmosphere. Thus the costs shown in Table A8 can be interpreted as the social costs of visibility reduction attributable to automobile SO_4 under each emission assumption. After discussions with the client, it was decided that the higher (more conservative) visibility costs computed using the logarithmic model should be used.

Sensitivity Analysis of Benefit Estimates

The fixed values assumed for many of the quantities on which the benefits estimates were based may be in error. Therefore, the analysts thought it useful to determine the sensitivity of the benefits estimates to variations in numerical assumptions. Table A9 shows the total health plus visibility benefits and the changes in computed benefits when several of the numeric assumptions were varied. In each case, the range of values over which the quantity was varied was selected to approximate the existing uncertainty on that quantity.

As indicated in Table A9, both the magnitudes of existing uncertainties

Table A9. Sensitivity of the estimated social benefit of complete desulfurization

	Nominal value	Sensitivity range	Social benefit (millions of dollars)	(millions of dollars)
Automobile emissions				
Catalyst deterioration rate	25%/yr	0—50%/yr	29.1—36.3	7.2
1974—80 gas mileage improvement	23%	0—40%	29.8—37.8	8.0
Fractions of cars containing oxidation catalysts	85%	50%—100%	28.6—32.5	3.9
Atmospheric transport and conversion				
Rate of conversion of SO_2 to SO_4	6%/hr summer 2% hr winter	1/3 to 2 times nominal	8.6—58.0	49.4
SO_2 deposition rate	2.88%/hr	1.44%—5.70%/hr	31.2—31.3	0.1
SO_4 deposition rate	1.44%/hr	0.11%—2.88%/hr	31.2—31.4	0.2
Reduced number of health effect cases				
Heart and lung	265 thousand	1/4—4 times nominal values	26.3—46.3	20.0
Asthma	90 thousand		30.4—34.4	4.0
Lower respiratory	18 thousand		30.4—34.4	4.0
Chronic respiratory	44 thousand		20.6—73.4	52.8
Pre-case costs				
Heart and lung	$19	$5—$100	27.5—52.7	25.2
Asthma	$12	$5—$100	30.6—39.1	8.5
Lower respiratory	$60	$10—$300	30.3—35.5	5.2
Chronic respiratory	$320	$25—$1,000	18.2—61.1	42.9
Improvement in visibility	.15 miles	1/4—4 times nominal	23.8—60.9	37.1
Visibility cost	$24.50/50% reduction	1/3—3 times nominal	24.6—51.0	26.4

Nominal social benefit = $31.2 million

and their impact on estimated benefits varied considerably. For example, there was a lot of uncertainty over the level of stationary sulfur oxide emissions in 1980. Independent of the deliberations over reducing automotive fuel sulfur, steps were being taken at the state and federal level that, depending on their effectivenesss, might reduce stationary sulfur emissions by as much as 50 percent.[11] Repeating the benefits analysis assuming a 50 percent reduction in stationary emissions (1980 emissions assumed above uniformly decreased by 50 percent) produced estimated total annual average sulfate concentrations of 12 to 16 $\mu g/m^3$, compared with the 15 to 25 $\mu g/m^3$ values shown in Figures A9 and A10. The computed morbidity benefits in this case were lower ($13 million compared with $21 million), because the incidence of exposures to concentrations above health-effects threshold values was reduced. However, the computed visibility benefits were higher ($15 million compared with $10 million), because the percentage improvement in visibility achieved by reducing automotive emissions was higher. Thus the net change in benefits obtained by varying the assumption for stationary emissions was not very great.

The uncertainties to which computed benefits were most sensitive were the conversion rate of SO_2 to SO_4, the dose-response relationships for cases of chronic respiratory disease and aggravation of heart and lung disease symptoms, the economic values assigned to these health effects, and the level and social value of visibility improvement obtained.

Cost-Benefit Comparison

Dividing estimated total morbidity and visibility benefits by the total number of vehicle miles estimated to be driven in 1980 produced estimated benefits of the desulfurization option of 63 cents per thousand vehicle miles driven. Costs of desulfurization, as indicated above were estimated to be 2 cents per gallon of gasoline produced. Using the base case assumptions for vehicle fuel economy, an average mileage for 1980 on-the-road vehicles was computed to be 16.6 miles per gallon. Assuming an energy penalty of desulfurization of 1 percent (due to reduced energy content of desulfurized gasoline), this resulted in an estimated incremental cost to the consumer of $1.22 per thousand miles driven. Thus, excluding the mortality benefits of desulfurization, a benefit-to-cost ratio of .52 was obtained.

Epilogue

The results of the analysis were explained to the client in the form of both a written report and an oral presentation. The analysts stressed that although economic costs were computed to be almost twice the dollar values that were assigned to benefits, this comparison ignores the impact of gasoline desulfurization on mortality and other effects, such as sulfur-related plant and material damage. They also reminded the client that there were important equity issues not addressed by the analysis. For example, those who currently enjoy the benefits of automotive travel are by and large not the very young, the sick, and the elderly who suffer the most from sulfur pollution. Furthermore, should desulfurization be required, the additional costs that individuals would be forced to pay would not generally be in proportion to the benefits that they would accrue. Although the analysts admitted that their analysis did not convincingly support or refute those arguing for desulfurization legislation, they claimed that the exercise was valuable in that it helped focus discussion away from the personalities and narrow interests of the principal stakeholders toward the economic, health, and other issues that were of most concern to Los Angeles residents by clarifying the magnitudes of the various costs and benefits associated with the decision and by indicating the major uncertainties and their significance.

Notes to Appendix A

1. The possibility that sulfur levels of 100 ppm or less could be obtained by an industry effort to allocate low sulfur crudes (those containing less than 0.5 percent sulfur) to Southern California was considered and then rejected. Reallocation was not perceived to be a viable long-term solution in view of the demand for low-sulfur crudes for residential fuel oil, diminishing supplies, and other factors.

2. Defining the modeling region in this way considerably simplified the analysis by permitting the direct use of results obtained by Cass (1978) and Roth *et al.* (1974). Cass developed a computationally sophisticated model for the formation and transport of sulfate in the Los Angeles air basin, using in part an inventory of emissions compiled by Roth. The square region defined here is approximately the same size as the region modeled by these authors and uses cell dimensions that are twice as large. Overlapping the region selected by these authors enabled the analysis to be completed within the time constraints imposed by the client by eliminating many data collection obstacles and by establishing a level of detail that was consistent with the available computational resources.

3. Least squares fitting assumed a relation of the form $Y = a(X_1 + X_2 + X_3) + b$, where Y = SO_4 concentrations predicted by the fitted model (determined by least squares to approximate measured SO_4 levels), X_1 = SO_4 concentrations contributed by mobile sources, X_2 = SO_4 concentrations contributed by stationary sources with low effective stack height, X_3 = SO_4 concentrations contributed by stationary sources with high effective stack height, b = background concentrations, and a = a multiplicative error factor. A best fit was obtained with $b = 3.3$ and $a = 1.2$ for the summer data and $b = 2.0$ and $a = 1.8$ for the winter data. A value $a \neq 1$ may be interpreted as an error factor for measured SO_4 values which might be attributed to conversion of SO_2 to SO_4 on the filter used in the measurement device. This error factor has been estimated to be as high as a factor of 2 (Bowman 1975).

4. Measured values are approximate, since they were estimated from figures appearing in Cass (1978) which summarize data from the Los Angeles Air Pollution Control District, the Community Health Environmental Surveillance System, and the National Air Surveillance Network.

5. Much of the data on which these health dose-response functions are based comes from a series of epidemiological studies conducted as part of the Community Health and Environmental Surveillance System (CHESS studies) (EPA 1974) conducted in the early 1970s. Unfortunately, these studies do not generally use precise or consistent definitions for the various health effects measured. A case of child's lower respiratory disease is defined as a physician-diagnosed case of bronchitis and croup. Measurements of the other health effects were, however, generally accomplished through self-administered questionnaires. Susceptible individuals were asked to keep diaries describing their state of health. Aggravation of heart and lung disease symptoms was defined as worsening of dyspnea, cough, or phlegm production or increased frequency or severity of angina. An asthma attack was defined as an attack of wheezing with shortness of breath. A case of chronic respiratory disease was defined as cough and phlegm on most days for at least three months each year. A health effect was recorded when an individual responded in the diary that he or she felt worse or much worse than usual. As a result of these and other methodological difficulties with the CHESS studies, the dose-response functions given here are *not* currently regarded as providing an adequate representation of the relationships between exposures to sulfur oxides and health.

6. More precisely, if $f(x;\, m_{ij},\, a_{ij})$ is the log-normal distribution estimated for cell i under emissions scenario j, and $y(x)$ is the dose response curve, then the percentage P of additional cases of the health effect attributable to daily levels of suspended sulfates is given by

$$P = \int_0^\infty y(x) f(x;\, m_{ij},\, a_{ij})\, dx$$

7. See, e.g., Eggleton (1969), White and Roberts (1977), Charlson *et al.* (1974).

8. Specifically, the frequency distribution derived for visibility V as a function of SO_4 removed had the form:

$$f(V) = \frac{1}{(2\pi)^{\frac{1}{2}} aV} \exp\left[-\frac{1}{2}\left(\frac{\ln V - m^2}{a}\right)\right]$$

with parameters

$$m = 1.47 + 0.47x + 0.10x^2 - \ln A$$

$$a = 0.624 - 0.043x$$

where x is the fraction of the base case SO_4 that is removed from the atmosphere under the assumed control policy, and A is the 1980 total particulate level divided by the average particulate level in downtown Los Angeles from 1965 to 1974.

9. For short-duration symptoms, the averages of the values tended to be larger than would be actual out-of-pocket expenses. E.g., the average willingness to pay to avoid severe (disabling) shortness of breath for one day was \$91; the income loss for a person of average income would be about \$75. The differences may reflect psychic costs of pain and suffering. For longer periods of illness — e.g., 90 days — the averaged values were much

less than estimated employment-related income losses for the disability. This may be attributed to the fact that most respondents had some form of insurance.

10. This is computed from an average of values given in Blank *et al.* (1977) and assumes 2.57 persons per household. The result is in fairly good agreement with other willingness-to-pay studies (e.g. Randall *et al.* 1974).

11. Preliminary results from the California Air Resources Board estimated 1980 sulfur emissions from stationary sources in the South Coast Air Basin at 302 tons per day. For comparison, the forecast by Hunter and Helgeson (1976) (used in this analysis) was for 743 tons per day for the same region. (The South Coast Air Basin overlaps but does not exactly equal the region considered in this analysis.)

References to Appendix A

Blank, F. M., *et al.* 1977. 'Valuation of Aesthetic Preferences: A Case Study of the Economic Value of Visibility.' Preliminary draft, University of Wyoming (quoted with the permission of R. d'Arge).

Bowman. 1975. Florida Department of Environmental Regulation. Personal communication.

Cass, G. R. 1975. 'Lead as a Tracer for Automotive Particulates: Projecting the Sulfate Air Quality Impact of Oxidation Catalyst Equipped Cars in Los Angeles.' Environmental Quality Laboratory, California Institute of Technology, Pasadena. May.

——. 1978. 'Methods for Sulfate Air Quality Management with Applications to Los Angeles.' Ph. D. Dissertation, California Institute of Technology, Pasadena.

Charlson, R. J., A. H. Vanderpol, D. S. Covert, A. P. Waggoner, and N. C. Ahlquist. 1974. '$H_2SO_4/(NH_4)_2SO_4$ Background Aerosol: Optical Detection in St. Louis Region.' *Atmospheric Environment* 8: 1257–1267.

Eggleton, A. E. J. 1969. 'The Chemical Composition of Atmospheric Aerosols on Teesside and Its Relation to Visibility.' *Atmospheric Environment* 3: 355–372.

Environmental Protection Agency. 1974. 'Health Consequences of Sulfur Oxides: A Report from CHESS, 1970–1971.' EPA–650/1–74–004, Office of Research and Development, National Environmental Research Center, Research Triangle Park, N.C. May.

Federal Power Commission. 1975. 'Power Generation: Conservation Health and Fuel Supply, Appendix 1. National Power Supply.' Report and Recommendations of the Task Force on Conservation and Fuel Supply to the Technical Advisory Committee on Conservation of Energy. GPO March.

Finklea, J. F., *et al.* 1975. 'Estimates of the Public Health Benefits and Risks Attributable to Equipping Light Duty Motor Vehicles with Oxidation Catalysts.' National Environmental Research Center, Office of Research and Development, Environmental Protection Agency, Research Triangle Park, N.C. February 1.

Horie, Y. and A. Chaplin. 1977. 'Population Exposure to Oxidants and Nitrogen Dioxide in Los Angeles — Volume III: Long-Term Trends. 1965–1974.' U.S. Environmental Protection Agency, Office of Air Quality Planning and Standards, Research Triangle Park, N.C. January.

Hunter, S. C., and N. L. Helgeson. 1976. 'Control of Oxides of Sulfur from Stationary Sources in the South Coast Air Basin of California.' Contract 1–ARB4–421, KVB Inc., Tustin, Calif. June.

Interdisciplinary Center for Aeronomy and Other Atmospheric Sciences. 1978. 'An Interdisciplinary Study of the Health, Social and Environmental Economics of Sulfur Oxide Pollution in Florida.' University of Florida, Gainesville.

Kawecki, J. M. 1978. 'Emission of Sulfur-Bearing Compounds from Motor Vehicles and Aircraft Engines: A Report to Congress.' EPA—600/9—78—028, Environmental Protection Technology Series, U.S. Environmental Protection Agency, Washington D.C. August.

Kellogg, W. W., *et al.* 1972. 'The Sulfur Cycle.' *Science* 175 (February 11): 587.

Los Angeles County Department of Health Services. 1976. 'Number and Rate of Deaths by Age Groups, Region, and District (Los Angeles County).' Los Angeles, Calif.

Merkhofer, M. W., R. J. Korsan, P. C. McNamee, and S. M. Olmsted. 1978. 'Cost/Benefit Analysis of Mobile Source SO_x Emission Control: An Analysis of the Benefits of Automobile SO_x Emission Reduction in the Los Angeles Basin.' Draft final report, SRI International, Menlo Park, Calif. May.

Middleton, W. E. D. 1951. 'Visibility in Meteorology.' In *Compendium of Meteorology*, pp. 91—97. Boston: American Meteorological Society.

Morgan, M. G., S. C. Morris, A. K. Meier, and D. L. Shenk. 1978. 'A Probabilistic Methodology for Estimating Air Pollution Health Effects from Coal-Fired Power Plants.' *Energy Systems and Policy* 2(3): 287--310.

National Center for Health Statistics. 1973. 'Prevalence of Selected Chronic Respiratory Conditions, United States — 1970.' *Vital and Health Statistics*. Series 10, No. 84, U.S. Department of Health, Education, and Welfare, Public Health Service, Rockville, Md. September.

——. 1977. 'Acute Conditions — Incidence and Associated Disability, United States — 1974—1975.' *Vital and Health Statistics*, Series 10, No. 114, U.S. Department of Health, Education, and Welfare, Public Health Service, Rockville, Md. February.

Nordsieck, R. A. 1975. 'Air Pollutant Emission Factor Estimates for California Motor Vehicles, 1967—2000.' California Department of Transportation, Los Angeles, Calif. January.

Orange County Services Agency. 1976. 'Vital Statistics Summary, Orange County, 1976.' Public Health and Medical Services, Santa Ana, Calif. July 1.

Randall, A. *et al.* 1974. 'Bidding Games for Valuation of Aesthetic and Environmental Improvements.' *Journal of Environmental Economics and Management* 1(2): 132—149.

Roth, P. M., *et al.* 1974. 'Mathematical Modeling of Photochemical Air Pollution — II: A Model and Inventory of Pollutant Enissions.' *Atmospheric Environment* 8: 97—130.

Schwing, R. C. 1976. 'Risks in Perspective — Longevity as an Alternative to "Lives Saved."' U.S. Research Publication GMR—2133, Research Laboratories, General Motors, Warren, Mich. February.

Southern California Association of Governments. 1976. 'SCAG—76 Growth Forecast Policy.' Los Angeles, Calif. January.

Trijonis, J. *et al.* 1975. 'An Implementation Plan for Suspended Particulate Matter in the Los Angeles Region.' Final Report, TRW Transportation and Environmental Operations, prepared for the U.S. Environmental Protection Agency, Region IX, San Francisco, Calf. March.

Turner, D. B. 1970. 'Workbook of Atmospheric Dispersion Estimates.' U.S. Environmental Protection Agency, Environmental Health Series (Air Pollution), Research Triangle Park, N.C..

White, W. H. and P. T. Roberts. 1977. 'On the Nature and Origins of Visibility-Reducing Aerosols in the Los Angeles Air Basin.' *Atmospheric Environment* 11: 803—812.

Appendix B

A Decision Analysis of Alternative Government Policies for Commercial Cryptography

This case study represents an application that derives its input data almost exclusively from the subjective judgments of experts. The analysis, which represents an application of social decision analysis, has been previously described by Merkhofer (1985).

Decision Situation

Envisioning a growing market for encryption, researchers are attempting to produce inexpensive cryptographic devices based on codes that are essentially unbreakable. Security agencies in the United States have urged that the private sector be restrained in its pursuit of this goal. They fear that foreign governments, terrorists, and others whose interests are hostile to the U.S. might begin using unbreakable commercially available U.S. codes, thereby eliminating important U.S. intelligence gathering and law enforcement intelligence sources. Should the U.S. government encourage or discourage private sector R & D into cryptography?

Background

The private sector became interested in cryptography through a series of interrelated events initiated by the federal government. Beginning in the 1960s, Congress passed several laws dealing with the proper handling of information (such as the Freedom of Information Act, the Foreign Corrupt Practices Act, and the Privacy Act of 1974). The White House responded by taking steps to require federal agencies to use encryption to protect nonclassified data,[1] and the National Bureau of Standards (NBS) established a Federal Data Encryption Standard (DES),[2] which specified the cryptographic algorithm for this purpose. Through standardization, the DES demonstrated that a mass market could enable the private sector to produce inexpensive cryptographic products that incorporate a relatively high level of encryption security.

Despite the introduction of low-cost DES products, the use of cryptography outside the government remained minimal. Many believed that the reason was a distinctly unfavorable characteristic possessed by the DES and all other existing cryptographic systems: keys (sequences of numbers or letters that specify how messages are encrypted and decrypted) must be distributed to and kept secret by all communicators. Because a separate key is required for each pair of potential communicators, distributing and maintaining the secrecy of keys in a large communication network (such as the telephone system) was seen as unworkable. If a key failed to reach a communicator, secure communication would be impossible. Furthermore, if a receiver of messages purposefully or accidentally revealed his key, then all communication based on that key could be revealed.

Then, in 1975, researchers developed the theory behind a new type of cryptographic system called public-key cryptography (PKC) (Diffie and Hellman 1976). PKC systems use separate but related keys for encrypting and decrypting messages. Decryption keys cannot be deduced from encryption keys. Thus encryption keys could be made public without compromising the security of the encrypted data. As pointed out by PKC developers, a user could place his personal encryption key in a public directory (like the phone book) along with his name and address. To send a message, a sender would simply look up the appropriate key from the directory and use it to encode the message. Only the intended recipient could decode the message because only he would have the decryption key that reverses the transformation produced by the encryption key (Hellman 1978). Because the proposed PKC systems were based on problems that are among the most complex known to mathematicians, the popular press quickly dubbed PKC the "unbreakable code" (Shapely 1978).

Spurred by the perceived competitive advantages of PKC, manufacturers of cryptographic products using traditional codes (especially DES) developed imaginative hardware and software to reduce the burdens of distributing and managing secret keys. Meanwhile, patents were secured to produce new systems based on PKC algorithms. Manufacturers optimistically forecast an explosive growth in the encryption market.

Needless to say, government agencies quickly became interested in PKC technology and the uncertain consequences of civilian sector development and use of encryption. In an unprecedented public address, the director of the National Security Agency (NSA) warned that unclassified cryptographic research might endanger national security by helping foreign governments decipher American codes or assisting them in masking their communications from U.S. intelligence (Inman 1979). The Department of Defense (DOD) urged that the private sector be restrained in cryptography research and development. Meanwhile, a report prepared for the Department of Commerce (DOC) recommended that government policy promoting commercial crytography be adopted (Walling *et al.* 1981).

To provide guidance in the formulation of cryptography policy and to help integrate the differing points of view expressed by DOC and DOD, a decision analysis of policy options was commissioned. The analysis was conducted in 1981.

Analysis

The key step in structuring the analysis was developing a logical framework for evaluating policy options. Figure B1 shows the logical framework established by the analysts with the aid of the clients of the study. Because directly relevant hard data were largely unavailable, the analysts were forced to base their evaluation almost entirely on subjective judgments solicited from experts. The logical framework provided a policy focus by indicating the information that was relevant to the evaluation and furnishing a structure for organizing that information.

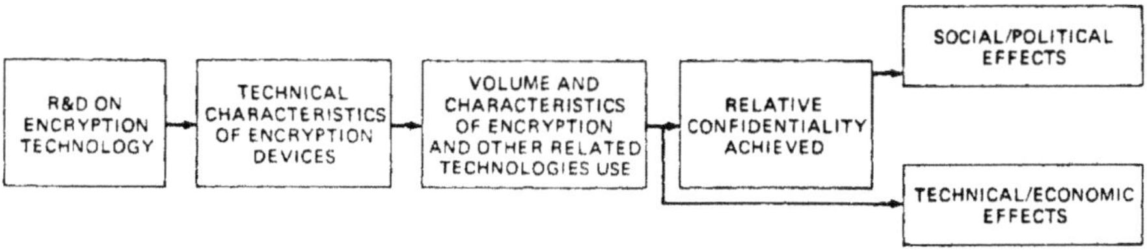

Fig. B1. Logical framework showing relationships influencing social impacts of public-key cryptography.

As illustrated in the logical framework, research in encryption and related areas was assumed to determine the technical characteristics of commercially available encryption devices and their costs. These technical characteristics, in turn, were assumed to influence the volume of use of encryption by application area and the level of use of related technologies, such as electronic message systems, that may be affected by encryption. Some societal impacts (labeled TECHNICAL/ECONOMIC EFFECTS in the diagram) were assumed to be a direct result of the use of encryption devices. Other impacts (labled SOCIAL/POLITICAL EFFECTS) were assumed to result from the increased confidentiality and security of data that are encrypted.

Individuals with expertise in each of the areas represented by the boxes in Figure B1 were identified by the analysts with the help of the study's clients. Researchers in crytology were chosen to provide judgments on likely developments in encryption theory; manufacturers were identified to provide data on the characteristics of future encryption devices; major users (such as communication system operators, financial institutions, and terminal manufacturers) were chosen as sources of information on future levels of encryption use; and sociologists, futurists, and other professionals concerned with the technical, economic, social and political future were selected to estimate cryptography impacts.

Specification and Characterization of Alternatives

To help identify available government actions and organize them into distinct policy options, the logical framework for the study (Figure B1) was applied. Any government action that alters one or more of the factors in the diagram was assumed to be a candidate for consideration as part of a policy option. Thus policy options were generated by identifying feasible government actions for altering each factor. Table B1 shows some of the actions that were so identified, with each listed according to the primary factor that it is meant to influence. Table B2 shows these same actions organized into three distinct policy options chosen by the analysts for further consideration.

The first policy option, "encourage PKC," consisted of actions designed to promote the development of PKC and commercial cryptography. The second option, "status quo," assumed the continuation of current policy. The third, "discourage PKC," consisted of actions designed to limit and control development and use of PKC and other crypto technologies. These options were defined somewhat arbitrarily, the intent being to make them sufficiently different that the course of the technology and its consequences would likely depend on which policy was instituted, while making them realistic enough to be of interest.[3]

Decision Criterion

For a criterion for evaluating policy options, the analysts chose the expected net present social value of the outcomes estimated to be produced by each option. Thus, as in most applications of social decision analysis, the evaluation consisted of estimating probability distributions for relevant outcome variables, assigning equivalent dollar values to those outcomes, discounting outcomes occurring in different time periods, and comparing the computed expected values.

Outcome Measures

Defining appropriate outcome variables for a decision analysis of alternative national crypto policies proved to be difficult. Table B3 summarizes the reasoning applied in arriving at a set of useful outcome variables. The first column in the table lists the major areas identified by experts as being potentially affected by PKC and civilian cryptography. Outcome variables for quantifying the level of impact in three of these areas are shown in the

Table B1. Sample government policy decisions and their primary points of impact

Research and development	Device characteristics	Applications and level-of-use	Confidentiality achieved	Impacts
Adjust project funding (amount, source, research area, security classification)	Intensify or relax patents and copyright restrictions	Require, regulate, or prohibit use	Strengthen privacy and other information-protection laws	Change legal penalties
Alter skilled labor pool (educational funding, licensing of individuals or facilities, postemployment restrictions, security clearance requirements)	Establish manufacturer licensing and product testing requirements	Subsidize or tax manufacturers or users	Encourage or discourage competing or associated technologies	Promote international treaties regarding transborder data flows
	Promulgate standards (algorithms, protocols, key management)	Provide government encryption services		
Control information dissemination (workshop and conference sponsorship and participation, publication restrictions)	Provide a government certification service	Encourage or restrict exports		
		Encourage or restrict imports		
		Use or prohibit use in government		

Table B2. Strategic options defined for quantitative analysis

	Representative actions		
Strategic option	Research and development	Device characteristics	Level of use
Encourage PKC	Government public crypto research funding doubled, PKC-related research emphasized	Provide compensation for profit loss resulting from patent secrecy orders	Limit application of export controls, provide favorable appeals procedure and approval criteria
	Non-DOD agency funding of crypto research encouraged	Establish NBS standards for PKC	Establish standards and conditions for approved government use of PKC
	Limit NSA review of proposals to "peer" capacity only	Offer NBS certification of devices and encourage nongovernment certification of algorithm security	
Status quo	Government funding of public crypto research held at a few million $/yr	Patent secrecy orders applied occasionally to protect national security	Export controls employed occasionally to protect national security
	NSF continues as major funding source for public crypto research		
	Prepublication review of private-sector crypto research by NSA remains voluntary		

Table B2 (*Continued*)

Strategic option	Representative actions		
	Research and development	Device characteristics	Level of use
	Government-funded crypto research directly impacting national security funded through and controlled by the DOD		
Discourage PKC	Reduce government funding of public crypto by one-half	Limit legal manufacture of leading-edge crypto devices to licensed firms	Prohibit export of leading-edge crypto devices
	Establish NSA as major source of government-funded crypto-related research		Regulate private-sector use of crypto devices
	Require prepublication review of all crypto-related research prepared under government support		Reiterate government faith in and commitment to DES
	License individuals and institutions for leading-edge crypto research		Require use of DES for protecting personal data held by private-sector organizations

Table B3. Some outcome areas, issues, and the outcome variables selected for quantitative analysis

Area	Relevant issues	Selected outcome variable
Personal privacy	Security of personal data and communications	Amount of unauthorized access to personal data or correspondence
Business security	Security of high-value corporate data and communications	Equivalent dollar value of losses sustained through wiretaps
National security	U.S. intelligence-gathering effectiveness	Fraction of foreign nations using "unbreakable" codes
	Private communication network security	
Social freedoms	Freedom to conduct research in, manufacture, and use technologies that promote personal privacy	
Economic	U.S. telecommunication/computer industry international competitiveness	
Political	Transborder data flows	
Legal	Ability to identify and use evidence for prosecution	

right-hand column. The middle column lists specific issues within each of the areas concerned.

As indicated in the table, outcome variables were selected to measure the impact of policy on the security of personal data and communications, the security of high-value corporate data and communications, and U.S. intelligence-gathering effectiveness. The quantitative variables defined for representing personal privacy and business activity were, respectively, the number of unauthorized accesses to personal data or correspondence and the equivalent dollar value of losses sustained through wiretaps. The quantitative measure selected for U.S. intelligence-gathering effectiveness was the fraction of foreign countries using codes that cannot be broken by U.S. intelligence-gathering agencies. The number of outcome variables was limited to three to reduce modeling complexity; it was regarded by the analysts to be more valuable in terms of likely insights to devote resources to quantifying considerations in these limited areas than to dilute the effort through an attempt to include a broader spectrum of considerations.[4]

Deterministic Analysis

Deterministic analysis consisted of conducting a highly simplified systems analysis. Two main steps were involved. First, the analysts developed a model to represent experts' judgments for the relationships between policy alternatives and outcome variables and to account for alternative social preferences for outcomes. Second, a sensitivity analysis was conducted to measure the relative importance of each variable in the model.

Decision model

Figure B2 shows the structure of the decision model, which was established in three steps. First, the analysts interviewed experts to identify the factors and interrelationships important to the estimation of decision outcomes. Then, based on their understanding of the experts' comments, the analysts proposed at tentative structure for the decision model. Finally, a panel of selected experts was convened which reviewed, modified, and ultimately approved Figure B2 as providing a suitable framework for quantitative analysis.

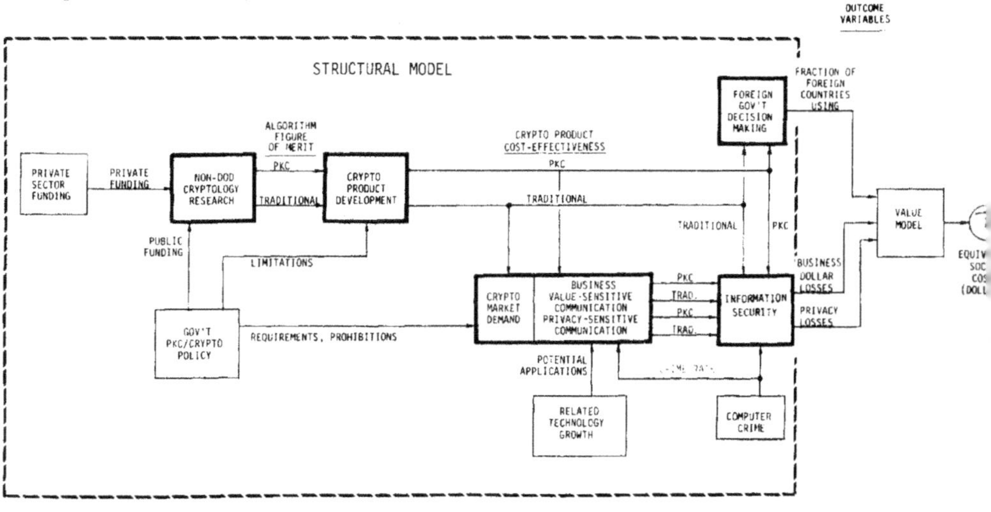

Fig. B2. Deterministic model structure.

The boxes drawn with heavy lines in the figure represent the major factors in the relationship between crypto research and impacts. (These five boxes correspond to the five elements of the logical framework shown in Figure B2). As represented by the arrows, each factor was assumed to affect, and be affected by, other factors.

Starting at the left of the figure, cryptology research was regarded as producing PKC and traditional crypto algorithms. The quality of these algorithms (as measured by "figures of merit") was presumed to be determined in part by the levels of private and public research funding. The figures of merit of available algorithms in turn were taken to influence the market attractiveness of the resulting crypto products, which might also be influenced by goverment requirements such as certifications or limitations on product development. The market attractiveness, together with government policies, was assumed to influence crypto demand, which itself was broken into two market segments: those applications that protect business value-sensitive information and those that protect public privacy-sensitive information. Demand was also assumed to be affected by the growth of potential application technologies, such as electronic mail and electronic funds transfer systems, and by the level of computer-related crime. The level of use of PKC and traditional crypto systems was presumed to influence the dollar and privacy losses suffered, whereas the fraction of foreign countries using "unbreakable codes" was assumed to be determined by the underlying security of the available crypto products. The value model was used to represent social preferences for trading off outcome variables against one another. The value model converts the three outcome variables to a common unit of measurement (dollars).

Having obtained agreement from the experts that Figure B2 provided an appropriate structure for the decision model, the analysts turned their attention to developing explicit, judgmental relationships between the inputs and outputs for each of the boxes shown. In some cases, experts suggested specific quantitative relationships. In others, the analysts developed tentative quantitative assumptions and then reviewed and modified the assumptions as required to obtain the experts' validation and acceptance. Assumptions represented in the resulting submodels are described in the paragraphs below.

Government policy and private-sector funding submodels

Interviews with experts in government policy revealed that a variety of existing laws and regulations could be invoked to influence the development of commercial cryptography. Through more careful channeling of research funds, for example, it was suggested that public-sector funding of research in areas directly relating to cryptography could be doubled or halved from current levels. Private-sector research funding, however, was judged to be difficult to control and likely to follow demand. The experts felt that it was relatively difficult to limit the dissemination of the results from any individual research project, but that export controls and patent secrecy orders were potentially effective if applied to prevent the commercial development of sophisticated cryptographic products. They also judged it to be relatively easy to promote or retard growth in the private-sector market for encryption products by applying existing laws and regulations and through other means of exerting government influence.

Reflecting these considerations, the government policy submodel was designed to estimate three outputs: public research funding level, product development limitations, and requirements for or limitations on use that influence the rate of growth of the encryption market. The values of these outputs were determined by the policy option selected. If the option was "encourage," public crypto research funding was assumed to be roughly doubled, and the assumed nominal growth rate of the encryption market was also doubled. If the "discourage" option was chosen, public funding was assumed to be halved, product development to be restricted in such a way that only crypto products that could be cryptanalyzed by U.S. government agencies would be produced for commerical use, and the nominal growth in the encryption market to be cut to zero. The private sector funding submodel was simply a variable that could be set to represent any arbitrary funding level. When the deterministic model was used to represent the dynamics of the encryption market, private sector funding in any given period was assumed to be proportional to crypto sales in the previous period.

Non-DOD cryptology research submodel

Because the quality of future commercially available codes was regarded as a critical uncertain factor, it was necessary to consider potential future quality differences in available PKC and traditional cryptographic algorithms. A major difficulty in this regard was the lack of practical measures of the quality of such an algorithm and the very limited understanding of how rapidly additional research might result in the improvement of existing codes. Consequently, the submodel developed to represent the relative quality of future PKC and traditional cryptoalgorithms was highly simplified.

The output of the cryptology research submodel was a numeric figure of merit meant to summarize the relative quality of available PKC and traditional codes. The higher the figure of merit, the more suitable the best code from that class of algorithms is for commercial products. Thus, to assess a figure of merit for PKC or traditional algorithms, a practical version of the algorithm useful for commercial applications must first be proposed. The primary consideration for determining the figure of merit was the ease or difficulty of breaking the algorithm given the best current method of cryptanalysis, but other relevant considerations included the length of the algorithm's key; the number of computational operations required for key generation, distribution, and management; and whether data expansion of cipher text occurs.[5]

The scale shown in Figure B3 was constructed to assess and interpret values for figures of merit. The DES was assumed currently to represent a

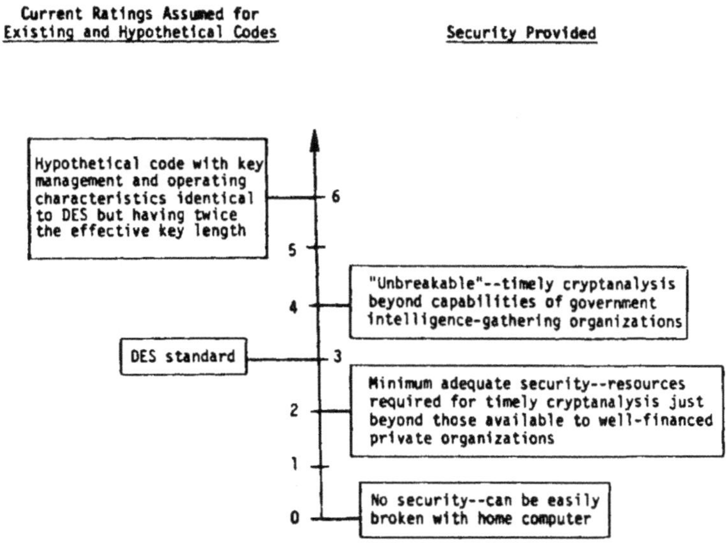

Fig. B3. Numerical scale for a code's figure of merit.

figure of merit of 3. If it were possible to double-encrypt with DES without any penalty in speed or key management, such an algorithm was assumed to have a figure of merit of 6.[6] Codes with figures of merit of 4 or above were assumed to be "unbreakable" in the sense that their use by foreign countries would diminish U.S. intelligence-gathering capability. Codes with figures of merit of 2 or below were assumed to provide inadequate security to protect high-value corporate information from sophisticated attacks by individuals or organizations.[7]

Note that with this formulation a code's figure of merit depends on the resources and techniques available to the cryptanalyst. As these improve — for example, through increases in computer processing speeds that shorten the time required for exhaustive key search — the figure of merit for a given implementation must fall. Thus a figure of merit that is now an adequate value of 2.5 may be reduced to an inadequate value of 1.5 as cryptanalysis capabilities improve.

For the purposes of the deterministic phase of the analysis, figures of merit were assessed directly through interviews with experts and provided as inputs. In the probabilistic phase described below a stochastic model was constructed to estimate future figures of merit for PKC and tradional algorithms based on extrapolations from current characteristics and accounting for private and public research funding levels.

Crypto product development submodel

This submodel was designed to represent judgment of manufacturers concerning the impact of price and the quality of an underlying crypto-algorithm on the market value of a crypto product. Marketing experts expressed the opinion that an inverse relationship exists between price and product attractiveness. It was also felt that the attractiveness of a product increased with the security of its underlying algorithm, but that there is a declining value to increasing security beyond some practical level.

To make such considerations explicit, the crypto product development submodel produced product attractiveness ratings for PKC and traditional crypto products. Each rating was defined to be a number reflecting a subjective judgment for the average attractiveness of a product based on a particular crypto algorithm; ratings were assumed to be a function of costs and the figure of merit of the algorithm. A rating of 0 was taken to represent a product that provides either no encryption security or one whose cost for the level of security provided is so high as to be uneconomical for all applications. Current crypto products based on the DES were assumed to possess a product attractiveness rating of 3. As noted in the description of the demand submodel below, two crypto products of identical attractiveness were assumed to achieve equal market penetration.

Figure B4 illustrates the form of the relationship assumed for product attractiveness rating.[8] As indicated, the attractiveness rating of a product

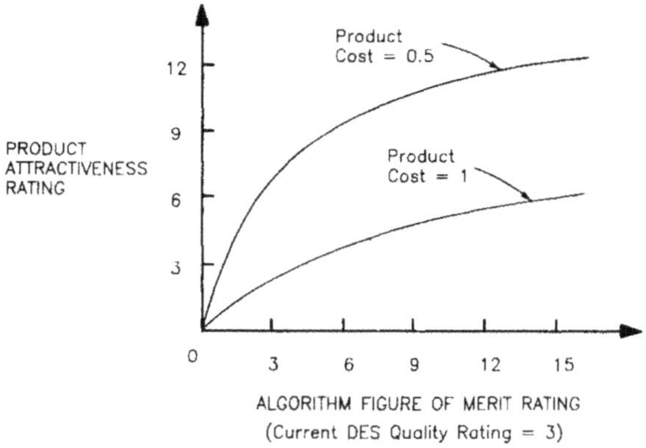

Fig. B4. Relationship assumed between algorithm figure of merit and crypto product attractiveness.

increases with the figure of merit of its underlying algorithm; however, doubling the figure of merit (e.g., by double encrypting) produces a less than doubling of product attractiveness. An inverse relationship between attractiveness and product cost was assumed. Thus doubling the cost of a product was assumed to halve its attractiveness. The figure of merit associated with a crypto product was assumed equal to the figure of merit of the underlying algorithm unless the government policy was to discourage PKC. In this case, the figure of merit of commercial products was assumed to be limited below a value of 4 so that the available products could be broken by U.S. intelligence agencies.

Crypto market demand, related technology growth, and computer crime submodels

The demand submodel measured level of use in two market segments: applications by business for protecting value-sensitive communications and applications to protect privacy-sensitive communications. Level of use was represented by an estimate of the fraction of the total number of messages in each market category that would be encrypted using each technology. The total number of messages was assumed to be determined by an estimate of the current number of messages sent within each market category and a growth rate. The fraction encrypted by each technology was determined by a nominal forecast and a factor that was applied to the nominal forecast to account for alternative assumptions for the market attractiveness of future crypto products.

The nominal forecast for the fraction of messages encrypted was developed using existing encryption market penetration forecasts (Kahn 1980; Diffie 1978; Carnegie-Mellon University 1980) and expert judgments concerning the impact of two critical variables: government policy (particularly the effect of legal requirements or prohibitions on use) and the level of computer-related crime. An "encourage" policy was assumed to be capable of doubling current encryption growth rate predictions. A "discourage" policy was assumed to be capable of holding the growth rate to zero. Because many experts felt that an unexpected increase in computer-related crime that could be prevented through encryption would spark more rapid demand growth, estimated encryption growth rates were assumed to double again if the level of computer crime were to increase significantly.

The factor applied to account for the effect of changes in future product attractiveness reflects the fact that the greater the attractiveness of PKC products relative to products based on traditional codes, the greater PKC's share of the market will be. Figure B5 shows the factor developed to reflect these considerations. The figure illustrates a case in which the attractiveness rating for PKC products is assumed to be 3. If the rating for traditional crypto products is also 3, each technology was assumed to make up exactly one-half the nominal demand forecast. If the PKC product attractiveness rating increases above 3, the fraction of the nominal forecast achieved by PKC was assumed to increase and the fraction achieved by traditional products was assumed to decrease. Notice that as the attractiveness rating for PKC products increase above that for traditional products, the fraction of nominal demand achieved by traditional products plus the fraction achieved by PKC products sums to greater than

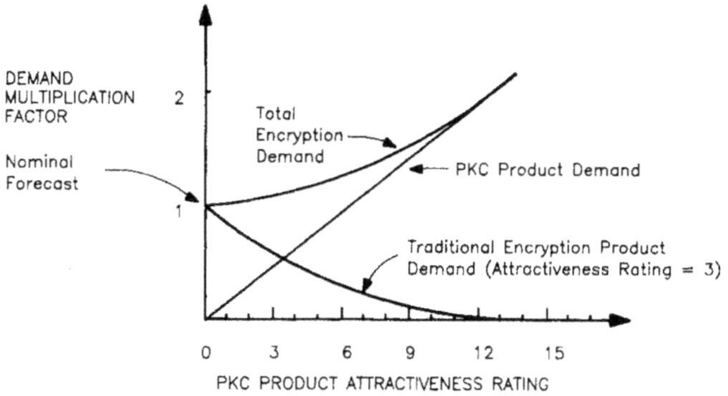

Fig. B5. Assumed effect of changes in crypto product attractiveness on forecasted encryption demand.

one. This assumption was designed to represent the judgment that a major improvement in one crypto technology not only would enable that technology to capture a larger fraction of the encryption market but would also produce additional demand as new customers used encryption in place of or in addition to the other means of protecting information.

Business security and privacy submodel

The business activity and privacy submodel produced as output an estimate of the dollar losses sustained by business as a result of disclosures of value-sensitive communications and an estimate of the number of unauthorized accesses to privacy-sensitive information occurring through successful wiretaps. For value-sensitive messages, encryption with a crypto system whose figure of merit is at least 2 was assumed to prevent all dollar losses that would have occurred through wiretapping without encryption. For privacy-sensitive messages, the assumption was that any encryption (figure of merit above 0) would prevent unauthorized access. The rate at which attempts are made to tamper with or obtain unauthorized access to messages was assumed to be proportional to the computer crime rate.

Foreign government decision-making submodel

The foreign government decision-making submodel estimated the fraction of foreign countries using codes that cannot be broken by U.S. intelligence agencies at future points in time (5-year intervals). To compute this estimate, the submodel took an initial fraction assumed to be using "unbreakable" codes and increased this by an amount determined by the higher of the two figures of merit associated with PKC and traditional codes. A figure of merit greater than 4 was assumed to be computationally unbreakable by the U.S. government (i.e., the time required to break the code is so great as to make the information obtained not worth the computational expenses). If the figure of merit for any commercially available crypto product is above 4, some foreign countries are likely to use that technology, and the fraction of foreign countries using "unbreakable" codes was assumed to increase. If the figures of merit of commercial products are below 4, there was assumed to be no change in the initial fraction. Figure B6 shows the form of the relationship for three different assumptions for the initial fraction of foreign countries using "unbreakable" codes.

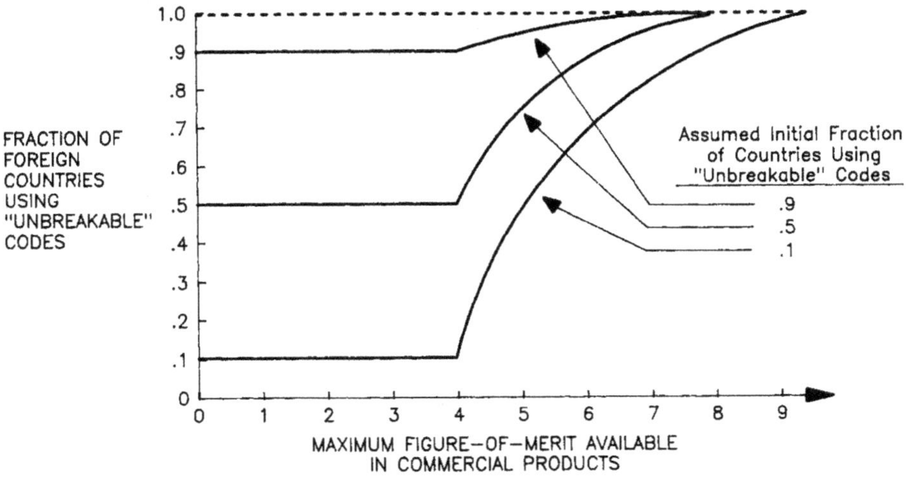

Fig. B6. Assumed relationships between fraction of foreign countries using "unbreakable" codes in 1986 and the figure of merit of commercial products in 1986.

Base-case model results and sensitivity analysis

The various submodels were implemented in computer code and base-case values assigned to all inputs. Table B4 summarizes some of the nominal input value assignments and base-case estimates obtained as a result of running the computer code assuming a 5-year horizon.

For the purpose of aggregating estimated outcomes into a single measure of equivalent social cost, nominal tradeoff weights were assigned as shown in the table. The nominal weights were \$50 per unauthorized access to privacy-sensitive information and \$100 million per foreign country using "unbreakable" codes. Since average dollar losses per computer-related crime were estimated to be \$500,000, the nominal weights imply the following assumption: for the purpose of selecting among alternative crypto policies, the loss of communications intelligence from one additional country using "unbreakable" codes produces a social cost comparable to 200 successful computer-related crimes or 2 million

Table B4. Base-case estimates produced through running the deterministic model with nominal values for independent variables (B = billion, M = million)

Sample nominal input values	Sample intermediate model variables	Outcome variables	Nominal tradeoff weights	Total equivalent social cost (through 1986)
Policy = status quo Initial PKC research funding Public = $4M/yr Private = $1M/yr	1987 potential encryption market privacy-sensitive applications = 3.5B messages/yr Value-sensitive applications = 1.75B messages/yr	Fraction of foreign nations using "unbreakable" codes = 0.5 (unchanged)	−$100M/additional country	$102M
Algorithm figures of merit PKC = 3 Traditional = 3 Initial product costs (normalized) PKC = 10 Traditional = 1	1987 market fraction encrypted by code PKC privacy-sensitive = 0.03 value-sensitive = 0.05 Traditional privacy-sensitive = 0.17 value-sensitive = 0.31	Number of unauthorized accesses through wiretaps = 13,980 messages (cumulative through 1986) Equivalent dollar value of business losses = $101M (cumulative through 1986)	−$50/access	
Initial potential encryption market Privacy-sensitive applications = 2B messages/yr Value-sensitive applications = 1B messages/yr				

Table B4 (*Continued*)

Sample nominal input values	Sample intermediate model variables	Outcome variables	Nominal tradeoff weights	Total equivalent social cost (through 1986)
Initial market fraction encrypted by code				
PKC				
privacy-sensitive = 0				
value-sensitive = 0				
Traditional				
privacy-sensitive = 0.005				
value-sensitive = 0.25				
Initial fraction of foreign countries using "unbreakable" codes = 0.5				
Fraction of computer crimes preventable through encryption = 0.01				

unauthorized accesses to personal privacy-sensitive messages. Because nominal tradeoff assignments were largely arbitrary, the policy implications of assuming both higher and lower relative values were also explored.

Table B5 shows the most sensitive variables identified through sensitivity analysis and their impacts. The sensitivity analysis consisted of varying each input variable across a range of values selected to approximate that variable's range of uncertainty, while holding all other variables constant at their nominal values. For example, in the base case, both PKC and traditional crypto products were assumed to have figures of merit of 3. As shown in Table B4 running the model with this assumption, base-case values for all other inputs, and the nominal tradeoff weights for outcome variables produced a total estimated social cost of $102 million for the period 1982—1986. In the sensitivity analysis, the model was run again under various assumptions for PKC and traditional code figures of merit between 0 and 5, while holding all other variables at their nominal values. Figure B7 shows the resulting variation in total social costs. The

Table B5. Sample sensitivity analysis results (Base-case equivalent social costs = $102 million)

Variable	Nominal value	Sensitivity range	Resulting variation in equivalent social costs (millions of dollars)
PKC/traditional algorithm figure of merit	3/3	0—5	4550
Computer-related crime rate	30/yr	10—150	938
Potential market growth rate	15%/yr	−5%—30%	412
Average cost per value-sensitive message wiretap	$500,000	$10K—$1M	395
Nominal forecast rate of encryption market share increase	10%/yr	0—30%	295
Ratio of PKC product costs to traditional product costs	10	1—20	210

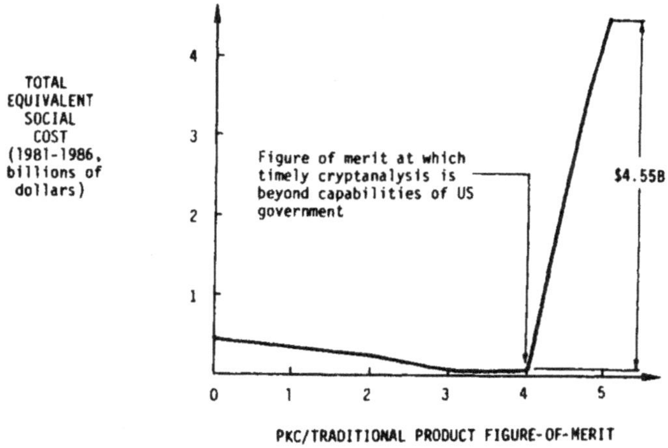

Fig. B7. Estimated total social cost as a function of commercial crypto technology figure of merit (with base-case assumptions).

difference between the highest and lowest values, $4,550 million, is the value noted in Table B5.

Those variables whose variations produced the greatest change in equivalent social costs were judged to represent critical uncertainties. As indicated by the table, the uncertainties identified as most critical were the figures of merit for PKC and traditional cryptoalgorithms, the incidence of computer-related crime, and market growth rates.

Probabilistic Analysis

Probabilistic analysis consisted of quantifying critical uncertainties and estimating the total uncertainty in net social costs under each policy alternative.

To represent dynamics and uncertainty in the decision process, a decision tree was used. Figure B8 shows the decision tree structure using a shorthand notation in which only the variables in the tree and the possible values for each variable are shown.[9] The structure of the tree indicates the various assumptions. As shown, time was divided into two 5-year intervals: 1982—1986 (short term) and 1987—1991 (long term). In 1982 the government was assumed to make an initial policy decision, choosing to encourage PKC and commercial cryptography, to continue the current policy (status quo), or to discourage PKC and commercial cryptopgraphy. By the end of 1986 it was assumed that the attractiveness of crypto products based on PKC would become known; PKC and traditional codes would each have some portion of the crypto market; and the approximate levels of each of the outcome variables (privacy violations, business losses

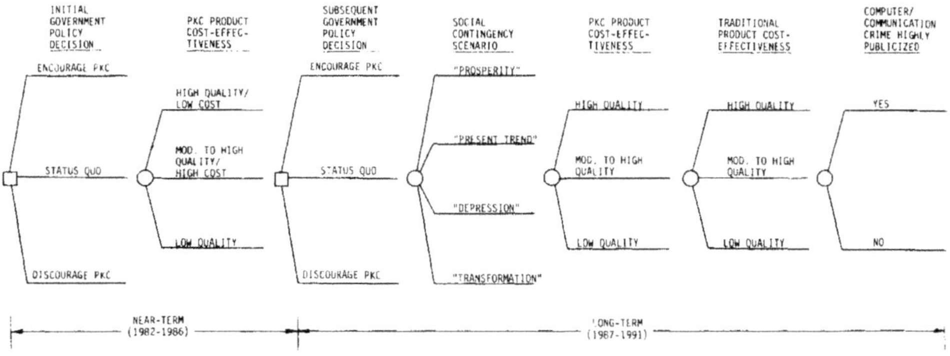

Fig. B8. Decision tree used to represent dynamics and uncertainty of decision process.

through wiretaps, and fraction of foreign countries using unbreakable codes) would become known. At this point the government was assumed to have the opportunity to revise crypto policy. The impact of the revised government policy was assumed to depend on several long-term uncertainties. As illustrated in the decision tree, these were a social contingency scenario, the longer-term attractiveness of both PKC and traditional products, and whether or not there is dramatic increase in the number of highly publicized computer-related crimes. Private research funding for each code in the period 1987—1991 was assumed to be proportional to the level of use achieved for that code during the period 1982—1986.

The manner in which uncertainty was represented and the process by which the tree was used to estimate overall uncertainty under each policy option is discussed in the section below.

Crypto product attractiveness

To represent uncertainty in crypto product attractiveness, a simple stochastic model was constructed. The attractiveness of a product was assumed to be determined by its underlying algorithm's figure of merit, which at any future time was assumed to be determined by three factors. First, research was assumed to produce random quantum gains in PKC and traditional code figures of merit as a result of technological advances (such as the development of a new algorithm). Second, the figure of merit of any given algorithm was assumed to decay at a fixed rate proportional to the increasing operating speeds of computers. As computers become capable of operating at higher speeds, breaking codes by exhaustive search[10] becomes easier, so figures of merit must decline over time. Third, mathematical breakthroughs occur that produce quantum decreases in the security of an algorithm. For example, publication of a symmetry property in the DES reduced the number of operations that must be performed in an exhaustive key search by a factor of 2. Mathematical advances can therefore be expected to produce random decrease in figures of merit.

Figures B9(a) and (b) illustrate sample behavior of a cryptographic algorithm's figure of merit over time as produced by the stochastic model. As illustrated, the probability of quantum gains in the figure of merit was assumed to be proportional to total research funding (private plus governmental). The probability of code improvement per unit of research funding, however, was considered to be higher for PKC than for traditional codes to reflect an assumption of greater opportunity due to the relatively unexplored nature of PKC. The heights (incremental increases and decreases) of gains and losses were assumed with equal likelihood to be anywhere between 0 and 100 percent of the current figures of merit, independent of funding. The probability of random drops in figures of merit was taken to be independent of funding level.

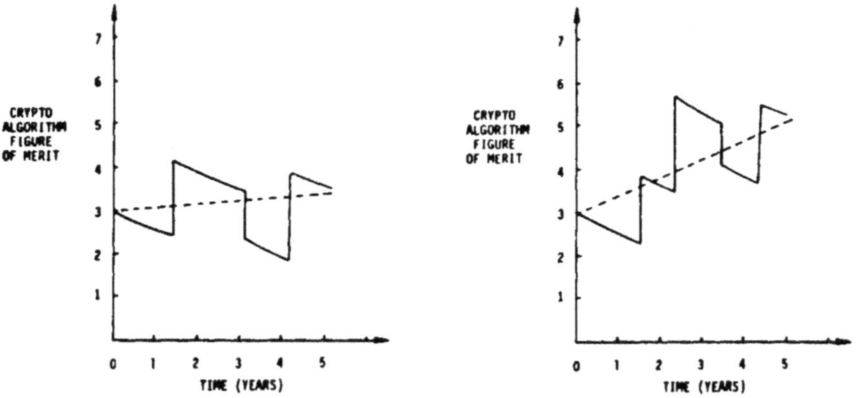

(a) Case 1: Total Research Funding of $5 million/yr

(b) Case 2: Total Research Funding of $10 million/yr

Fig. B9. Sample behavior of stochastic model used to represent dynamics of uncertainty in crypto algorithm figure of merit.

Crypto product costs were assumed to decline at 15 percent per year. PKC products were assumed in the base case to cost ten times as much as traditional products. Mass production of PKC was assumed to permit a cost comparable to that of traditional (DES) cryptosystems. Therefore, PKC costs were assumed to drop to the same level as traditional product costs when and if the demand for PKC products exceeded a specified minimum value (50 percent of the current demand for traditional products).

The stochastic model for the behavior of algorithm figures of merit and the model for decline in costs were used to obtain high, medium, and low estimates for future product attractiveness ratings and corresponding probabilities for each attractiveness level. For example, with an initial policy of status quo (implying an assumed total annual funding level of $5 million), near-term (1986) PKC product figure of merit was estimated to be 4.5 with probability 0.1, 3.4 with probability 0.8, and 1.1 with probability 0.1. These estimates were reviewed with cryptology researchers and judged to be reasonable.

Computer-related crime rate

Many have suggested that the number of crimes involving computers might increase and that should this occur, the commercial demand for cryptography would grow dramatically. Whether such an increase will occur within the next ten years, however, is highly uncertain.

To represent and quantify subjective uncertainty over future levels of computer-related crime, the following uncertain variable was defined: the total number of computer-related crimes involving losses in excess of $1 million reported in the *New York Times* during the period 1987 through 1991. Figure B10(a) show a cumulative subjective probability distribution

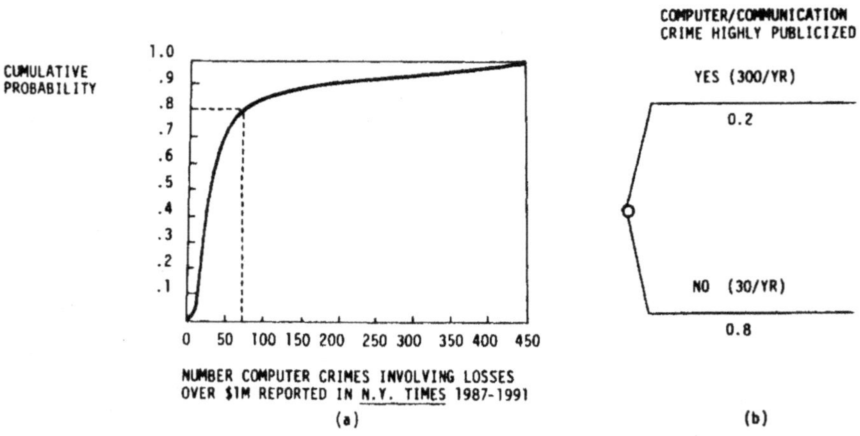

Fig. B10. Probability distribution illustrating uncertainty over number of publicized computer crimes (a) and discrete approximation used in decision tree (b).

for the value of this variable encoded from one individual knowledgable in issues of computer-related crime. The curve shows the probability that the actual number of reported crimes will be less than any value along the X axis. For example, the probability that the number is less than 75 (dashed line) is 0.8. This distribution was used to obtain probabilities for high and low values for the computer crime rate as illustrated in Figure B10(b). These values were subsequently used in the probabilistic analysis.

Social contingency scenario

For the purpose of estimating alternative market growth rates, alternative social contingency scenarios were taken from long-range planning forecasts produced by SRI International's Center for the Study of Social Policy (See Schwartz and Mitchell 1976 for an explanation of methodology). These forecasts recognize four alternative futures for the United States: prosperity, present trend, depression, and transformation. Prosperity and present-trend scenarios assume essentially linear extensions of the present, with prosperity being based on a condition in which fears concerning resource and food shortages, pollution, alienation, and system dysfunctioning are far overstated. A depression scenario involves a failure of the free enterprise system, bringing about a prolonged economic collapse and turn to an increasingly authoritarian form of government. A transformation scenario is characterized by a fundamental shift in values away from conventional industrial views and production and a low economic growth philosophy.

To simplify the analysis, depression and transformation scenarios were assumed to have the same basic effect on commercial cryptography. Prosperity was assumed to imply the high values used for market growth rates in the sensitivity analysis (Table B5), present trend to imply the base-case values, and depression or transformation to imply the low value. Based on discussions with futurists, a probability of 0.3 was assumed for prosperity, 0.5 for present trend, and 0.2 for depression or transformation.

Tree analysis

Each path through the decision tree of Figure B8 (obtained by selecting one of the branches emanating from each of the seven nodes) describes a possible scenario. For example, selecting the second from the topmost branch from each node produces the following scenario: initial status quo policy; near-term PKC products turn out to have a moderate quality level and high cost; selection of a long-term status quo policy; followed by a social contingency characterized by the continuation of present trends; long-term PKC products of moderate quality; long-term traditional crypto products of moderate quality; and no increase in the number of computer-related crimes. A total of 1,458 unique scenarios can be constructed in this way.

Analysis of the decision tree consisted of running the deterministic model using input values corresponding to each of the 1,458 scenarios through the tree. For each scenario, the deterministic model was run twice, first to simulate the period 1982—1986 and then the period 1987—1991, with the outputs of the first period providing initial conditions for the second. Table B6 shows the values of the three outcome variables that were generated for selected scenarios.

Table B6. Values of outcome variables for selected paths through decision tree

	Scenario (path through decision tree)						
No.	Initial Policy	PKC Attrac-tiveness	Subsequent Policy	Social Contingency	PKC Attrac-tiveness	Traditional Attrac-tiveness	Crime Rate
1	Enc.	Nom.	Nom.	Nom.	Nom.	Nom.	High
2.	S.Q.	Nom.	Nom.	Nom.	Nom.	Nom.	High
3.	Disc.	Nom.	Nom.	Nom.	Nom.	Nom.	High
4	S.Q.	V.Good	Nom.	Nom.	Nom.	Nom.	High
5.	S.Q.	Poor	Nom.	Nom.	Nom.	Nom.	High
6	Enc.	Nom.	Nom.	Nom.	Nom.	Nom.	Low
7.	S.Q.	Nom.	Nom.	Nom.	Nom.	Nom.	Low
8.	Disc.	Nom.	Nom.	Nom.	Nom.	Nom.	Low
9.	S.Q.	V.Good	S.Q.	Nom.	Nom.	Nom.	Low
10.	S.Q.	V.Good	Disc.	Nom.	Nom.	Nom.	Low
11.	S.Q.	Poor	S.Q.	Nom.	Nom.	Nom.	Low
12.	S.Q.	Poor	Enc.	Nom.	Nom.	Nom.	Low

Total number of unauthorized access to privacy-sensitive information (thousands, 1982—1991)

Total equivalent dollar losses of value-sensitive information (billions of S. 1982—1991)

Fraction of foreign countries using "unbreakable" codes (1991)

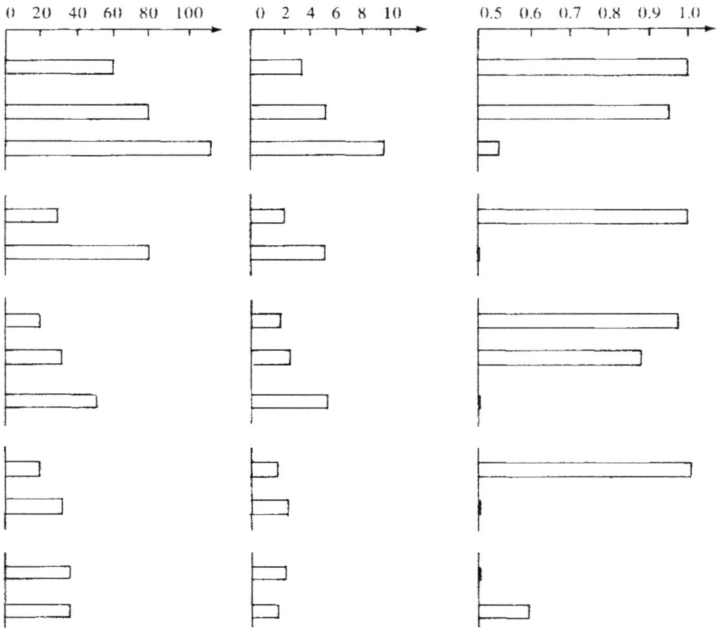

An examination of Table B6 clarifies the decision model and provides some insights into possible future behavior of the commercial crypto environment. The first three scenarios each involve high computer crime rates and differ only in the initial government policy. As the policy is made less favorable to PKC, the number of unauthorized accesses and dollar losses increases. This occurs because the less encouraging policies produce a lower level of use both through a direct effect on market growth rates and through an indirect effect of lowered product attractiveness brought about by reduced research funding. Notice that by 1991 nearly all foreign countries are estimated to be using codes that cannot be broken by U.S. intelligence agencies under either the encourage or status quo policies. This is because simulation under these scenarios produced figures of merit for PKC products that are well above 4, the threshold representing a code that is just incapable of being broken. Specifically, the 1991 calculated figures of merit were 6.3 and 12.9 for PKC compared with 3.1 and 3.7 for traditional products under status quo (Scenario 2) and encourage (Scenario 1), respectively. Higher figures of merit were estimated for PKC than for traditional codes because, as noted earlier, the likelihood of improvement through research was assumed higher for PKC. Traditional code figures of merit are strongly tied to DES and other standards that might be promulgated by the NBS. Such standards were assumed to be established in such a way as to encourage the use of private-key codes that do not pose a national security threat. Reflecting these assumptions, figures of merit for traditional codes remain below 4.

Public-key cryptographic codes with high figures of merit might be produced to ensure that encrypted information remains secure against advancements in cryptanalysis for longer periods of time; but the higher the figure of merit, the more likely the code is to attract international attention and to be used by other nations. Only the "discourage" option prevents the commercial sector from producing PKC codes whose quality is so high that they result in significant foreign use. The 1991 figures of merit were 4.0 for PKC and 2.7 for traditional products under the discourage policy (Scenario 3). With the "discourage" policy, products with figures of merit greater than 4 are prohibited during the 1982—1986 period.

Scenarios 4 and 5 represent the cases where PKC products are initially significantly more and less attractive than currently anticipated under the status quo government policy.[11] As illustrated in the table, uncertainty over PKC market attractiveness was found to have a major influence on privacy and dollar losses (because the model predicts a significant market for PKC under the high and moderate initial quality scenarios — about 55 percent (high quality case) and 39 percent (moderate quality case) of value-sensitive messages by 1991). If near-term PKC product attractiveness is worse than expected (Scenario 5), the ultimate level of PKC use was estimated to be lower (about 17 percent of value-sensitive messages); and even with the nominal assumption for subsequent long-term PKC cost-effectiveness, no PKC products with high figures of merit were estimated to be produced. As a result, and because under normal assumptions traditional products will not be unbreakable, the fraction of foreign countries using unbreakable codes remains at the initial value of 0.5. Thus, if PKC turns out not to be cost-effective during the first five years, the model shows PKC products not improving during the second five years to the point that they obtain significant use by foreign nations.

Scenarios 6—8 are identical to Scenarios 1—3 except that crime rate is low (about one-tenth the value assumed in the high crime case and approximately equal to current levels). The behavior of the estimated outputs is similar, except that the fraction of foreign countries using unbreakable codes was found to be slightly lower in each case. This is because a lower crime rate reduces demand and research into cryptography, resulting in a slower rate of improvement in the quality of the codes. The number of privacy violations and dollar losses was estimated to be much more than one-tenth the value computed in the high crime case, because there is less use of encryption under the low crime scenarios (about 62 percent of value-sensitive messages in 1991 under Scenario 7 compared with 93 percent under Scenario 2).

Scenarios 9—12 illustrate the predicted limited ability of a subsequent policy decision either to retard a very good technology or to improve upon a poor one. In Scenarios 9 and 10 the initial attractiveness of PKC products is higher than currently anticipated, the difference in the scenarios being that in Scenario 10 the initial status quo policy is shifted in 1987 to a discourage policy. The effect of the policy shift in this case appears favorable; privacy and value losses were found to increase only slightly, and the problem of foreign countries using the codes is almost completely eliminated. Scenarios 11 and 12 show the opposite case in which PKC products have a level of attractiveness that is much poorer than anticipated. Here, shifting to an encourage policy in 1987 was estimated to produce only a small effect on privacy and value losses and to have the detrimental effect of producing a code of high enough quality to result in its being unbreakable by U.S. intelligence and used by a significant number of additional foreign countries. The reason for this is that it is difficult to increase commercial usage by much during the second five-year period if a code is poor during the initial five years. According to the model, it is much less effective to use government policy to improve upon a poor cryptotechnology than to limit production and use of an extremely good technology. In the latter case, adequate security for commercial applications can be obtained without a code that is so secure that it is computationally unbreakable by sophisticated cryptanalysis organizations.

Solution of the decision tree

A decision tree can be solved to obtain the decision strategy that minimizes expected net social costs (where expected cost is the average over all possible costs weighted by their probabilities) provided that the outcome variables are expressed in the common unit of dollars. The nominal tradeoff weights given in Table B5 were used for this purpose: \$100 million equivalent social cost per foreign country using an unbreakable code and \$50 equivalent social cost per unauthorized access to privacy-sensitive information. As noted previously, the selection of these tradeoff weights was meant only to serve as an example for analysis. To aggregate outcomes occurring at different points in time, a discount factor of 5 percent per year was used. A relatively low discount rate was selected because all costs were expressed in constant dollars and therefore do not require an adjustment for inflation.

Figures B11 and B12 show the results of solving the decision tree under two different asumptions. In Figure B11 the initial fraction of

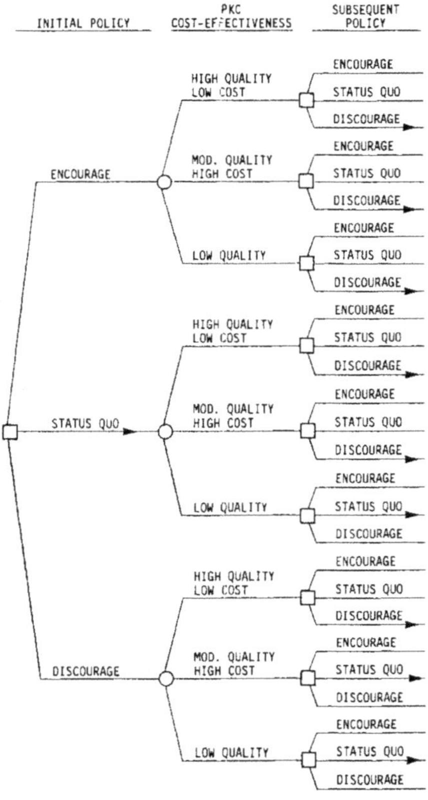

Fig. B11. Decision tree showing strategy minimizing social cost assuming 50 percent of foreign nations currently use "unbreakable" codes.

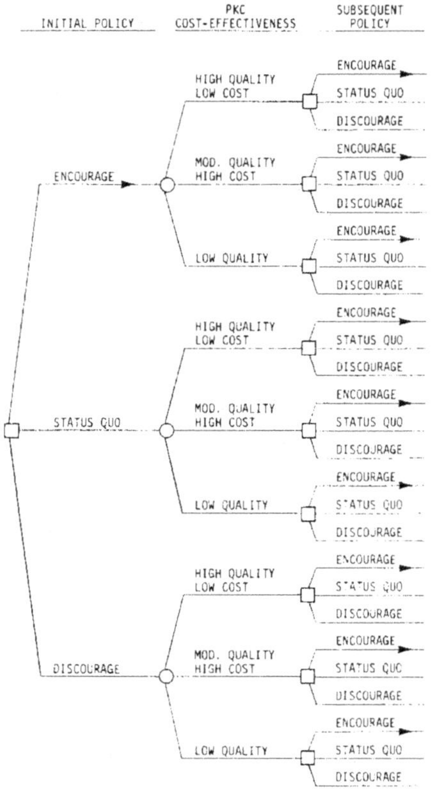

Fig. B12. Decision tree showing strategy minimizing social cost assuming 90 percent of foreign nations currently use "unbreakable" codes.

foreign countries using codes that U.S. intelligence cannot break was assumed to be 0.5. In Figure B12, the fraction was assumed to be 0.9. Arrowheads drawn on branches emanating from decision nodes indicate the alternatives that minimize expected net social costs (given the events represented by the branches leading to the nodes). Thus, for example, if the initial fraction was 0.5, the best strategy was "status quo" followed by "discourage," unless PKC turns out to be less attractive than anticipated, in which case a status quo policy should be maintained. The difference between the results in Figures B11 and B12 points out how crucial is the question of how much intelligence information the U.S. is currently able to obtain through cryptanalysis. If 90 percent of the foreign countries already use codes that the United States cannot break (or will by 1991, regardless of U.S. commercial cryptography), then, assuming the nominal tradeoff weights, the optimal strategy is uniformly to encourage PKC.

Figure B13 shows another result of analysis: probability distributions on discounted net social cost computed from the decision tree for the case

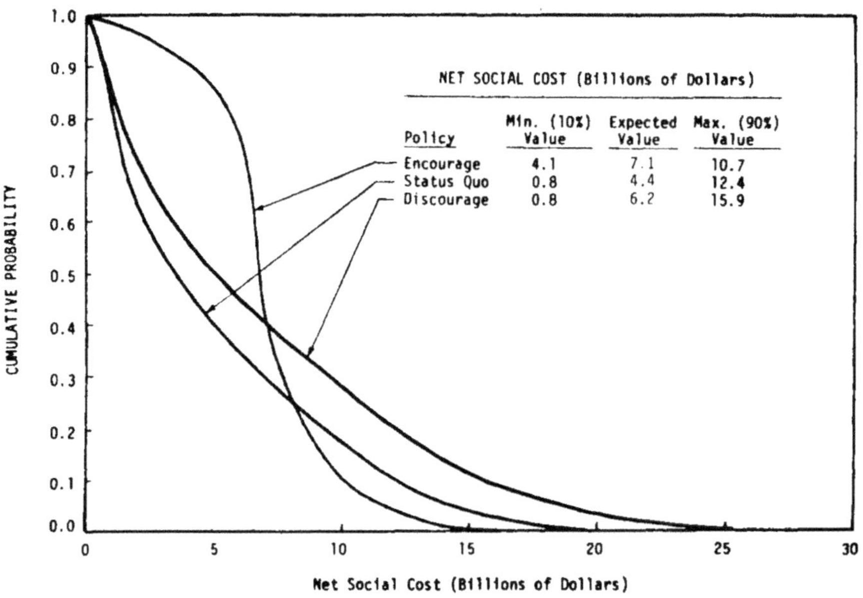

Fig. B13. Estimated probability that discounted net social cost will be greater than or equal to various values under each policy.

where the initial fraction of foreign countries using unbreakable codes is assumed to be 0.5.[12] The curves are complementary cumulative distributions: they show the probability that the actual value of social costs will be greater than or equal to any value on the X-axis. They also provide a quantitative estimation of the risks associated with each initial policy option. Notice that the curve associated with the discourage alternative lies to the right of the status quo curve, indicating that the probability of exceeding any given social cost was estimated to be higher under the discourage alternative than under status quo. For example, the probability of producing net social costs greater than $15 billion is shown to be roughly 0.04 under status quo and 0.12 under discourage. Thus status quo appears uniformly preferable to a discourage policy under the assumptions of the analysis. On the other hand, the encourage and status quo curves cross. Status quo, which has a lower expected cost than the encourage policy ($4.4 billion compared with $7.1 billion), produces a higher probability of very large costs. For example, the probability of social cost greater than $15 billion is nearly zero with the encourage policy. The reason for this is that the highest uncertain social costs are produced for paths through the tree in which the attractiveness ratings of PKC and traditional codes are very low (i.e., the codes are broken) and computer-related crime is very high. With the nominal tradeoff weights assigned ($50 per unauthorized access and $100 million per additional

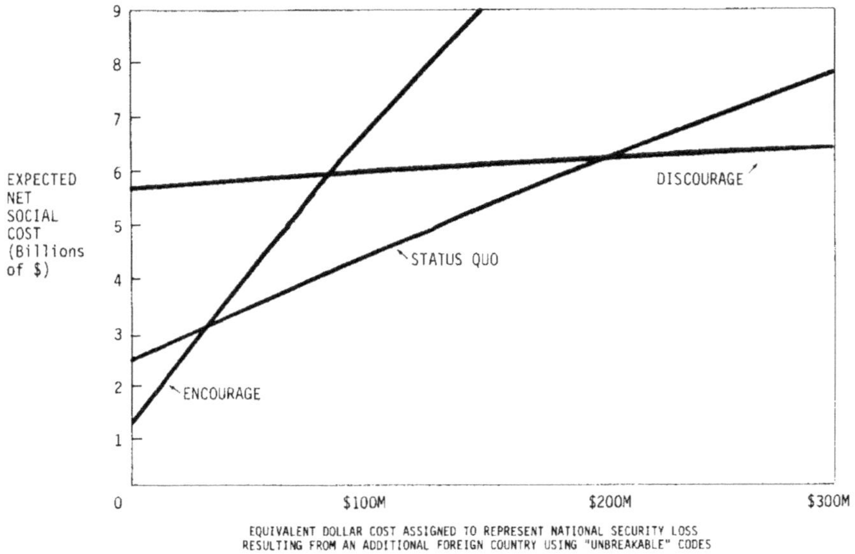

Fig. B14. Policy diagram showing how initial crypto policy decision depends on social cost assigned to increased security of foreign communications.

foreign country using "unbreakable" codes), these costs are higher than those generated from national security losses. The encourage policy therefore reduces risk by making the highest possible losses from computer-related crime less likely. Thus encouraging PKC produces a higher expected cost than does status quo (because of likely security losses resulting from use by other countries), but less risk of very large social costs (as a result of, e.g., higher than expected computer-related crime).

Figure B14 is a policy diagram that suggests how the initial crypto policy decision might depend on the social cost ascribed to increased security of foreign communications. The curves show how the estimated expected discounted net social cost of each policy option depends on the dollar tradeoff value assigned per additional foreign country using "unbreakable" codes. To obtain these curves, the analysis was repeated assuming a variety of values between $0 and $300 million. As illustrated, the policy minimizing expected social costs depends on the value chosen. For value assignments between about $30 and $200 million, the status quo policy produces minimum expected social costs. Below $30 million, expected social costs are minimized with the encourage policy; above $200 million, costs are minimized with the discourage policy. The quantitative analysis thus enabled the decision among alternative crypto policies to be related to a more fundamental social preference question; namely, the relative social cost of being unable to cryptanalyze foreign communications compared with the social cost of privacy and business losses resulting from insecure private-sector communications.

Informational Analysis

The fundamental idea of informational analysis is that of placing a monetary value on gathering additional information. A conceptual aid to computing this value is the concept of clairvoyance. Suppose someone existed who knew in advance just what value of a particular uncertainty would result in a decision problem. A standard decision analysis calculation is to estimate how much such information would be worth. The approach is to determine by how much expected costs could be reduced if the information were known prior to the decision. This reduction, then, would be the maximum amount an individual would be willing to pay a clairvoyant to obtain the information before having to make the decision. Computing the value of clairvoyance is useful because it places an upper bound on what it might be worth to attempt to gather better information on a given uncertainty before making a decision. Practical information-gathering schemes such as surveys and computer modeling are costly and yield less than perfect information. The value of perfect information can thus be used to help guide information-gathering decisions.

Another useful concept is that of wizardry. A wizard is defined as someone who can set an uncertain quantity to any specified value. Because in many situations a decision-maker may have some ability to alter probabilities of uncertain events, it is of value to know how much this ability is worth. The value of wizardry may be calculated by setting uncertainties to specific values and observing the effect on calculated expected social cost. The difference between the costs assuming the specified value for the variable and those assuming uncertainty yields the value of wizardry.

Table B7 shows the values of clairvoyance and wizardry that were

Table B7. Expected values of resolving uncertainty on and controlling computer-related crime rate

	Billions of dollars
Expected cost without information	4.4
Expected cost with perfect information on crime rate	4.2
Expected cost with crime rate low	1.7
Expected value of clairvoyance (4.4—4.2) (value of resolving uncertainty on computer-related crime rate)	0.2
Expected value of wizardry (4.4—1.7) (value of guaranteeing a low level of computer-related crime)	2.7

calculated for the most sensitive uncertainty — computer crime rate. Recall (Figure B10) that the decision model assumed that there is a 20 percent chance that the rate of major, publicly reported, computer-related crimes will increase by a factor of 10 within the next decade. As indicated, knowing whether this will happen is worth, according to the decision model, $200 million. Being able to guarantee a low crime rate was estimated to be worth much more — $2.7 billion.

The reason for these high values is the potential impact of crime-rate information on the policy decision. With uncertainty over whether computer-related crime will increase, the best strategy is status quo. If it were known that the crime rate would increase, the decision tree indicates the best strategy to be to encourage commercial cryptography. National security costs would likely be higher, but personal privacy and business losses would be minimized. If the computer-related crime rate were known to remain low, the best strategy would be to discourage cryptography, thereby minimizing national security costs. Being able to adjust policy in this way in accordance with knowledge over future computer crime reduces the expected social costs calculated from the decision tree by $200 million. On the other hand, ensuring that crime rate will remain low allows the discourage option to be applied and guarantees that both personal privacy, business, and national security losses will be low. The result in this case is the much larger $2.7 billion reduction in net social costs.

Epilogue

The analysts described their analysis in a series of presentations to the sponsors, the various experts who participated in the effort, and to representatives from the DOC, DOD, and other interested parties. The analysts stressed that because many of the quantitative assumptions were necessarily speculative due to the lack of information, the specific numeric results could not be relied upon. The analysis was described as providing a highly simplified quantitative illustration of considerations and conclusions that had evolved more generally from the interactions and investigations of the analysts and experts during the course of the project. They suggested that the analysis provided value by ensuring a highly disciplined exploration of issues and by generating insights that might not otherwise have been recognized.

For their primary conclusions, the analysts listed thirteen qualitative consensus observations put forth by the principal participating experts. An example was the conclusion that the sensitivity of the decision to the national security value derived from communications intelligence (which could not be adequately explored because of the unclassified nature of the study) and the desirability of coupling commerical crypto policy with efforts to prevent computer-related crime. The general reaction of those briefed on the analysis was typified by that of a representative from the president's Office of Science and Technology Policy who had responsibility for integrating the differing opinions previously advanced by DOC and DOD. He commented that the value of the analysis lay not so much in its specific conclusions, which he suggested were fairly intuitive, but in its clarification of the logic behind those conclusions.

Notes to Appendix B

1. E.g., Circular No. A—71, 'Responsibilities for the Administration and Management of Automatic Data Processing Activities,' Executive Office of the President, Bureau of the Budget, Washington, D.C., March 6, 1965.

2. 'The Data Encryption Standard,' Federal Information Processing Standards Publication 46, USA, January 1977.

3. Obviously a great many other options are available for a national crypto policy, and specification of a complete national policy would be much more detailed than the set of actions listed in the table. For the purposes of the quantitative analysis, however, it was assumed that policy alternatives were limited to a choice among these three action sets.

4. Because not all of the relevant issues were represented by outcome variables in the analysis, quantitative results do not reflect a completely comprehensive consideration of issues. Major issues listed in the table that were not addressed include the national security benefit of improved private-sector communication network security, freedom of choice, and increased cost-effectiveness of secure communication. Each of these issues tends to argue for policies that encourage, or at least do not discourage, private-sector cryptography. This might seem to provide a systematic bias against the "encourage" option. On the other hand, there may be other issues arguing for controlled civilian crypto development beyond limiting the fraction of foreign nations whose codes we cannot break. The analysts believed that the function of the quantitative analysis was not to identify an optimal policy but to provide information and understanding through the analysis of a highly simplified quantitative approximation of the relations among major factors.

5. Because the figure of merit represents characteristics of the mathematical encryption algorithm only, it did not account for the economic costs of implementation. Costs were accounted for in the product development submodel.

6. The only currently known method for breaking the DES or proposed PKC systems is to try all possible keys until the correct one is found. Thus the time required to break a code is determined by the total number of possible keys, which in turn is determined by the length of the key. IBM reports that the length of the DES key, 56 bits, was selected based on tradeoffs of exhaustive search difficulty, computational time, and engineering considerations (Kolata 1977). Double encryption can be used effectively to double the length of the DES key, but double encryption also increases computational operations and reduces speed.

7. Thus an assessment of a figure of merit of 2.5 for PKC, e.g., would represent the judgment that the best current PKC algorithm is, overall, of somewhat less value in

commercial applications than the DES but capable of providing adequate security at present to protect sensitive corporate data. An assessment of PKC figure of merit of 4.5, on the other hand, would signify the judgment that PKC is of greater value than DES; i.e., commercially practicable design characteristics may be set so that breaking the code is beyond the current resources of the most sophisticated cryptanalysis organizations.

8. In deriving this functional form, the following kinds of questions were used. Suppose as a potential user of cryptography you have a choice between two products. The products have identical operating characteristics, but one is 25 percent more expensive and has an exhaustive-search work factor equivalent to twice the key length of the other one. Is the additional security worth the added cost? Obviously the answer depends on whether the security provided by the less expensive product is regarded by the user as adequate. As shown in Figure B4, a doubling of effective key length without operating penalities (which doubles the figure of merit) produces nearly a doubling in the product's attractiveness rating when the code's security is marginal (figure of merit around 2), but only a small change when the security provided is already likely to be more than adequate for many years (figure of merit around 6).

9. To display the complete tree, each node and its branches in Figure B8 would be duplicated at the end of each branch emanating from the node to the left of it.

10 See n. 6 regarding exhaustive search.

11. The assumptions for these cases were obtained using the crypto product development submodel and the stochastic model for code figure of merit. Both PKC and traditional cryptographic algorithms were assumed to have initial (1981) figures of merit of 3. As shown in Figure B9, these figures were assumed to change over the next five years in a random way that depends on research funding level. Resulting 1986 values were grouped into the 80 percent middle values and the 10 percent upper and lower extremes. The average PKC figure of merit values in the upper, middle, and lower groups (4.5, 3.1., and 1.1) were taken as the high (Scenario 4), nominal (Scenario 2), and the low (Scenario 5), cases. As noted previously, PKC product costs were assumed to be ten times that of traditional code products during the first 5-year period.

12. To obtain these results the net discounted social cost was computed for each path through the decision tree (assuming contingency decisions as shown in Figure B13). The probability of each path was then obtained by multiplying the probabilities corresponding to the outcomes associated with the branches along that path. The probabilities for the high, nominal, and low figure-of-merit branches in each case were assumed to be 0.1, 0.8, and 0.1, respectively, with corresponding numerical values determined by the stochastic model for figure of merit, as described in the previous note. The probability of high crime rate was taken as 0.2, as shown in Figure B12. Knowing the net discounted social cost and the probability of each path, the probabilities of all net social costs greater than or equal to each given amount were identified and summed to obtain the complementary cumulative probability curves.

References to Appendix B

Carnegie-Mellon University. 1980. 'An Assessment of Civil Sector Users of Digital Data Encryption.' Department of Engineering and Public Policy, Department of Social Sciences and School of Urban and Public Affairs, Pittsburgh, Penn, November.

Diffie, W. 1978. 'The Outlook for Computer Security.' *Mini-Micro Systems* October: 42—44.

—— and M. E. Hellman. 1976. 'New Directions in Cryptography.' *IEEE Transactions on Information Theory* IT—22(6): 644—654.

Hellman, M. E. 1978. 'An Overview of Public Key Cryptography.' *IEEE Communications Society Magazine* 16(November): 24—32.

Inman, B. R. 1979. Speech before the Armed Forces Communications and Electronics Association, January 11.

Kahn, D. 1980. 'The Market for Encryption.' *Cryptologia* 4(1): 42—44.

Kolata, G. B. 1977. 'Computer Encryption and the National Security Agency Connection.' *Science* 197(July 29): 438—440.

Merkhofer, M. W. 1985. 'A Decision-Analytic Technology Assessment of Public Key Cryptography.' In R. L. Schultz, ed. *Applications of Management Science.* Vol. IV. Greenwich, Conn.: JAI Press.

Schwartz, P. and A. Mitchell. 1976. 'The Art of Exploratory Planning: Identifying the Corporate Long-Range Course.' Research Report 582, Business Intelligence Program, SRI International, Menlo Park, Calif. December.

Shapely, D. 1978. 'The New Unbreakable Codes — Will They Put NSA out of Business?" *Washington Post,* Outlook, Sec. Bl. July 9.

Walling, V. C., Jr., D. B. Parker, and C. C. Wood. 1981. 'Impacts of Federal Policy Options for Non-Military Cryptography.' SRI Project 1663 Final Report, SRI International, Menlo Park, Calif.

Appendix C

An Application Of Applied Social Welfare Theory to a Space-Mission Planning Decision

This case study describes one of the very few applications of applied social welfare theory so far conducted to support an actual decision. Dyer and Miles (1976, 1977a), the analysts involved in this study, required a decision-aiding approach that permitted the active participation of the principal stakeholders in the decision process. The description provided here closely follows the documentation provided by Dyer and Miles.

Decision Situation

As part of the Mariner Program, the National Aeronautics and Space Administration (NASA) plans to launch two identical spacecraft toward Saturn for the purpose of conducting experiments designed to gather information on the Jupiter and Saturn planetary systems. The quality of the scientific information expected from these experiments depends on the precise trajectories that are selected for each spacecraft. Furthermore, different trajectories would be optimal for different experiments. Given that the scientists associated with the various experiments differ in their preferences, which particular pair of trajectories should be selected?

Background

Approximately once every twenty years, Earth, Jupiter, and Saturn are arranged in such a way as to allow a spacecraft to travel efficiently from Earth to Saturn by "swinging around" the strong gravitational pull provided by Jupiter. The most recent of those opportunities occurred in the 1976—1980 time period and was seized upon by the 1977 Mariner Jupiter/Saturn project, which was renamed Voyager.[1] The objectives for this ambitious mission were to investigate Jupiter and its largest satellites, Callisto, Ganymede, Europa, and Io; to proceed on to Saturn and to explore several of its moons, such as Titan and Iapetues; and to investigate the interplanetary medium and its interactions with the solar wind.

Figure C1 illustrates the configuration of the twin Voyager spacecraft,

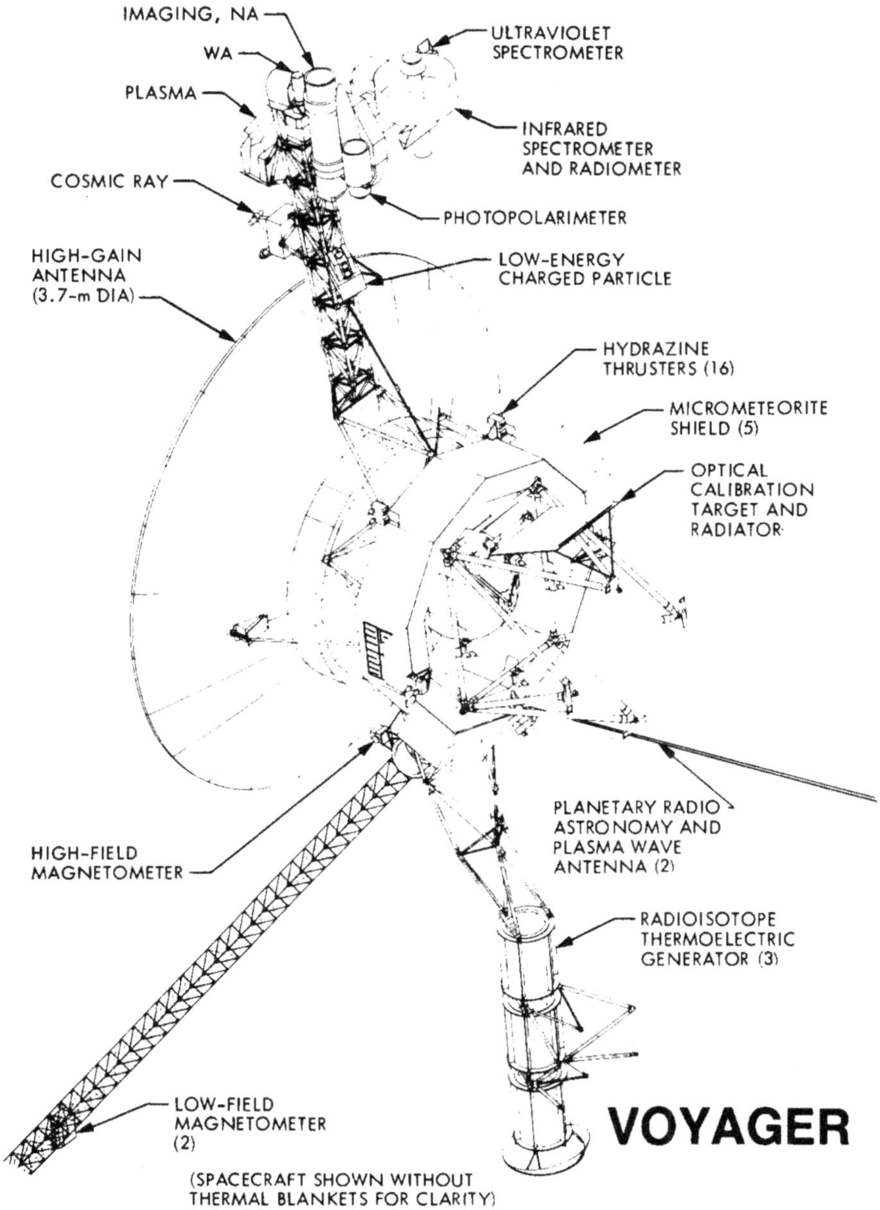

Fig. C1. The Voyager Spacecraft.

which were designed for long life, long-range communications, and precision navigation. Electric power was provided by independent radioisotope thermoelectric generators. On board each spacecraft were instruments for eleven science investigations involving measurements covering visual, infrared, and ultraviolet regions of the spectrum as well as other remote-sensing studies of the planets and satellites. The intended experiments consisted of studies of radio emissions, magnetic fields, cosmic rays, and lower-energy particles; studies of plasma (ionized gases) waves and particles; and experiments using the spacecraft radios. The mission plan called for the spacecraft to be launched on a pair of dates selected in August and September of 1977, to reach Jupiter in 1979, to encounter Saturn in late 1980 or in 1981, and then to leave the solar system.

Although the need to rely on Jupiter's gravitational pull en route to Saturn constrained the spacecraft arrival geometry at Jupiter, a trajectory speedup or slowdown could be used to time the Jupiter arrival so as to synchronize the spacecraft for Jupiter satellite encounters. Similarly, trajectory variations at Saturn could be used to determine the geometry of an encounter with Titan. The configurations of the Jupiter and Titan encounters would produce a major effect on the planned scientific experiments. Consequently the scientists participating in the mission were expected to have specific preferences for various trajectories made possible by midcourse propulsive maneuvers.

Management responsibility for Voyager was assigned by NASA to the Jet Propulsion Laboratory (JPL). The eighty participating scientists were divided by specialty into eleven science teams, each responsible for one of the instrument packages that formed the payload of the spacecraft. Table C1 lists the science teams and describes the measurements with which each was concerned. The team leaders of each science team collectively formed a Science Steering Group (SSG). An important function of the SSG was to advise the project manager on the preferences of the participating scientists over alternative trajectory pairs. It was the goal of the project manager to select a pair of trajectories that best met the preferences of the participating scientists.

An important decision that had to be made early in the mission planning process was the selection of a "standard trajectory pair"; that is, a tentative flight plan for each spacecraft. Although flexibility was to be maintained to permit trajectory changes up to the time of launch, the standard trajectory pair represented an important commitment because it provided a basis for designing the mission systems and represented an initial bargaining position for later selection of actual flight trajectories. Thus the selection of a standard trajectory pair was viewed as a significant milestone, both by the project manager and by the scientists.

To help select the standard trajectory pair, an analysis was commissioned. The clients for the study established four objectives for the decision-aiding approach:

1. It should produce a trajectory pair compatible with science requirements and mission constraints.

Table C1. The Voyager science investigations

Science team	Abbreviation	Primary measurements
Radio science	RSS	Physical properties of atmospheres and ionospheres. Planet and satellite masses, densities, and gravity fields. Structure of Saturn rings
Infrared radiation	IRIS	Energy balance of planets. Atmospheric composition and temperature fields. Composition and physical characteristics of satellite surfaces and Saturn rings
Imaging science	ISS	High-resolution imaging of planets and satellites at large phase angles. Atmospheric dynamics and surface structure
Photopolarimetry	PPS	Methane, ammonia, molecular hydrogen, and aerosols in atmospheres. Composition and physical characteristics of satellite surfaces and Saturn rings
Ultraviolet spectroscopy	UVS	Atmospheric composition including the hydrogen-to-helium ratio. Thermal structure of upper atmospheres. Hydrogen and helium in interplanetary and interstellar space
Cosmic ray particles	CRS	Energy spectra and isotopic composition of cosmic ray particles and trapped planetary energetic particles
Low-energy charged particles	LECP	Energy spectra and isotopic composition of cosmic ray particles and trapped planetary energetic particles
Magnetic fields	MAG	Planetary and interplanetary magnetic fields
Plasma particles	PLS	Energy spectra of solar wind electrons and ions, low-energy charged particles in planetary environments, and ionized interstellar hydrogen
Planetary radio astronomy	PRA	Planetary radio emissions and plasma resonances in planetary magnetospheres
Ultraviolet photometry*	UVP	Deuterium-to-hydrogen ratio of atmospheres. Structure of upper atmospheres. Physical structure and chemical composition of Saturn rings. Hydrogen Lyman alpha emission in interplanetary space

* Not selected by NASA at the time of the study

2. It should be compatible with the existing project management structure and science interfaces.
3. It should not create dissension among the SSG members.
4. It should be conceptually simple.

The analysis was conducted in 1973, at which time ten of the eleven project teams had been selected.

Analysis

Rather than argue for any single approach, the analysis team formulated the trajectory selection problem in four ways: as a problem of collective choice based on rank sum (ordinal) judgments, as a bargaining problem, as a social welfare maximization problem, and as a decision for a supradecision-maker concerned with maximizing group preferences. The analysts regarded the purpose of their study to provide information useful to the decision process. Because no single formulation was regarded as compatible with the wide range of views held by all participants, the analysts believed that attacking the problem from different perspectives would provide the greatest insight.

Specification and Characterization of Alternatives

In principle, an infinite number of trajectories was feasible. Because it was computationally practical to analyze at most only a small number of alternatives, it was necessary to generate a few potentially favorable options that spanned the range of trajectories and that met the minimal requirements of each science team. This was accomplished through an iterative process involving the JPL engineers, who were familiar with the characteristics of and constraints on feasible trajectories, and the science teams, who understood the implications of trajectory characteristics on the science experiments.

The first step in alternative generation consisted of having the individual science teams document their investigation objectives and, based on the performance characteristics of their instruments, establish the most important scientific criteria for the trajectories. These were

At Jupiter:

1. Penetration of the Jupiter magnetic flux tube associated with the satellite Io with a range to Io of less than 40,000 kilometers
2. Passing within 50,000 kilometers of at least one Galilean satellite other than Io
3. Occultation of the sun and Earth by Jupiter as viewed from the spacecraft (i.e., orientation of the spacecraft so that Jupiter eclipses the sun and Earth)

At Saturn:

1. Passing within 50,000 kilometers of Titan with occultation of both the sun an Earth by Titan
2. Multiple satellite encounters at less than 100,000 kilometers

3. Occultation of the sun and Earth by Saturn.
4. Occultation of the sun and Earth by the rings of Saturn
5. Escape from the solar system in the direction of the sun's motion through interstellar space

The priority ranking of these criteria varied from team to team, and the teams differed in their preferences for making each spacecraft's trajectory similar in order to achieve the most important criteria in a redundant manner (and thereby protect against failures), versus making them complementary to achieve the maximum number of criteria.

The scientific criteria were provided to the JPL engineers, who then developed candidate trajectories that were consistent with the capabilities and constraints of the spacecraft and launch vehicle. In the vicinity of Jupiter it was necessary that the trajectories avoid the most intense regions of the planet's radiation belts. Jupiter arrival times were constrained within a period bounded at the early end by launch vehicle capability and at the far end by the need to limit total flight time because of reliability risks and the overhead expense of maintaining an operations staff during the long interplanetary cruise phase of the mission. Within these constraints the engineers generated feasible trajectories by considering free-flight trajectories that departed from Earth on each of the most probable launch days (the opening of the launch period and approximately 11 days thereafter) and that covered every feasible Titan arrival opportunity at Saturn from November 1980 to September 1981. The trajectories were then varied plus or minus a few days at Jupiter to obtain the most preferred geometries with respect to Jupiter satellite encounters. In addition, every feasible Iapetus arrival opportunity was covered, as well as several multiple satellite opportunities at Saturn that could be retargeted to Titan as late as 90 days prior to Saturn arrival.

In this manner 105 distinct trajectories were generated. Candidate trajectory pairs were then developed by picking combinations that included one trajectory corresponding to the opening of the launch period and one to be launched about eleven days later. A total of 2,624 pairs were thus assembled from the 105 single trajectories. This number was then reduced by roughly a factor of 2 by imposing the constraint that the Jupiter encounter dates for the two trajectories be separated by at least one month. This constraint was necessary to ensure that the data retrieval capabilities of mission operations not be overloaded during the encounter periods.

To reduce the number of remaining trajectories to a size suitable for analysis, the JPL engineers and science teams repeatedly screened, eliminated, and added to the initial options. While overseeing this exercise, the analysts urged that trajectory pairs spanning the widest range of alternatives be retained. Ultimately the list was reduced to 32 potentially favorable trajectory pairs which were numbered 1 through 36 with four deletions.[2]

Decision Criteria

Four separate decision criteria were developed corresponding to the four alternative formulations employed in the analysis.

Rank sum collective choice formulation

The rank sum rule is one of the oldest and most widely used decision criteria for preference aggregation. If each science team assigns the number 1 to its most preferred trajectory pair, the number 2 to its second most preferred, and so on, the rank sum decision criterion is to choose the alternative whose average ranking across all teams is the lowest number.

Young (1974) has shown that four axioms are sufficient to justify the use of the rank sum rule:

1. *Faithful.* If the group consists of only a single individual, the choice rule must choose that individual's most preferred alternative.
2. *Consistent.* If the choice rule chooses the same alternative for two different groups, it must choose that alternative when the two groups are aggregated.
3. *Neutral.* The names of the alternatives must make no difference.
4. *Cancellation.* The preference of any one individual can be cancelled by contrary preferences of any other individual.

Like all social choice rules based only on ordinal preferences, the rank sum rule fails to satisfy Arrow's axioms for group decision making (see Chapter 2).[3]

Nash bargaining model formulation

The bargaining model formulation involved expressing the trajectory selection process as an n-person, cooperative game defined in a way that would allow a modification of the solution method provided by Nash (1950). Nash investigated a two-person game defined as follows. Two individuals may barter goods but have no money to facilitate the exchange (so they are unable to make side payments). The players may agree to any one of a finite number of specific trades (reapportionments of their joint goods), $t^1, t^2, \ldots, t^m$, or may agree to randomizations between trades such that the specific choice between alternative trades is determined by the outcome of a specified lottery.[4] The status quo may be regarded as the special "trade," denoted t^0, in which no actual exchange occurs. The players are assumed rational in the von Neumann-Morgenstern sense, so that their preferences for trades may be expressed by respective utility functions u_1 and u_2. Thus the bargaining problem may be characterized by the feasible set of actual and randomized trades and the utilities u_1^0 and u_2^0 assigned by the players to the initial bargaining point t^0.

Nash's "solution" to this problem consists of a process for obtaining a "fair" payoff, u_1^+, u_2^+, corresponding to one or more of the feasible trades. The solution process consists of two steps:

1. Selecting the origins for the players' utility functions in such a way that for each player the utility of status quo is zero (i.e., $u_1^0 = 0$, $u_2^0 = 0$)
2. Finding the unique point (u_1^+, u_2^+) that is the maximum of all of the possible products $u_1 u_2$ associated with feasible trades

The reason Nash's process is of interest is that it can be shown to produce a bargaining outcome that is

1. Invariant with respect to utility transformations (the solution does not change depending on the arbitrary origins and ranges established for the players original utility functions).
2. Pareto optimal (the solution is at least as good as the status quo, and there is no other outcome simultaneously preferred by both parties).
3. Independent of irrelevant alternatives (the solution is insensitive to the introduction of additional feasible trades that do not jointly dominate the calculated solution).
4. Symmetric (if the bargaining game is such that the players are placed in entirely symmetric roles, the solution provides them with identical utility).

Harsanyi (1959) was able to generalize Nash's result to the case of n players; that is, if there are n players, the unique payoff vector (specified in terms of the utility functions of the n players) that satisfies Nash's four axioms of fairness will be that trade from the set of feasible trades (actual plus randomized trades) that maximizes the product of the players' utility functions.

In expressing the trajectory selection as a Nash bargaining problem, the science teams were regarded as the players (so that $n = 10$). Because the preferences of the scientists within each team were relatively homogeneous, the analysts felt that it was reasonable to treat the science teams as individual entities. The actual trades were assumed to consist of the set t^1, . . . , t^{36} of candidate trajectory pairs. Since all science teams must fly the same mission, the set of available strategies was taken to be identical for each team. The utility functions for each team were assumed to be scaled so that $u_i = 0$, $i = 1, \ldots, 10$, if a trajectory is selected which yields no useful information for team i. Such a situation could arise, for example, if the trajectory pair were chosen in such a way that the instruments associated with an experiment could not function, or if the mission were cancelled. The status quo (situation in which no bargain takes place) was assumed to be tantamount to cancelling the mission with the result that no useful information would be obtained by any science team. Thus, $u_i^0 = 0$, for all $i = 1, \ldots, 10$.[5]

With these assumptions the trajectory decision was formulated in terms that fit the Nash model. The analysts, however, were unwilling to apply the Nash solution directly because of concern that a randomized solution (one that called for picking a trajectory by some random process) would be unacceptable regardless of its "fairness." They decided, therefore, to restrict the set of feasible solutions of payoff vectors to those that result from actual trajectories. The decision criteria thus consisted of finding the trajectory pair t^k from the candidates $t^1, \ldots, t^{36}$ that maximized the product of the science teams utilities:

$$u_1(t^k)u_2(t^k)\ldots u_{10}(t^k)$$

This modified multiplicative choice rule for the Nash bargaining problem does not necessarily produce a unique solution. Furthermore, it may be shown to violate one of the Nash axioms (symmetry). Nevertheless, the analysts observed that their modified multiplicative rule could be justified through an argument posed by Zeuthen (1930) and elaborated on by Harsanyi (1956).[6]

Social welfare formulation

In the social welfare formulation, trajectory selection was viewed as a problem of aggregating the preferences of the science teams. The analysis involved defining an n-person social welfare function over the set of possible trajectories and then determining the specific trajectory pair that maximized this function. The analysts chose to use an additive form for the function. Thus the decision criterion for this formulation was

$$w(t^k) = w_1 u_1(t^k) + w_2 u_2(t^k) + \cdots + w_{10} u_{10}(t^k)$$

where the t^k, $k = 1, \ldots, 36$, are the alternative trajectory pairs, the $u_i(\cdot)$ are the utility functions associated with the science teams, and the w_i are weighting factors. The conditions given by Harsanyi (1955, 1975) and Keeney and Kirkwood (1975) (see Chapter 2) were used to justify this form.[7]

If the individual utility functions $u_i(\cdot)$ are scaled so that values represent approximately equal preference differences for each team, the magnitudes of the weighting factors can be interpreted as indicating the relative importance of the various scientific experiments. Thus the analysts sought to impose interteam normalization by scaling each u_i so that very good and very bad trajectories that produce similar consequences for each team are assigned identical utility values. A "no data" trajectory t^0 would produce an identical worse-case outcome for each team, and it was therefore assumed that

$$u_i(t^0) = 0$$

for all i. To complete the scaling, it was assumed that the best trajectory t_i^+ for each team i would produce consequences for that team that were as good as the best trajectories for any other team. Therefore

$$u_i(t_i^+) = 1$$

for all i.

Supradecision-Maker Formulation

In the supradecision-maker formulation, trajectory selection was viewed as the responsibility of the project manager rather than of the science teams. The project manager, however, was interpreted to be a "benevolent dictator": although trajectory selection was considered a single-decision-maker problem involving multiple attributes, the decision maker was assumed to view the appropriate attributes for evaluating the options to be the preferences of the science teams. Following the usual practice in applications of multiattribute utility theory, the analysts chose a form for the project manager's utility function that was a linear combination of utility functions defined on the individual attributes (in this case, the utilities assigned by the science teams). Thus

$$U_{PM}[u_1(t^k) \cdots u_{10}(t^k)] = w_1 f_1[u_1(t^k)] + w_2 f_2[u_2(t^k)] + + \cdots + w_{10} f_{10}[u_{10}(t^k)]$$

where the w_i are weighting factors reflecting the relative importance of the science teams' preferences, and the $f(\cdot)$ are transformations applied to the team utility functions. These transformations permit the project manager to scale the team utility functions uniformly from 0 to 1 and permit him to make corrections if he feels that reported utilities contain errors or otherwise should be adjusted to make them more consistent with his own beliefs.

The analysts appealed to Fishburn (1970) for their justification of the linear form for the decision maker's utility function. Fishburn showed that a multiattribute utility function may be evaluated by an additive utility function if the desirability of any lottery from the viewpoint of the decision maker depends only on the marginal probability distributions of the u_i and not on their joint probability distributions.[8]

Derivation of Science Team Preferences

Three of the problem formulations (the bargaining, social welfare, and supradecision-maker formulations) required that the preferences of the individual science teams over candidate trajectory pairs be quantified in the form of cardinal utility functions. The analysts developed a two-phase process for assessing these utility functions.

In the initial phase the science teams were called on to determine their individual rankings of the 32 candidate trajectory pairs from best to worst. Each team i was then asked to consider each trajectory pair t^k in comparison with its most- and least-preferred trajectories and to assign a probability number p_i^k such that the team was just indifferent between receiving the trajectory t^k versus a lottery that would yield its most-preferred trajectory pair with probability p_i^k and its least-preferred trajectory pair with probability $1 - p_i^k$. In this manner each science team i generated 32 probability numbers p_i^k.

The second phase involved a comparison of each team's least-preferred trajectory pair, $t^{\text{least}(i)}$, with a lottery involving a uniformly worst possible trajectory and its most-preferred trajectory pair. The worst possible trajectory was taken to be a no-data trajectory pair called the "Atlantic Ocean Special" (in remembrance of the flight of Mariner 8, which terminated abruptly in the Atlantic Ocean). Each team i was asked to assign a probability number p_i^{least} such that the team was just indifferent between receiving its least-preferred trajectory pair versus a lottery that would yield its most-preferred lottery with probability p_i^{least} and the no-data trajectory pair with probability $1 - p_i^{\text{least}}$.

Figure C2 summarizes the comparisons made in each phase of the assessment. By scaling the utility functions so that the utility assigned by each team for its most-preferred trajectory pair was 1 (as required by the social welfare function formulation) and so that all teams assigned to the no-data trajectory pair a utility of 0 (as required in the bargaining model and social welfare formulations), science team utilities could be calculated as

$$u_i[t^{\text{least}(i)}] = p_i^{\text{least}}$$

$$u_i(t^k) = p_i^k + (1 - p_i^k)p_i^{\text{least}}$$

The analysts delivered detailed written and verbal instructions for the assessment process and allowed the science teams approximately one month to apply the procedure and report their results. Unfortunately, the utility values initially received indicated problems with the second phase of the process (which was intended to normalize the utility functions for interteam comparison). In reviewing the utility values assigned by the science teams to their least-preferred trajectories, the JPL project scientist was convinced that some science teams had assigned values that were

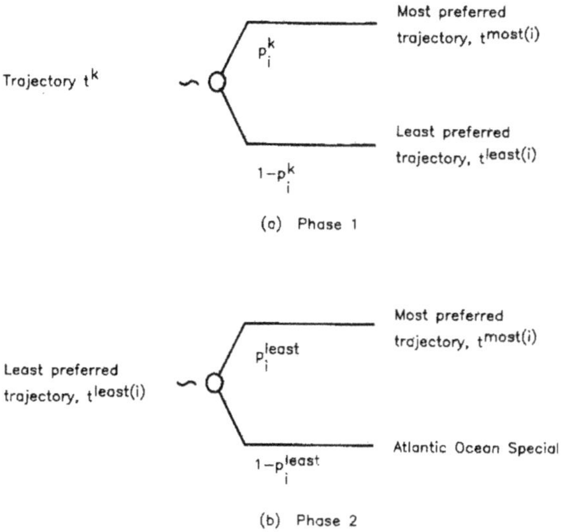

Fig. C2. Indifference judgments obtained in the assessment of science team utility functions.

much too high while others had assigned values that were much too low. The high values appeared to be due to the strong risk aversion exhibited by some of the teams to a "no-data" outcome. One science team, for example, responded initially that it would accept a probability of no less than 0.001 of obtaining the no-data outcome, so that their utility function values for candidate trajectory pairs covered "the remarkable range of 1.0 to 0.999." A representative of another team that had also assigned a high value to the team's least-preferred trajectory expressed the opinion: "the utility values indicate, in my opinion, that the team lacks a gambling nature, not that the pairs are of approximately equal value to our experiment . . ." (Dyer and Miles 1976). Utility values that were too low appeared to be the result of teams deliberately biasing their assessments downward. Some science teams apparently recognized that they could maximize their influence on the collective choice process by obtaining the maximum possible spread in their range of utilities. The analysts agreed that in retrospect it would probably have been better to have the project scientist or project manager conduct the Phase 2 evaluations of the assessments rather than asking the science teams to determine, in effect, their own handicaps.

Table C2 shows the final results obtained following efforts by the analysts to renegotiate more reasonable values for teams reporting either extremely high or extremely low utility values for their least-preferred trajectory pairs. Noting that these final values for least-preferred pairs covered the broad range from 0.101 to 0.800, the analysts recognized that

Table C2. The science team ordinal rankings and cardinal utility function values

Trajectory pair	Radio science (RSS)		Infrared radiation (IRIS)		Imaging science (ISS)		Photo-polarimetry (PPS)		Ultraviolet spectroscopy (UVS)		Cosmic ray particles (CRS)		Low-energy charged particles (LECP)		Magnetic fields (MAG)		Plasma particles (PLS)		Planetary radio astronomy (PRA)	
	Rank	Utility	Rank	Utility	Rank	Utility	Rank	Utility	Rank	Utility	Rank	Utility	Rank	Utility	Rank	Utility	Rank	Utility	Rank	Utility
1	6.0	0.772	28.0	0.871	13.5	0.850	22.0	0.550	17.0	0.820	24.5	0.850	14.0	0.700	15.0	0.500	26.5	0.600	19.0	0.546
2	6.0	0.722	24.5	0.875	22.0	0.750	25.0	0.530	22.0	0.790	5.5	0.910	7.0	0.830	19.0	0.480	11.0	0.750	18.0	0.547
3	20.5	0.600	14.5	0.925	4.0	0.960	28.5	0.520	5.0	0.910	28.5	0.840	28.0	0.450	27.5	0.350	7.0	0.900	24.0	0.109
4	20.5	0.600	6.0	0.948	26.0	0.700	20.0	0.570	2.0	0.970	28.5	0.840	27.0	0.480	26.0	0.351	7.0	0.900	25.0	0.108
5	20.5	0.600	9.0	0.943	28.5	0.670	17.0	0.600	4.0	0.920	13.0	0.880	6.0	0.900	6.0	0.800	1.5	1.000	6.0	0.600
7	20.5	0.600	32.0	0.750	32.0	0.600	32.0	0.500	18.0	0.810	32.0	0.800	30.0	0.430	25.0	0.360	26.5	0.600	26.0	0.107
8	20.5	0.600	20.0	0.886	10.0	0.900	5.0	0.780	13.5	0.830	21.0	0.860	11.0	0.740	7.0	0.700	17.5	0.700	9.0	0.556
9	20.5	0.600	12.0	0.930	16.0	0.800	13.0	0.650	26.0	0.760	9.5	0.890	19.0	0.610	16.0	0.490	17.5	0.700	13.0	0.552
10	3.0	0.887	11.0	0.935	24.0	0.720	8.5	0.720	19.5	0.800	2.0	0.960	20.0	0.600	24.0	0.400	17.5	0.700	7.0	0.558
11	20.5	0.600	13.0	0.928	7.0	0.920	3.5	0.810	28.5	0.720	5.5	0.910	26.0	0.490	31.0	0.260	32.0	0.500	27.0	0.106
13	8.0	0.652	18.0	0.915	22.0	0.750	28.5	0.520	13.5	0.830	31.0	0.810	31.0	0.320	30.0	0.280	26.5	0.600	28.0	0.105
15	3.0	0.887	29.5	0.869	16.0	0.800	18.5	0.580	9.0	0.840	17.5	0.870	16.0	0.650	20.0	0.430	11.0	0.750	22.0	0.543
17	20.5	0.600	4.0	0.966	13.5	0.850	6.5	0.750	13.5	0.830	24.5	0.850	9.0	0.770	14.0	0.510	17.5	0.700	8.0	0.557
18	20.5	0.600	8.0	0.944	10.0	0.900	15.0	0.630	22.0	0.790	13.0	0.880	24.0	0.560	17.5	0.485	26.5	0.600	29.0	0.104
19	20.5	0.600	26.5	0.873	10.0	0.900	28.5	0.520	26.0	0.760	13.0	0.880	25.0	0.500	29.0	0.300	26.5	0.600	30.0	0.103
20	6.0	0.772	10.0	0.937	31.0	0.620	22.0	0.550	32.0	0.630	21.0	0.860	22.0	0.580	27.5	0.350	26.5	0.600	16.0	0.549
21	20.5	0.600	14.5	0.925	3.0	0.970	31.0	0.510	22.0	0.790	28.5	0.840	29.0	0.440	11.0	0.575	7.0	0.900	31.0	0.102
22	20.5	0.600	29.5	0.869	10.0	0.900	18.5	0.580	31.0	0.640	5.5	0.910	17.0	0.630	23.0	0.405	26.5	0.600	23.0	0.542
23	20.5	0.600	24.5	0.875	16.0	0.800	25.0	0.530	30.0	0.670	5.5	0.910	8.0	0.780	22.0	0.408	26.5	0.600	20.0	0.545
24	3.0	0.887	17.0	0.919	26.0	0.700	28.5	0.520	1.0	1.000	28.5	0.840	32.0	0.300	32.0	0.250	26.5	0.600	32.0	0.101

Table C2. (*Continued*)

Trajectory pair	Radio science (RSS)		Infrared radiation (IRIS)		Imaging science (ISS)		Photo-polarimetry (PPS)		Ultraviolet spectroscopy (UVS)		Cosmic ray particles (CRS)		Low-energy charged particles (LECP)		Magnetic fields (MAG)		Plasma particles (PLS)		Planetary radio astronomy (PRA)	
	Rank	Utility	Rank	Utility	Rank	Utility	Rank	Utility	Rank	Utility	Rank	Utility	Rank	Utility	Rank	Utility	Rank	Utility	Rank	Utility
25	20.5	0.600	7.0	0.947	28.5	0.670	25.0	0.530	9.0	0.840	1.0	0.970	5.0	0.960	5.0	0.950	11.0	0.750	1.0	1.000
26	20.5	0.600	2.0	0.996	10.0	0.900	11.0	0.700	7.0	0.850	17.5	0.870	3.0	0.980	1.0	1.000	1.5	1.000	2.0	0.990
27	20.5	0.600	1.0	1.000	30.0	0.650	16.0	0.610	3.0	0.930	17.5	0.870	1.0	1.000	4.0	0.970	4.0	0.950	3.0	0.980
28	1.0	1.000	23.0	0.876	26.0	0.700	14.0	0.640	13.5	0.830	24.5	0.850	13.0	0.720	17.5	0.485	11.0	0.750	11.0	0.554
29	20.5	0.600	5.0	0.957	19.0	0.770	6.5	0.750	9.0	0.840	3.0	0.930	2.0	0.990	2.0	0.990	4.0	0.950	4.0	0.970
30	20.5	0.600	22.0	0.878	1.0	1.000	12.0	0.690	26.0	0.760	17.5	0.870	10.0	0.750	21.0	0.420	17.5	0.700	12.0	0.553
31	20.5	0.600	3.0	0.987	5.0	0.940	8.5	0.720	6.0	0.860	8.0	0.900	4.0	0.975	3.0	0.980	4.0	0.950	5.0	0.960
32	20.5	0.600	19.0	0.890	19.0	0.770	22.0	0.550	28.5	0.720	13.0	0.880	18.0	0.620	13.0	0.570	17.5	0.700	14.0	0.551
33	20.5	0.600	26.5	0.873	6.0	0.930	10.0	0.710	19.5	0.800	9.5	0.890	21.0	0.590	12.0	0.573	17.5	0.700	15.0	0.550
34	20.5	0.600	16.0	0.921	19.0	0.770	3.5	0.810	13.5	0.830	21.0	0.860	15.0	0.670	8.0	0.600	11.0	0.750	17.0	0.548
35	20.5	0.600	21.0	0.883	2.0	0.980	2.0	0.820	13.5	0.830	13.0	0.880	12.0	0.730	9.0	0.595	17.5	0.700	10.0	0.555
36	20.5	0.600	31.0	0.813	22.0	0.750	1.0	1.000	24.0	0.770	24.0	0.850	23.0	0.570	10.0	0.594	26.5	0.600	21.0	0.544

Source: Dyer and Miles (1977b)

the results probably still contained distortion due to gaming and differing attitudes toward risk aversion.

A statistical analysis of data reported by the science teams was conducted to investigate the extent to which teams with related experimental objectives agreed or disagreed in their rankings. Two groups of teams were formed during the evaluation process, one consisting of the fields and particles science investigations (those denoted CRS, LECP, MAG, PLS, and PRA in Table C1) and the other of the platform-mounted science investigations (IRIS, ISS, PPS, and UVS). To quantify the level of agreement among the science teams within these groups, Kendall's coefficient of concordance (see, e.g., Siegel 1956) was computed. The coefficient, denoted W, measures degree of agreement on a scale from 0.0 to 1.0, where 0.0 implies complete disagreement and 1.0 implies complete agreement. Using the ordinal rankings of the science teams as the basis for the calculation, a value $W = 0.30$ (significant at the 0.001 level) was computed. If only the rankings for teams in the fields and particles group were considered, a value $W = 0.63$ (significant at the 0.001 level) was obtained. For the platform-mounted science group, a value $W = 0.33$ (which was not significant at the 0.05 level) was obtained. A relatively high degree of agreement could be expected for the science teams in the fields and particles group because their science objectives would be mutually related by their common interest in investigations of cosmic phenomena governed by Maxwell's equations of electromagnetism. The experiments by teams in the platform-mounted science groups were not so closely related.

Collective Choice Analysis

The utility values for the candidate trajectory pairs reported by the science teams in Table C2 provided the basic input data for applying the decision criteria corresponding to the various problem formulations. Figure C3 summarizes the complete set of results. The analysis involved in each case and the interpretation of the results are discussed below.

Rank sum choice formulation

Figure C3(a) shows the rankings and values obtained by applying the rank sum choice rule to the ordinal rankings of trajectory pairs produced by various science teams. The figure shows the rank sum values for the top ten trajectory pairs linearly scaled onto the range [1/32, 1], with a value of 1 being most preferred. This scaling provided for approximate comparability with the values generated by the other choice rules. As illustrated, trajectory pair number 31 achieved the highest rank sum.

Nash bargaining model formulation

Figure C3(b) shows the rankings and values obtained by applying the

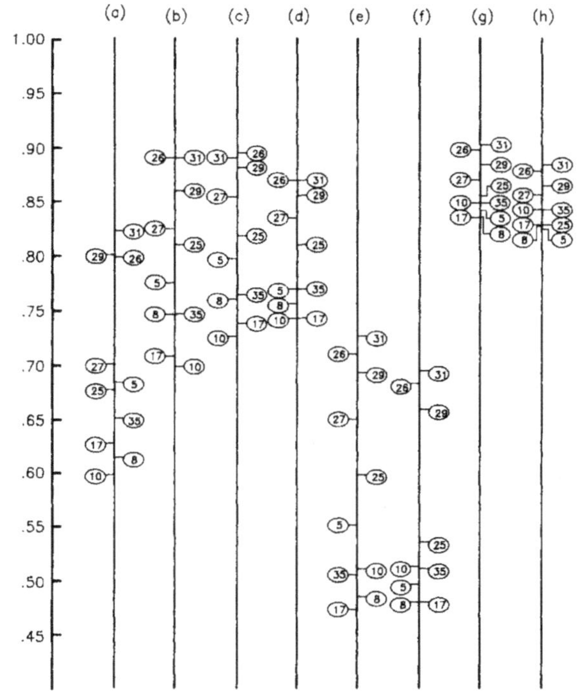

Fig. C3. Collective choice rankings and values.

modified multiplicative choice rule to the science team utility values, assuming that the no-data trajectory was the status quo alternative. The figure includes only those ten trajectories ranked highest by the rank sum rule. As indicated, trajectory pairs 31 and 26 tied for the highest ranking.

Social welfare formulation

Application of the additive, linear, social welfare function,

$$W = w_1 u_1(t^k) + w_2 u_2(t^k) + \cdots + w_{10} u_{10}(t^k)$$

required specification of the weights w_i. Because an attempt was made to scale the utility functions to represent approximately equal differences in the strengths of preferences across the science teams, these weights might be interpreted as measures of the relative importance of one team's preferences over another. Thus if one team's experiment was generally regarded to be more important than another's, that team could be assigned a higher weight.

Neither NASA nor the scientists would claim that the different science experiments were equally important. Nevertheless, the analysts and the JPL project scientist believed that assigning explicit, numeric weighting factors that quantified such differences would be extremely divisive and therefore would run counter to the objectives specified for the decision process. Consequently the analysts chose to use equal values of $w_i = 1$ for all science teams for the base-case calculations. For sensitivity analysis, however, the analysts also considered a case in which equal weights of $w_i = 2$ were assigned to the encounter-oriented science teams (RSS, IRIS, ISS, PPS, UVS, and PRA), and weights $w_i = 1$ were assigned to all other science teams (CRS, LECP, MAG, and PLS). Both sets of weighting factors were discussed with and readily accepted by the science steering group. Figures C3(c) and (d) show the results for the social welfare function for the base case and sensitivity case, respectively. In each case the values have been scaled onto the range [0.0, 1.0] to simplify comparison with the values produced by the other formulations. As illustrated, trajectory 26 obtained the highest rating under equal weights, while trajectory 31 was rated highest when encounter-oriented experiments were weighted more heavily.

Supradecision-maker formulation

With the supradecision-maker decision criterion,

$$U_{PM} = w_1 f_1[u_1(t^k)] + w_2 f_2[u_2(t^k)] + \cdots + w_{10} f_{10}[u_{10}(t^k)]$$

the project manager was given the opportunity to alter the preferences expressed by the science teams either by adjusting the weights w_i or by imposing nonlinear transformations on the science team utilities through the functions f_i. This freedom permitted conducting the analysis under several different assumptions to allow checking the sensitivity of the results.

The project scientist, who was assigned the task of manipulating the u_i values by the project manager, suggested two transformations designed to eliminate possible biases due to the unrealistically high and low utility values assigned by some teams to their least-preferred trajectory pairs. The first involved specifying a common value of 0.0 to every least-preferred trajectory pair. This had the effect of transforming each team's utility function from the original range reported in Table C2, $[p_i^{\text{least}}, 1.0]$ to the range [0.0, 1.0]. The implied assumption in this case was that each team's least-preferred alternative was equally bad. The second transformation consisted of linearly transforming each team's utility function into either the range [0.6, 1.0] or the range [0.8, 1.0], depending on the project scientist's assessment of the appropriateness of the least-preferred trajectory pair for the team's science investigation. The range [0.6, 1.0] was assigned to the science teams whose science investigations emphasized data collected during encounters, and the range [0.8, 1.0] was assigned to the teams with both cruise and encounter objectives. As in the social welfare function formulation, the analysis considered a case in which all of the weights w_i were set equal to 1 and a case in which the w_i were set equal to 2 for the encounter-oriented science teams.

The six farthest-right scales in Figure C3 show the rankings and values obtained for each case considered. When no transformation is applied to the team utility functions, the supradecision-maker criterion function is identical to the additive social welfare function. Therefore, if the project scientist does not adjust the science team preferences, the results for the equal weight and unequal weight cases are identical to those shown by the scales (c) and (d) in Figure C3. The scales (e) and (f) show the results for the equal and unequal weights, when the utilities of all teams are transformed so that the utility of the least-preferred trajectory pair is 0. The scales (g) and (h) show the similar results obtained when the project scientist determined a value of 0.6 or 0.8 for each team's least-preferred trajectory pair. As in the other cases, the values assigned by the decision criteria have been scaled so that a value of 1 would be produced if all teams preferred the same trajectory pair. Each analysis in the supradecision-maker formulation gave trajectory 31 the top rank.

Comparison of Results

A comparison of the rankings obtained by the collective choice rules shows substantial agreement among the various analyses. This agreement was partially due to the process used by NASA in selecting the experiments for the mission (because compatibility with general trajectory characteristics developed in the mission definition phase was a selection criterion), and partly a result of the statistical nature of the collective choice rules (which may be expected to be highly correlated, because sums and products of random variables are highly correlated, even if the random variables are independent and uniformly distributed over their domain). Although there are differences in the extact order of ranking, all of the choice rules list 31, 29, 26, and 27 as the top four trajectory pairs, and all agree that trajectory pairs 31, 29, and 26 are preferable to trajectory pair 27. Thus if the science teams were able to express accurately their preferences over the candidate trajectory pairs, and if the choice rules were appropriate, then the selected trajectory pair should be one of the three (31, 29, 26).

Table C3 abstracts from Table C2 the ordinal rankings for these three, highest-scoring, trajectory pairs. The table thus shows the ordinal rankings for the three pairs as originally provided by the individual science teams. Despite the apparent agreement suggested by the collective choice rules, the individual science teams differed significantly in their preferences for these trajectories. Only one science team (MAG) was in total agreement that 31, 29, and 26 were the top three trajectory pairs, and only one other team (PLS) had selected even one of these three as its most-preferred trajectory pair. Furthermore, one of the science teams (RSS) gave its lowest possible ranking (20.5) to all three of these trajectories. Based on

Table C3. Science team ordinal rankings for preferred trajectory pairs

Trajectory pairs	Rank sum	Radio science (RSS)	Infrared radiation (IRIS)	Imaging science (ISS)	Photo-polarimetry (PPS)	Ultraviolet spectroscopy (UVS)	Cosmic ray particles (CRS)	Low-energy charged particles (LECP)	Magnetic fields (MAG)	Plasma particles (PLS)	Planetary radio astronomy (PRA)
31	1	20.5	3	5	8	6	8	4	3	4	5
29	2	20.5	5	19	6	9	3	2	2	4	4
26	3	20.5	2	10	11	7	17.5	3	1	1.5	2

this disagreement, the analysts recognized that even though all of their approaches produced the same top three alternatives, the actual selecton of the trajectory pair for the mission would still be controversial.

Epilogue

A special two-day meeting of the SSG was convened for the purpose of recommending a standard trajectory pair for the space mission. After an initial unsuccessful attempt to obtain immediate consensus over a trajectory pair, the SSG requested a formal presentation of the collective choice analysis. The analysts described the rationale underlying each of the collective choice rules and the results of their applications.

In the ensuing discussion the majority of the SSG expressed a preference for one or more of the three pairs ranked highest (31, 29, 26) by the collective choice rules. The only significant opposition to this view was presented by the team leader of Radio Science Investigation, whose dislike for these three trajectory pairs (and the 24 other trajectories evaluated by his team as "least preferable") was due to a desire for a trajectory pair with at least one trajectory component that would be occulted at Saturn along the major axis of Saturn's rings (so as to produce a complete radio attenuation profile of the ring structure). Unfortunately, an optimum ring occultation and a good Titan encounter were mutually exclusive requirements. The SSG indicated its general concern that any improvement to the ring occultation not be obtained at the expense of degrading the secondary satellite encounters.

The team leader of the Imaging Science Investigation team expressed a dislike for trajectory pair 29 (to which the team had assigned an ordinal ranking of 19 and a utility function value of 0.770) but indicated that either trajectory pair 31 or 26 would be acceptable (utility values of 0.940 and 0.900). The team leader for Radio Science stated that his team would have a preference for trajectory pair 26 over 31 (one member of the SSG immediately observed that the Radio Science team should have expressed this preference in their ordinal rankings and utility function values). The team leader of the Cosmic Ray Investigation team stated that they could accept trajectory pair 26, which, even though ordinally ranked at 17.5, was given a utility function value of 0.870 (compared with 0.900 for trajectory pair 31).

On the basis of this discussion, trajectory pair 26 was tentatively selected as that preferred. The JPL trajectory analysts worked that evening to improve the geometry of pair 26 for the RSS team without degrading the features of the trajectory pair that appealed to the other science teams. These recommended revisions were presented to the SSG the following morning and met with approval. The modified version of trajectory pair 26 was subsequently approved by the project manager and became the

Voyager "Standard Trajectories." The two were labeled JSI and JSG, where JS stands for Jupiter/Saturn, and I and G stand for Jupiter's satellites Io and Ganymede, which were to be encountered on the corresponding trajectories (Beerer *et al.* 1974).

Notes to Appendix C

1. Because the mission was based on Mariner spacecraft design, it was initially named Mariner Jupiter/Saturn, or MJS; in 1977 the name was changed to Voyager. For a detailed description of the mission see National Aeronautics and Space Administration (1982).

2. Alternatives numbered 6, 12, 14 and 16 were voluntarily dropped by the science teams as being of clearly less interest.

3. In this application the rankings obtained were dependent on irrelevant alternatives: the options ranked 1 and 2 could be reversed if the option ranked 10 was eliminated.

4. A randomized trade might consist, e.g., of a 50—50 chance of actual trades t^i and t^k.

5. The Nash solution can be sensitive to the condition assumed for the status quo, so it is important to consider the reasonableness of equating failure to reach agreement with cancellation of the mission. In reality, if no agreement was reached, the project manager would resolve the conflict, e.g., by imposing a trajectory on the science teams or by requiring a redesign of the mission that would result in the elimination of one or more of the disagreeing teams in order to obtain a more compatible set of experiments. While recognizing this limitation, the analysts were unable to test how seriously the simplifying assumption regarding the status quo would distort the solution obtained from the Nash model.

6. Zeuthen argued that if there are two players, and player 1 is demanding a trade A that produces payoffs to player 1 and player 2 of u_1^A and u_2^A, respectively, while player 2 wants trade B that produces respective payoffs of u_1^B and u_2^B, player 1 should concede (i.e., accept or suggest another trade) if his "relative loss" in the concession is less than that for player 2; that is, if

$$\frac{u_1^A - u_1^B}{u_1^A} < \frac{u_2^B - u_2^A}{u_2^B}$$

This is mathematically equivalent to the criterion function

$$u_1^A u_2^A - u_1^B u_2^B$$

which implies that one of the parties will concede whenever

$$u_1 u_2 - u_1^0 u_2^0 > 0$$

(the modified Nash solution). Harsanyi subsequently showed that this same choice rule could be derived by assuming five behavioral axioms:

1. *Symmetry*: the players follow identical rules of behavior.
2. *Perfect knowledge*: each player can correctly estimate the probability that the other will reject an alternative.
3. *Monotonicity*: the probability of Player i ($i = 1$ or 2) rejecting A and insisting on B is a monotone, nondecreasing function of the utility difference $u_i^B - u_i^A$.

4. *Expected utility maximization*: unless players agree to make simultaneous concessions, each will make a concession if and only if the alternative gives him a higher expected utility than the uncertain outcome of refusing to make the concession.
5. *Efficiency*: the players will make simultaneous concessions if the new alternative gives both players higher expected utilities.

7. As described in Chapter 2, both Harsanyi (1955, 1975) and Keeney and Kirkwood (1975) justified an additive form for a social welfare function by assuming that the von Neumann-Morgenstern axioms apply to the individual (team) utility functions and to the collective choice rule. Harsanyi shows that the following additional assumption is then sufficient to require that the social welfare function have the additive form: if two alternatives are indifferent from the standpoint of each individual, then they are indifferent for the group as a whole. Keeney and Kirkwood obtained the same result by replacing Harsanyi's axiom with three of their own: (1) if all members but one are indifferent to all possible consequences, then the preferences of the group will be those of the one; (2) if all members but two are indifferent to all possible consequences, then the preferences of the group will be governed only by the preferences of those two; and (3) if all members of the group have the same utility function over the consequences, then the collective choice rule should always yield the same decisions as the common utility function. Both the Harsanyi and the Keeney and Kirkwood axioms might be questioned. Suppose, e.g., that each science team was indifferent between trajectories that yielded exactly the same probabilities of total success and total failure for their individual experiment. As a group, however, the science teams might prefer trajectories for which the failure probabilities for experiments were nearly independent (so that the probability of all failing simultaneously would be very remote) to trajectories for which dependencies implied that whatever caused one experiment to fail would likely cause the others to fail as well. Because the major risks of failure affected all experiments equally (e.g., see comments later in Appendix C regarding the "Atlantic Ocean Special") and because there appeared to be little differentiation in the nature of the dependencies among different trajectories, the analysts regarded the Harsanyi and Keeney and Kirkwood axioms as reasonable.

8. As might be expected, given the similarity between the above criterion function and that for the social welfare formulation, the assumptions underlying the supradecision-maker formulation are very similar to those noted in the social welfare function formulation. If the marginal probability distributions for the u_i are the same for any two lotteries, then the expected values of the u_i must be the same, implying that the science teams would be indifferent between those lotteries. This condition would be satisfied if the decision maker subscribed to the following: if two alternatives are indifferent from the standpoint of each individual, then they are indifferent for the group as a whole. This is the assumption used by Harsanyi in the justification of the additive form for the social welfare function, as discussed in the previous note and in Chapter 2.

References to Appendix C

Beerer, J. G., P. A. Penzo, and R. A. Wallace. 1974. 'Mariner Jupiter/Saturn 1977 Standard Trajectories.' Project Document 618—113, Jet Propulsion Laboratory, Pasadena, Calif.

Dyer, J. S. and R. F. Miles, Jr. 1976. 'An Actual Application of Collective Choice Theory to the Selection of Trajectories for the Mariner Jupiter/Saturn Project.' *Operations Reseach* 24: 220—244.

—— and ——. 1977a. 'A Critique and Application of Collective Choice Theory:

Trajectory Selection for the Mariner Jupiter/Saturn 1977 Project.' JPL Internal Document 900—777, Jet Propulsion Laboratory, Pasadena, Calif, February.

—— and ——. 1977b. 'Alternative Formulations for a Trajectory Selection Problem: The Mariner Jupiter/Saturn 1977 Project.' In E. E. Bell, R. L. Keeney, and H. Raiffa, eds. *Conflicting Objectives in Decisions.* New York: Wiley.

Fishburn, P. C. 1970. *Utility Theory for Decision Making.* New York: Wiley.

Harsanyi, J. C. 1955. 'Cardinal Welfare, Individualistic Ethics, and Interpersonal Comparisons of Utility.' *Journal of Political Economy* 63: 309—321.

——. 1956. 'Approach to the Bargaining Problem before and after the Theory of Games: A Critical Discussion of Zeuthen's, Hicks', and Nash's Theories.' *Econometrica* 24: 144—157.

——. 1959. 'A Bargaining Model for the Cooperative n-Person Game.' In A. W. Tucker and R. D. Luce, eds. *Contributions to the Theory of Games* 40(4): 325—355.

——. 1975. 'Nonlinear Social Welfare Functions.' *Theory and Decision* 6: 311— 332.

Keeney, R. L. and C. W. Kirkwood. 1975. 'Group Decision Making Using Cardinal Social Welfare Functions.' *Management Science* 22: 430—437.

Nash, J. F., Jr. 1950. 'The Bargaining Problem.' *Economica* 18 (April): 155—162.

National Aeronautics and Space Administration. 1982. *The Voyager Flights to Jupiter and Saturn.* NASA Document EP-191, GPO Document 1982-584-903. Washington, D.C.: GPO.

Siegel, S. 1956. *Nonparametric Statistics.* New York: McGraw-Hill.

Young, H. P. 1974. 'An Axiomatization of Borda's Rule.' *Journal of Economic Theory* 9: 43—52.

Zeuthen, F. 1930. *Problems of Monopoly and Economic Warfare.* London: Routledge.

Index of Names

Index of Subjects

GPSR Compliance

The European Union's (EU) General Product Safety Regulation (GPSR) is a set of rules that requires consumer products to be safe and our obligations to ensure this.

If you have any concerns about our products, you can contact us on ProductSafety@springernature.com

In case Publisher is established outside the EU, the EU authorized representative is:

Springer Nature Customer Service Center GmbH
Europaplatz 3
69115 Heidelberg, Germany

Batch number: 10761599

Printed by Printforce, the Netherlands